TECHNICAL EKPHRASIS IN GREEK AND ROMAN SCIENCE AND LITERATURE

Ekphrasis is familiar as a rhetorical tool for inducing *enargeia*, the vivid sense that a reader or listener is actually in the presence of the objects described. This book focuses on the ekphrastic techniques used in ancient Greek and Roman literature to describe technological artifacts. Since ancient literary discourse on technology extended beyond technical texts, this book explores "technical ekphrasis" in a wide range of genres, including history, poetry, and philosophy as well as mechanical, scientific, and mathematical works. Technical authors like Philo of Byzantium, Vitruvius, Hero of Alexandria, and Claudius Ptolemy are put into dialogue with close contemporaries in other genres, like Diodorus Siculus, Cicero, Ovid, and Aelius Theon. The treatment of "technical ekphrasis" here covers the techniques of description, the interaction of verbal and visual elements, the role of instructions, and the balance between describing the artifact's material qualities and the other bodies of knowledge it evokes.

COURTNEY ROBY is Assistant Professor in the Department of Classics at Cornell University. Her research interests include the literary techniques used in scientific and technical texts from the ancient Mediterranean, the interaction of verbal and visual elements in those texts, and the role of the imagination in ancient scientific discourse.

TECHNICAL EKPHRASIS IN GREEK AND ROMAN SCIENCE AND LITERATURE

The Written Machine between Alexandria and Rome

COURTNEY ROBY

Cornell University

CAMBRIDGE
UNIVERSITY PRESS

CAMBRIDGE
UNIVERSITY PRESS

Shaftesbury Road, Cambridge CB2 8EA, United Kingdom

One Liberty Plaza, 20th Floor, New York, NY 10006, USA

477 Williamstown Road, Port Melbourne, VIC 3207, Australia

314–321, 3rd Floor, Plot 3, Splendor Forum, Jasola District Centre, New Delhi – 110025, India

103 Penang Road, #05–06/07, Visioncrest Commercial, Singapore 238467

Cambridge University Press is part of Cambridge University Press & Assessment, a department of the University of Cambridge.

We share the University's mission to contribute to society through the pursuit of education, learning and research at the highest international levels of excellence.

www.cambridge.org
Information on this title: www.cambridge.org/9781107077300

First published 2016

A catalogue record for this publication is available from the British Library

Library of Congress Cataloging-in-Publication data
Names: Roby, Courtney, 1978– author.
Title: Technical ekphrasis in Greek and Roman science and literature : the written machine between Alexandria and Rome / Courtney Roby.
Description: Cambridge ; New York : Cambridge University Press, [2016] | Includes bibliographical references and index.
Identifiers: LCCN 2015041128 | ISBN 9781107077300 (hardback)
Subjects: LCSH: Ekphrasis. | Rhetoric, Ancient. | Literature and science–Greece. | Literature and science–Rome. | Science–Greece–History–To 1500. | Science–Rome–History–To 1500. | BISAC: PHILOSOPHY / History & Surveys / Ancient & Classical.
Classification: LCC PN56.E45 R639 2016 | DDC 808/.0481–dc23
LC record available at http://lccn.loc.gov/2015041128

ISBN 978-1-107-07730-0 Hardback

Contents

Figures

Acknowledgments

This book could not have been written without a great deal of help. I am most deeply indebted to Reviel Netz, who gave advice on the project from its very beginning. The work was also shaped in its early years by Alessandro Barchiesi, Maud Gleason, Natasha Peponi, and Jessica Riskin; Serafina Cuomo also provided helpful comments on an early draft. For suggestions that strengthened the project in later stages of its development, I am grateful to colleagues at Cornell, particularly Hayden Pelliccia, Verity Platt, and Jeff Rusten. Funding was provided by the Ric Weiland Graduate Fellowship, the Geballe Dissertation Fellowship at the Stanford Humanities Center, and the Himan Brown Charitable Trust. For shepherding the project through the publication process and suggesting a myriad of improvements, many thanks to Michael Sharp and to the anonymous readers of Cambridge University Press. Finally, my heartfelt gratitude to Luis, and to the other friends and family who have been so patient and encouraging throughout the craft process: *modo nobis stet illud, una vivere in studiis nostris.*

Introduction: The Written Machine

How can an artifact most effectively be represented in words? Should an author strive to provide the maximum number of details about the object's physical attributes, or can the reader be made to understand the object more deeply through an appeal to less immediately visible attributes, such as how it was made or its effect on the viewer? Frontinus gives an answer to these questions in his work on the aqueducts that watered Rome:

> It did not seem irrelevant to me to include the lengths of the rivers and the path of each in the appearance of the works. For since the greatest part of this office is the safekeeping of these, it is only right that their overseer should know which require the greater expenditures. Indeed, it was not enough for my concern to put them individually before the eyes, but I also took care to write about the forms of their paths, so that from these it would become clear where the valleys were and how large, where rivers were crossed, where hollows in the sides of mountains demand assiduous care in protecting and fortifying the stream. From this that utility appears, that we can have the thing as though it were in view, and consider it as though standing right beside it.[1]

Here, indeed, is one answer, but it is not the only one. Frontinus' cavalcade of structural and geographic details, dimensions, and carrying capacities contrasts strongly with Cicero's description of a miniature astronomical model attributed to Archimedes.[2] Cicero outlines the artifact's history (it was looted after the siege of Syracuse), the philosophical insight it inspires (that Archimedes could have designed a model of the cosmos suggests that the real thing was likewise created by a divine mind), and its intercultural interpretability (even barbarians could understand the perfect reason that produced it). He does not, however, provide anything in the way of technical detail beyond noting that the device imitates the motions of the sun,

[1] Frontinus, *De aquae ductu urbis Romae* 1.17.

[2] Cicero, *De re publica* 1.21, *De natura deorum* 2.88, *Tusculanae disputationes* 1.63.

moon, and five planets. Instead, the reader learns what it is like to see the *sphaera*, to undergo the revelatory experience of seeing the cosmos, as it were, from outside. Cicero still encourages us to think as though we were in the presence of the object, but his descriptive strategy is completely different. This book is about what those differences are, and what they mean.

Describing an artifact in text places it at a busy crossroads of connections between the object and its textual representation, and between the text's author and audience. The meetings that occur at these crossroads take a variety of forms. The author may communicate closely with his audience, structuring the text like an interaction between mentor and student, or he may take a more remote stance, delivering information without interposing himself in the text. The object may be brought before the reader's eyes in vivid historical detail, or abstracted from its context of use. Its material aspects may be scrutinized deeply, or kept at arm's length, the artifact's physical properties overshadowed by discussion of the philosophical reflections it induces in its viewers, the bodies of theoretical knowledge to which it is connected, and so forth.

Many of these descriptive possibilities might apply to a whole range of subjects, from paintings to trees to human beings, and even to less concrete subjects like events or emotional states. Indeed, ancient rhetorical manuals indicate that all of these, and others besides, were considered fair game for the kind of vivid description often labeled "ekphrasis."[3] The unifying factor was not subject matter but effect on the reader: an ekphrastic description is supposed to create such a vivid mental image that the audience feels almost as if the object is present. For example, the rhetorician Aelius Theon defines ekphrasis as "descriptive speech vividly (ἐναργῶς) bringing into view the thing shown."[4] So when Frontinus promises to describe the water system in such detail that the reader has it, as it were, in his own sight, he is really promising his reader an ekphrasis.

As Theon's definition makes clear, ekphrastic descriptions carry the power of *enargeia*, the clarity of imaginary vision sparked in the reader's

[3] I will use "ekphrasis" throughout this work as a convenient catch-all for a literary phenomenon that went by many names, though the term only appeared in the first century CE. However, it dominates the modern theoretical discourse of the textual representation of objects (particularly works of art), and as I will be referring to some of these modern theories it will help to maintain a consistent terminology, while noting both the utility and the shortcomings of these recent approaches. The subjects explicitly associated with ekphrasis in ancient rhetorical manuals are listed at Webb 2009, 213–14.

[4] Theon, *Progymnasmata* 118.7–8. On Theon's dates see ibid., 14. The later rhetorical authors ps.-Hermogenes and Aphthonius provide nearly word-for-word identical definitions (ps.-Hermogenes, *Progymnasmata* 10.1; Aphthonius, *Progymnasmata* 10.36.22).

mind by skillfully deployed words, so that he effectively becomes an eyewitness to the things being described.[5] Its cognitive effect on the audience is central to *enargeia*; as Leach observes, "*enargeia* differs from pictorial verisimilitude in the complexity of the receptive act it demands from the spectator."[6] *Enargeia* is most often referred to by ancient rhetorical theorists as a visual phenomenon, though it is sometimes described as being able to invoke other senses as well.[7]

This book will focus on a particular category of ekphrastic subject: the "written machine," technological artifacts represented in text so that the reader can develop such a vivid mental image that he comes into virtual contact with the thing itself. The phrase *graphomenon organon* is used by Hero in his *Dioptra* to refer to an odometer he describes in precise structural, mechanical, mathematical, and graphical detail.[8] Unlike its English translation, the Greek term captures the idea of drawing as well as writing. A focus on the "technical ekphrasis" emerges quite naturally from the combination of an attribute of technological artifacts and an attribute of ekphrasis: machines embody knowledge that can be decoded, and ekphraseis are designed to perform that decoding. The complexly layered systems of knowledge that will be relevant here include, to begin with, the mechanical and physical principles that inform their design and the craft expertise that shapes their construction.[9] Some devices (for example, astronomical, mathematical, or surveying instruments) are designed to help the user acquire new knowledge, and the demands and resources of these external bodies of knowledge help to shape both the artifact and its description.[10]

Ekphrasis leads a double life much like that of the mechanical artifacts. To be sure, ekphrasis usually communicates the material details, as it were the sensible qualities of its object. However, composing an ekphrasis is at the same time an interpretive activity: as Baxandall and others have noted about works of art, seeing is "theory-laden."[11] Goldhill likewise observes,

[5] On *enargeia* more broadly see Elsner 2001, 2–3; Webb 2006, 114–15; Goldhill 2007, 3–7; Squire 2009, 143–6; Webb 2009, 21–2, 27, 51–4, 70–4, 87–130; Platt 2011, 7, 173–8, 216–17, 230–4.

[6] Leach 1988, 7.

[7] For the emphasis on the visual see, among others, Theon, *Progymnasmata* 118.7–8; ps.-Hermogenes *Progymnasmata* 10.2, 10.25. For the use of other senses, see for example Cicero, *Part. or.* 6.20.

[8] Hero, *Dioptra* 34.9.

[9] For modern perspectives on what types of knowledge these are, see for example Bunge 1979; Mendelsohn and Elkana 1981, 19–27; Dipert 1993, 87–95; Faulkner 1994; Houkes 2009; Nightingale 2009; Thomasson 2009, 194–8; Hodder 2012, 10–14.

[10] On the types of knowledge encoded in technological artifacts, see Mitcham 1994, 192–6; Thomasson 2009, 194–8.

[11] Baxandall 1985, 107. Baxandall is, of course, preceded in this observation by historians of science as early as Popper 1935; reproduced in Popper 1959; the same English formulation is found in Hanson

in accounting for the power of ekphrasis to produce and shape a viewing subject, that "the critical gaze … is committed to a value-laden view of things."[12] These intensely descriptive passages create cultural context for the object described, which may take the form of praise or blame for an artifact, historical information about its provenance, or references to the mythical narratives to which a work of art alludes or prior works by which it was influenced.[13] The very act of selecting certain attributes to expand upon in the ekphrasis is a value-laden act insofar as it forces the audience to encounter that subject in a particular way.[14]

The author of a technical ekphrasis has a rich set of attributes to choose from. For example, a theatrical automaton can be portrayed in text from the perspective of a designer, in deep structural and functional detail. Alternatively, an author can describe what a mystified outsider would see, "black-boxing" its functional details and describing only what is apparent on the surface to the layman. Another approach allows the viewer's experience, rather than information about the artifact itself, to dominate the ekphrasis, shifting the focus of description still further from the mechanism. One text might describe a catapult in meticulous mathematical terms, slowly aggregating details over dozens of pages, while another hops briskly from one model of ballistic device to another, sketching out a whole field of machines through quick vignettes. Some authors explicitly confine themselves to describing the artifact itself, while others allow their analysis to range more freely into associated bodies of knowledge like material physics, history, or philosophy. The artifact itself can be presented from a historical distance, or it can come together before the reader's eyes through a description of its construction process; the reader can even be implicated in this process through the directive guidance that building instructions provide. Any of these approaches might either make use of purely visual descriptive terms, or invoke tactile and auditory information as well; diagrams and other visual elements can be integrated directly into the text to enhance the immediacy of the encounter still further.

A concrete example will show how a few of these possibilities play out. Hero of Alexandria's *On Automaton-making* (*Peri automatopoiētikēs*)

1958, 19. For the time being, however, let us keep the focus on seeing *artifacts*, for which Baxandall's testimony is especially relevant.

12 Goldhill 2007, 2. *Contra*, see Heath 1987, 98–111; Heath's argument is effectively countered at Fowler 1991, 28.

13 Subjects of ekphrasis other than artifacts have their own expansive contextual information: in the case of an ekphrasis on a battle, for example, Theon lists expenditures, fears, damage to man and landscape, victory, and trophies among the features to be described, which in turn invest the ekphrasis with cultural, historical, and ethical value.

14 On the expansiveness of ekphrasis, see Webb 2009, 52–5, 74–6.

describes the structure and function of two types of theatrical automaton. For the most part, the text presents the artifact as under construction. Consider, for example, the structure and function of the pneumatic subassembly that causes milk to flow from the thyrsus held by a miniature Dionysus: "And from a vessel ΝΟ, let a tube ΠΡΣΤ run to another tube ΥΦ, which is tightly enclosed by a tube ΓΔ, connected from the parts underneath to the surface on which the shrine rests."[15] This is just one part of one subassembly of a very complicated device; the letter labels are essential for organizing all that descriptive information, as well as for linking it to the accompanying lettered diagrams. Hero draws these textual tools from an established tradition of geometrical prose, and he often assimilates the components of the automaton to their geometrical idealizations: for example, two different-sized wheels are assimilated to sections of a cone. The imperative forms Hero uses to "construct" the device in the reader's imagination are likewise characteristic of geometrical prose. Other components are described in material rather than mathematical detail: the grains of millet draining out of the counterweight that operates the device are "light, small, and adherent";[16] the cord that the counterweight draws is smeared with wax and resin to keep it from stretching and contracting;[17] pivot points have to be smoothed and oiled to ensure proper rotation.

Because automata are designed to mystify an audience, Hero takes into account the perspective of the audience as well as the designer. He does this both by describing decorative elements like the figures, backdrops, and garlands, and by indicating what is hidden from the viewer. The mechanisms for driving the moving backdrop of a sailing scene, igniting a flame, and lifting up the figure of Athena are composed mainly of elements the audience doesn't know are there. Hero advises adding a plank that will cast a shadow over all the working parts,

> so that none of them will be seen from in front of the theater. So that the plank will not be placed there for no apparent reason, a pediment is placed on it as on a shrine; the wings projecting on each side of the plank should be made black or blue.[18]

The reader thus gets to encounter the artifact during construction and performance, from the inside and the outside, verbally and visually, as a mathematical ideal and a material reality. Each of these descriptive approaches links the artifact to different bodies of knowledge: geometry, craftsmanship,

[15] Hero, *Peri automatopoiētikēs* 13.3.
[16] Hero, *Peri automatopoiētikēs* 9.4.
[17] Hero, *Peri automatopoiētikēs* 5.1.
[18] Hero, *Peri automatopoiētikēs* 28.1–3.

stagecraft, and so forth. Indeed, Hero says automaton-making encompasses every part (μέρος) of mechanics, indicating that he conceives of mechanics as a system of interconnected subdisciplines. The automaton is described as a crafted artifact of a particular form, displayed for a particular audience with a specified level of knowledge compared to the designer. At the same time, it is described as a member of a more general class of mechanical devices, connected to other devices that operate by the same basic theoretical principles and material components. This text, like many of Hero's works, highlights mechanical systematization and interdisciplinary connections, and Hero congratulates himself for its broad accessibility. However, he sacrifices neither material detail nor mathematical accuracy: he could have chosen to describe only the exterior of the machine, or to omit the geometrical explanations, but he did not. Hero creates for his reader a certain picture of this artifact and what ought to be known about it, carefully choosing the elements that feature in his description.

Ekphrasis in antiquity was a broadly inclusive textual device, a fact largely ignored by modern critics who prefer to associate ekphrasis peculiarly with art objects. This claim has already been thoroughly dismantled by Webb, who provides a useful capsule history of how the error developed over the last two centuries.[19] Koelb draws further attention to Denniston's article on ekphrasis in the 1949 edition of the *Oxford Classical Dictionary*, which groundlessly defines ekphrasis specifically as a "rhetorical description of a work of art."[20] This error was more widely propagated when Leo Spitzer drew on Denniston's article for the definition of ekphrasis used in his influential essay on Keats's "Ode on a Grecian Urn." Modern critical studies of ekphrasis, building on this store of work, inherited these aesthetic preoccupations. In particular, the Horatian *ut pictura poesis* was taken to indicate a special, even exclusive, relationship between ekphrasis and works of art.[21] For some critics, even this limitation is not enough. Heffernan, for example, goes further: "ekphrasis is the verbal

[19] For the development of this misconception over the nineteenth and twentieth centuries and its implications for modern critical theory on ekphrasis, see Webb 2009, 28–36. For a concise summary of modern critical approaches to ekphrasis, see Squire 2009, 140–2.

[20] Koelb 2007, 2. Koelb goes on to specify that "like Spitzer, Denniston curiously cites no classical authority for his definition of ecphrasis; or rather, he misattributes it to J. W. H. Atkins's article titled 'Progymnasmata' in the same volume. Atkins translates ecphrasis – correctly – as 'description,' not description of art in particular" (734).

[21] Horace, *Ars poetica* 361, itself owing much to Simonides' characterization of painting as silent poetry and poetry as speaking picture, which survives only as a quotation from Plutarch (*De gloria Atheniensium* 346F4). An example of this attitude occurs in the introduction to Robillard and Jongeneel 1998, which brings in the Horatian tag immediately and uses it as a jumping-off point to limit ekphrasis to "artworks."

representation of visual representation."[22] Mitchell's essay on ekphrasis relies on the same formulation.[23]

The special association of ekphrasis with representational art is quite deeply rooted in the modern critical landscape, and has yielded much insightful work. But for a discussion of ekphrasis in ancient literature, this perspective will not do. Both in the theory and practice of ancient rhetoric, ekphrasis is applied to people, events, and objects of all kinds, and not particularly associated with works of art. The varieties of ekphrasis are sometimes categorized in the rhetorical manuals by how they expand on their source material, whether that means a narrative of circumstances, an account of an event's causes, and so forth. Many technical ekphraseis belong to the category Theon calls the "ekphrasis of the *tropos*."[24] This kind of ekphrasis focuses attention on the *way* something gets done, and in fact Theon uses the construction of siege fortifications as an illustrative example. The "technical ekphrasis" is thus evidently not only theoretically permissible within the broad boundaries of ancient ekphrasis, but actually recognized in ancient rhetorical literature.

Yet there is undoubtedly something special about descriptions of artworks, particularly representational art, that drives critics like Heffernan to identify them as the core targets of ekphrasis. Ekphraseis on works of art attract special critical attention because they can be mined for cultural significance, and this in turn is possible because designing and crafting a work of art requires encoding cultural knowledge in some plastic form.[25] In the specific context of Hellenistic epigram, for example, Gutzwiller denotes epigrams on artworks as an "identifiable category" among these epigrams.[26] Platt echoes this phrase in explaining why her own treatment focuses on this type of object, while explicitly acknowledging the breadth of subjects for ekphrasis used in ancient rhetoric.[27] Artworks are cultural artifacts; they are physical objects shaped by culturally significant beliefs, stories, or aesthetic principles. Ekphraseis on these objects are then

22 Heffernan 1993, 3; formulated slightly differently in Heffernan 1991, 299: "ekphrasis is the verbal representation of graphic representation."

23 Mitchell 1994, 152.

24 Theon, *Progymnasmata* 68.7–10; on this form of ekphrasis see Webb 2009, 69–70.

25 These processes of encoding and decoding are complex and ambiguous; it is not my intention here to suggest otherwise. Discussion of the connections between ekphrasis and the interpretation of works of art can be found, for example, in Thomas 2001; Mitchell 2005; Panofsky and Lavin 1995, particularly Summers, "Meaning in the Visual Arts as a Humanistic Discipline," and Bredekamp, "Words, Images, Ellipses"; Squire 2009, where ch. 1 treats these questions particularly explicitly.

26 Gutzwiller 2002, 85 n. 1.

27 Platt 2011, 173 n. 9.

culturally charged; the author chooses for some reason to focus on certain aspects of the artwork, which link the ekphrasis and the object to the surrounding culture in a particular way. Squire cites the fact that ekphrasis is particularly important "as an embodiment of … cultural theories of representation and textuality" as motivation for the relatively large number of critical explorations of ekphrasis on art objects.[28] Mitchell argues that while ekphrasis is not some kind of "key to the difference between ordinary and literary language," it does serve to "[focus] the interarticulation of perceptual, semiotic, and social contradictions within verbal representation."[29]

The role played by ekphrasis in mediating between objects and texts is rapidly becoming better understood, thanks to a number of recent studies of the theory and practice of ekphrasis, its development within the context of ancient rhetoric, and the value of ekphrasis to the study of art history. Webb's groundbreaking study engages ancient theories of ekphrasis on a particularly broad range of objects.[30] Several other recent analyses have focused on particular types of object, most often painted or sculpted art objects, though unlike the works critiqued earlier, they do not pretend that ekphrasis is relevant to art alone.[31] Many of them address the possible tension between objects and their associated texts and visual representations, and the different ways the reader can be implicated in the text–object relationship.[32]

Because human culture is encoded in technological artifacts as well as works of art, technical ekphraseis are also freighted with culturally bound knowledge, and they may make this knowledge explicit in different ways and to different degrees. In practice, basic technologies like the loom and plow embody knowledge that often remains tacit, while "high technologies" like the catapult are more often explicitly discussed in text.[33] Texts might focus on the mechanical principles of a catapult design, the craft knowledge that goes into its construction, or the historical events in

[28] Squire 2009, 141. [29] Mitchell 1994, 180. [30] Webb 2009.

[31] For example, Boehm and Pfotenhauer 1995; Yacobi 1995; Newby and Leader-Newby 2007; Squire 2009; Squire 2010; Baumann 2011; Platt 2011.

[32] On tensions between text and object, see for example Goldhill 2001; Platt 2007; Squire 2009, 15–23, 92–111, 146–93, 247–8. On the reader's engagement with the text and object, see Goldhill and Osborne 1994; Webb 2006; Tueller 2008; Newby 2009.

[33] Knowledge about how to design and build more basic technologies was presumably more often transmitted through direct contact with experienced craftsmen, as these artifacts continued to propagate despite the infrequency of detailed descriptions of them in text. On textual and non-textual transmission of knowledge about technology, see Meißner 1999, 50–4, 62–88, 123–46; Meißner 2003.

which it participated. Ekphraseis on technological artifacts decode and transmit this knowledge, linking the object into the surrounding culture. These cultural links may emphasize the object's significance for its immediate scientific environment, or they may establish connections that range beyond the scientific and technological to philosophical, aesthetic, or political concerns.

Technology and culture

Technology is complexly bound to its culture of creation. Though the focus here will remain on how information about technological artifacts is propagated in literature, we should not lose sight of the social context in which their construction and use are inevitably embedded. Ellul focuses on the social context for the particular case of "technology" in more or less the sense we use it today, while Ihde's meditation on the links between technology and the "lifeworld" effectively explores the interconnections between technologies, nature, and humans, for all that it often sharply dichotomizes them.[34] Bauer and Herder suggest the need to study "socio-technical systems" as combinations of technological artifacts and purposive actors that can only be understood through their interactions.[35] Meskell calls for a re-evaluation of the role of "material *habitus*" in archaeology, linking object analysis more closely to studies of the social relations those objects activate.[36] Hodder focuses on the "dependencies" between things and people, and the "entanglements" that emerge from these dependencies, shaped by material affordances, changing human needs, and the chaos of contingency.[37] All this requires an infrastructure of communication: the combined textual and verbal methods that record and spread the working principles of the artifacts and their manufacture among practitioners, and mediations between technological discourse and the larger culture that establish the utility of technologies and motivate their further propagation and development.

The multifarious possibilities for the relationship between an artifact, its contexts of manufacture and use, and the texts that carry knowledge about the artifact can be clarified by considering a familiar example: the Ikea construction manual. The marvelously simple, wordless instructions with which Ikea provides the amateur assembler of furniture are entirely

[34] Ellul 1990; Ihde 1990. [35] Bauer and Herder 2009, 601. [36] Meskell 2005, 1–7.
[37] Hodder 2012.

dependent on a complicated infrastructure of manufacture and distribution which, for example, allows me to make sense of what is meant by a part called "102509." I can only do this because that system has already provided me with a sealed, labeled bag of just as many 102509s as I will need to assemble the desk of which they are a part. In the absence of such a system, the difficulties of description and instruction of course have to be shifted to the reader or builder; Hero of Alexandria cannot "black-box" them like Ikea does. Many of the differences between an "Ikeastic"[38] and an "ekphrastic" description of a technological artifact are due precisely to this difference in infrastructure.

Like works of art, technological artifacts depend on human intentionality for their existence, not only causally, but also in the sense that Thomasson calls "conceptual" or "existential" dependence.[39] That is, unlike natural entities, the underlying concept of an artifact is of something intentionally created by humans. Technological artifacts become the focus of diverse literary endeavors because they are important to their host culture – perhaps because they are useful for solving practical problems, or because they display special inventiveness and cleverness, or for a host of other reasons. The ancient Mediterranean world was pervaded by a socio-technical system of uses for these artifacts (including military sieges, waterways that required bridging and dredging, and the complex social and technological machinery of theaters) that motivated their production and development. Hero, for example, cites a very ancient theatrical tradition for the automata he describes, though he acknowledges that they have increased greatly in complexity by his own time. Ancient technology was linked, as intimately as its modern counterpart, to the discourse that supported its propagation.[40] A study of that discourse should therefore take into account not only descriptions of decontextualized artifacts, but descriptions of artifacts in production and use as well (when these are given ekphrastic clarity); these are all aspects of technology, properly considered.

The link between technological artifacts, the social context of their development, and the literature in which knowledge about them was

[38] Thanks to Reviel Netz for inventing this adjective. A related point about the dependence of Ikea furniture on its individual pieces is made briefly, though without reference to communicating information, at Hodder 2012, 40.

[39] Thomasson 2009, 194–5.

[40] Meißner's insistence that technology must involve the specification of its own domain, as well as articulating responses for practitioners who find themselves faced with a problem that lies outside the realm of effectiveness of a particular technology, is a particularly forceful statement of this linkage (Meißner 1999, 50).

disseminated has been perhaps most eloquently described by Shapin and Schaffer, who unpacked the idea of "literary technology" in the seventeenth-century context of Boyle's experiments with the air pump. Boyle's argument that experimental work was the appropriate way to acquire natural philosophical knowledge required him to combine the "material technology" of the apparatus itself with a "literary technology" that propagated knowledge about the experimental phenomena it demonstrated, and a "social technology" for validating the knowledge claims made by experimental philosophers.[41] Shapin and Schaffer further emphasize that "all three are *knowledge-producing* tools." Something analogous to this "literary technology," which included tools like letter labels and diagrams, was likewise used by ancient authors to bring their readers into cognitive contact with technological artifacts.

The connection between technological know-how and the surrounding culture is a two-way street. Just as the technical knowledge that goes into creating mechanical artifacts informs the surrounding culture, cultural priorities shape the types of technology that are produced and the uses it is put to.[42] Sometimes the proliferation of new variations on existing technologies is privileged; sometimes the surviving texts indicate a preference for technologies that can be implemented quickly to serve an immediate need, even if they are not quite on the cutting edge.[43] Cultural preferences and exigencies also affect the degree of esteem afforded the various activities involved in the production of technological artifacts. Some authors foreground the mathematical aspects of technological design, emphasizing its closeness to geometrical work. Others privilege the novel mechanical or structural results obtained in design and construction, stressing the craftsmanship required to produce the material objects rather than the abstract principles of their operation. Still others highlight the use and refinement of technology in the field.

Shifts in focus between the mathematical and material, the theoretical and the practical, the artifact and its context of production or use, suggest the many conceptual "layers" that comprise a technological artifact, and the many ways an author might choose among them. Cuomo invokes these variations by contrasting the mathematician Pappus of Alexandria to Hero: while the latter constantly reminds his reader of the materiality

[41] Shapin and Schaffer 1985, 25.

[42] DeLaine 1996; Fleury 1996; Meißner 1999, 37–50, 89–123, 139–42; Chouquer and Favory 2001, 64–71, 95–110, 264–76; Meißner 2003; Rihll 2007; Cuomo 2007, 41–76; Jaeger 2008, 48–127; Berryman 2009, 105–220.

[43] See, for example, Meißner 1999, 165–6, 248–50, 252–4, 277–83, 312–17.

of the objects he describes, "Pappus' mechanics … is made to operate within a mathematical universe of reference."[44] Even in his more theoretical texts such as the *Mechanics*, Hero clothes the mathematical "skeleton" of mechanical principles in the "flesh" of a concrete physical environment. In framing the inclined-plane problem, for example, Hero emphasizes the challenges posed by friction and material irregularity, whereas Pappus downplays these components in favor of the problem's geometrical aspects.

However, Hero's decision to model the wheels of his automaton as slices of a cone highlights that this contrast is more a local than a global preference for one descriptive approach over another. Hero sometimes detaches his automaton from its material reality when he wants to use mathematical tools of diagrammatic illustration or proof on it, as when he assimilates the wheels of the automaton to sections of a cone. Underlining such differences between texts tells us something about their different cultural contexts of production. So, for example, Schiefsky argues that Hero's presentation of mechanical principles in his *Mechanics* is deeply informed by recent and contemporary mechanical practices.[45]

Texts written by mechanical laymen can reveal just as much about how cultural priorities influence technological discourse. Plutarch, for example, brings some Platonic sympathies to bear in devaluing Archimedes' mechanical work in its original Hellenistic setting: he claims the preparations for the siege of Syracuse were not seen by Archimedes as serious work, but only as a geometrical game.[46] He presents geometry as nobler intellectual work than mechanics, so much so that Archimedes' ventures into designing and constructing mechanical devices, even those that served to defend against the invading Roman navy, can only be explained as a game. Plutarch then proceeds to an archaeology of mechanics, revealing that from its very beginnings mechanics was but a handmaiden of geometry, and as such was reviled by Plato for moving geometry away from the abstract and toward the material, and by relying on tools and materials that required manual labor.[47]

Plutarch's contrast between his own culture's high regard for the mechanical arts and what he describes as its original debased status highlight how perceptions of technology's cultural status change over time, while his editorializing claims about the relative virtues of mathematics

[44] Cuomo 2000, 116.
[45] Schiefsky 2008.
[46] γεωμετρίας δὲ παιζούσης; Plutarch, *Marcellus* 14.7–8.
[47] Plutarch, *Marcellus* 14.9–11.

and mechanics testify to the fact that no culture was monolithic in its attitude toward any given scientific activity.[48] The range of philosophical backgrounds and professional allegiances of technological practitioners, as well as of the non-technical authors who represented their work in other forms of literature, ensures that any conceptual construction such as "Hellenistic science" conceals a variety of viewpoints on the goals and values of scientific activity in that era.[49] However, I will nevertheless attempt to identify some shifts, for example, between what may be construed as "Hellenistic mechanical literature" and "late ancient Roman mechanical literature," while remaining cognizant of the complexities hiding in each of those constructions.

While literary representations of technology vary as they cross cultural and generic boundaries, at the same time they create stability and continuity for the technical knowledge they help to propagate. If an author conforms to contemporary literary preferences in describing an artifact, he maximizes his chances of increasing popular awareness of it. Beyond enabling the construction of the artifacts themselves, texts describing technological artifacts may also celebrate the wonder they create or the power they embody, which in turn plays a vital role in fueling interest and investment in technology. Texts describing technology are very often more mobile than the artifacts themselves, so that they can spread technology from one place to another.[50] Since texts can bear information not just about a single artifact, but about the development and improvements made to a type of artifact over time, they can contextualize the current state of the art by describing particular stops along the road. Hero's *Belopoeica* provides this kind of historical view of the long-term development of ballistic technology, while Oribasius does the same for medical technologies in his *Collectiones medicae*.

48 τὴν γὰρ ἀγαπωμένην ταύτην καὶ περιβόητον ὀργανικήν; Plutarch, *Marcellus* 14.9.1.

49 The question of the resemblance between philosophical schools and professional associations in antiquity is tricky territory, and the difficulty is multiplied by the relative obscurity of the biographical details of most mechanicians. For comparison with the professional relationships between Roman surveyors, see Cuomo 2007, chs. 4–5. Stückelberger 1994, 99 identifies Hero as being of the "school" of Ctesibius and Philo. While the influence of Ctesibius on Philo is undeniable, and while Hero certainly refers to the example set by Philo in another field in his *Peri automatopoiētikēs*, the identification of a "school" is more dubious. A detailed (if unskeptical) account of the evidence for and persons affiliated with the mechanical "school of Alexandria" may be found in Gille 1980, 54–102.

50 Technological knowledge is of course often propagated by its practitioners, who are likewise mobile. Often both forms of propagation work in tandem: Philo of Byzantium, for example, seems to have journeyed, perhaps by way of Rhodes, to Alexandria where he studied alongside the protégés of Ctesibius, developing knowledge which he then puts into a text dedicated to a recipient elsewhere. On the sketchy details of Philo's biography see Philo 1974, 13–18.

The power of technical ekphrasis to delineate and propagate technological knowledge does not confine authors to their native technological cultures; indeed, a vital role in self-reflective scientific development is played by texts that celebrate (explicitly or implicitly) the technological prowess of others.[51] Ovid describes the flying machine of Daedalus in some detail in his *Metamorphoses*, bringing his reader into contact not only with the idea of a marvelous piece of technology, but also with the world of Hellenistic poetry, in which scientific discoveries and marvels play an especially important role.[52] Ovid's invocation of the "science" of Daedalus' invention signals an encounter between Roman and Hellenistic literary cultures just as much as it does an encounter between poetic and scientific genres. In this case the machine is of course particularly fantastical, but Claudian's late ancient epigram on the artistry of Archimedes' celestial model is similarly layered with the strangeness of the technical artifact and the remoteness of the culture that generated it.[53] Textual representations of mechanical technology create a different type of engagement with the knowledge the artifact encodes than a direct encounter with the artifact itself would: they allow the reader to see the technology through a particular set of eyes, to see how it is "read" within the context of a culture.

Throughout these variations of culture and genre, the fundamental task of the technical ekphrasis remains the same in some essential ways: to bring the reader into imaginary contact with the artifact. An ekphrasis is, at its most basic structural level, a mapping between an object and a text, a verbal encoding of the attributes of an object so the reader can recreate it in his imagination. Mitchell, though focused on ekphrasis as verbal representation of visual representation, evokes a similar process of "conversion" of an object to verbal representation, and "reconversion" into an image in the reader's mind.[54]

The work that follows will consider in detail how this recreation happens, rhetorically and cognitively, as well as the variety of possible relationships between text, object, author, and reader. The object may be real, theoretical, or imaginary; contemporary or historical; it may be familiar or something neither author nor reader has ever experienced. The text may

[51] Journalistic coverage of developments like Sputnik that helped fuel the space race between the U.S. and U.S.S.R. is perhaps a modern example of this.

[52] Ovid, *Metamorphoses* 8.183–95. Appropriate caution about runaway claims about the prominence of science in Hellenistic poetry is provided by Asper 2009, 15–16.

[53] Claudian, *Minora* 51.

[54] Mitchell 1994, 152, 164.

offer a scientific description of the object's functional attributes, a poetic or philosophical meditation on the experience of seeing it, or an historical analysis of its development. The reader may be addressed directly by the text to imagine the object or even to create a new one for himself. This latter possibility is a special feature of texts that provide instructions for building machines. The guidance these texts offer gives new life to the author–reader relationship, turning the author into a hands-on mentor who speaks directly to the reader about what "you" and "we" are doing, encouraging the reader to experience the text actively rather than passively, to imagine a brand new artifact coming together.

Technical ekphraseis, particularly in instructional texts, can also reach beyond the verbal to the pictorial, including diagrams which depict the object in whole or in part, in abstract geometrical form or as the constructed artifact would appear to a viewer. Diagrams serve as a kind of middle ground between the object itself and its more abstract verbal depiction in the text; they may even be used as templates for construction, further bridging the gap between the imagined and constructed object.

This book will follow these issues through a group of texts that range broadly in time, culture, and genre. It might be asked what advantages such an approach offers over, for example, an investigation limited to the work of Vitruvius or Philo. Such tightly focused studies are invaluable to the study of ancient science and technology, and this book is itself of course deeply indebted to many of them, from Drachmann's study of the pneumatic works of Philo and Hero to Creese's work on the monochord (and so many in between). This is a different kind of enterprise, a study of the fundamentally *literary* problem of how ancient authors grappled with the challenges of representing many different varieties of mechanical technology in textual form. Discovering the particular features of Hellenistic technical ekphrasis (for example) requires a broad-ranging investigation that allows it to be compared to the way later literary cultures (which were nevertheless part of a contiguous literary tradition) undertook substantially the same task.

By "substantially the same task" I mean that from Hellenistic Greece to late antiquity, designers and craftsmen worked on a constantly developing family of technologies, the branches of which (such as pneumatic devices, traction devices, and ballistic devices) remained fairly stable, as did their means of production in the absence of radical modern developments like assembly lines and division of labor. The knowledge-propagation infrastructure required to develop these technologies also changed relatively slowly, as mechanicians passed down their knowledge through text or through some kind of comparatively informal apprenticeship process, in the absence of the formal technical schools that would only appear in Europe in the sixteenth century.[55]

Nor did the "literary technologies" that helped to propagate mechanical knowledge change drastically. The texts bearing these descriptions continued to be hand copied on papyrus or parchment, without the potential for wide distribution that the printing press made possible, which was such a vital part of the scientific developments of the seventeenth century.[56] The technology of diagrams had not yet undergone radical changes such as the emergence of combined orthographic projections and mechanical reproduction.[57] The rise of the codex effected certain changes in the prevailing style of illustrations, but the surviving manuscripts show that the forms of illustration Weitzmann characterizes as "the papyrus style" often persist in codex copies of mechanical works.[58] The result is a combination of mechanical and literary technologies that can with some justification be called a continuous tradition of technical ekphrasis, which began in earnest in Hellenistic Greece and did not lapse until late antiquity.

Some remarks on what this book is not attempting to do are in order here. It is not an account of technological products and practices, but of how technological artifacts (which I will also refer to as mechanical artifacts) were represented in Greek and Roman literature. Its territory is thus even narrower than that established by Nicolet for a volume on Roman technical literature, in which he declares that the volume's focus excludes "les textes proprement scientifiques" such as mathematical texts, and even "scientific" texts with a more markedly "technical" character, as well as medicine, veterinary medicine, and pharmacology (which, he notes quite rightly, have their own copious secondary literature).[59] "Scientific" subjects will not be the primary concern here either, though occasional reference will be made to some of them when a comparative approach may be illuminating. For example, the mathematical problem of doubling the cube was treated by many different authors, including Hero, Apollonius, and

[55] This is not, of course, to say that there were no changes in technical education from the Hellenistic era to late antiquity; on this subject see Meißner 2003, 160–1. There was a change in the social *structures* of information transmission (say, from intra- to extra-familial), but while texts seem to have become more common vehicles for transmitting technical knowledge, no great shifts in the available literary and instructional technologies themselves are evident.

[56] On this claim in particular, see Hall 1983.

[57] On the revolutionary changes brought about by the advent of printing technologies enabling "exactly repeatable pictorial statements," especially in technical illustration, see Ivins 1969. For a cautionary note on how "exactly repeatable" they were, see Kusukawa 2011, 64–81.

[58] On these changes see, canonically, Weitzmann 1947, 47–57, 81–112.

[59] Nicolet 1996, 7.

Diocles. Eratosthenes' solution pays particular attention to the mechanical tool known as the mesolabe, which was developed to solve this problem. His response to the problem, which presents this mechanical tool through fable, epigram, and geometrical proof, will therefore be treated here, whereas the purely mathematical approaches will not.

This project is thus situated on the sidelines of the vigorous discussion of the literary aspects of Greek "scientific knowledge" more generally considered that has been taking place over the last few decades,[60] and comparable considerations on the Roman side which are now emerging.[61] However, mechanical artifact texts do often exhibit some of the features that such studies point out in texts from more centrally "scientific" disciplines: prefaces defending the value of the knowledge contained in the text, authors who insert themselves as experts into the text itself, and sometimes a concern for problems of natural, logical, and ethical philosophy.

However, the overlap is far from complete. Mechanical artifacts also engage with a set of concerns that have no analogue in many "scientific" texts: the craftsman's tacit knowledge, rules of thumb for defining appropriate building materials and dimensions, assessment of individual machines without seeking to generate universal principles from particular observations, and an emphasis on the material, hands-on aspects of the craft process. Ancient mechanical artifacts are thus connected to a set of bodies of knowledge that corresponds only partially to the set associated with medicine, or that associated with mathematics.

The following chapters fall into three pairs. The first pair establishes the background of the problem, and might be skimmed or skipped by readers primarily interested in the nuts and bolts of technical ekphrasis. Chapter 1 introduces the textual passages as a group, discussing the range of genres in which the passages are found (as well as some of the problems of defining such a range), and the types of mechanical artifacts found in ancient literature. Chapter 2 provides a synoptic overview of some of the changes in technical practices that in turn effected changes in the textual representation of these artifacts from culture to culture.

The second pair explores the textual tools available to ancient authors in shaping the reader's encounter with mechanical artifacts. Chapter 3 connects the strategies used in ekphraseis on technical artifacts with

[60] Highlights of this tradition include Fuhrmann 1960; Deinlein 1975; Lloyd 1987; Sorabji 1988; Taub 1993; Vázquez Buján 1994; Clarke 1999; Netz 1999; Cuomo 2000; Tybjerg 2003; Asper 2007; Lehoux 2007; Berryman 2009; Taub and Doody 2009.

[61] For example, Fleury 1990; Beagon 1992; Nicolet and Gros 1996; Jaeger 2008; König and Whitmarsh 2007b. Lehoux 2012 offers some tantalizing suggestions of where such studies might go.

developments in the surrounding rhetorical and literary culture. Chapter 4 expands this discussion to include visual representations that sometimes supplement the verbal description.

The final two chapters inquire more deeply into the special characteristics of technical ekphraseis, treating these artifacts not just as physical objects, but as embodiments of knowledge. Chapter 5 explores the relationship between description and instruction, devoting particular attention to the verbal techniques that create different relationships between author and reader, ranging from a sense of a shared, hands-on workspace to a disengaged approach where the artifact seems to come together without human intervention. Chapter 6 examines structural techniques that foreground the working relationship between knowledge and machine, whether focused on the different bodies of knowledge required to design and construct a machine or the external knowledge to which certain types of artifact (including astronomical, harmonic, surveying, and calculating devices) provide access.

CHAPTER 1

Genres: Texts and Artifacts

Introduction

In the analysis of technical ekphrasis that follows, I have aimed to capture the varieties of description rather than collecting every ancient description of a machine, something more like a specimen cabinet than a warehouse. Major points of variation include the degree of technical detail, the proportion of the text the passage takes up, and even the physical relationship of the text to the object described. A single text may contain several passages of technical ekphrasis. For example, Biton's *Kataskeuai polemikōn organōn kai katapaltikōn* describes several defensive and offensive siege engines, one right after another. In cases such as this, I have not included every passage from the text, but rather chosen a range that demonstrates the spectrum of descriptive techniques. The resulting collection explores how engagement with the knowledge embodied by technological artifacts was reflected throughout the literary world of the ancient Mediterranean, over a wide range of genres and time periods. The earliest authors whose descriptions of technological artifacts survive are probably Philo and Biton, both of whom are thought to have written in the third or second centuries BCE, though considerable uncertainty remains in these authors' dates.[1] The latest whose date we can be sure of is Cassiodorus, who wrote in the sixth century CE.

[1] Marsden argues that the king Attalus to whom Biton addresses his work is Attalus I because he does not include any descriptions of the torsion engines that appeared after (but did not replace) the earlier tension engines (Marsden 1971, 5–6, 78). However, we have no reason to think that it was standard practice to describe only state-of-the-art machines; Hero's own *Belopoeica*, for example, describes engines available in Ctesibius' time. Lewis argues that the addressee was Attalus II (Lewis 1999). This claim rests on the basis of some technical terminology, the absence of torsion engines, and Biton's reference to the "enemies" of Pergamum, which Lewis identifies as Prusias II of Bithynia, who invaded Pergamum in 156 BCE. Rihll completes the set of possibilities by arguing, largely based on the likely identities of the exemplary designers that Biton lists, that the work is most likely addressed to Attalus III (Rihll 2007, 164–9). Meißner writes first (161) of Attalus I, and later (247) of II and III (Meißner 1999).

Some come from texts that might be labeled "technical," but this classification raises many questions. A "technical" text may provide some historical background for the machines it describes, but this will not be its primary concern. Even texts offering instructions that could guide the reader in constructing a new artifact may look back to antecedent models, but this does not impel us to classify such texts as histories. Consider, for example, Biton's description of a certain catapult:

> This is the stone-thrower constructed in Rhodes by Charon of Magnesia. It has straight beams AB. And so let these have a length of 6 feet, and a width and depth of ½ foot. Then [let there be] two parallel beams ΓΓ joined to the first, having length 3½ feet and width 1 foot. These will be plated with iron or bronze at the ends of the beams, over a length not less than 2 feet, so that they will not break under force.[2]

Biton matches each device he describes with a famous antecedent, but it is the physical, structural detail rather than the historical background that takes center stage. The artifacts' mechanical characteristics organize the text, while references to the place or personnel of their invention serve only as a quick introductory statement. At the end of this passage (and, formulaically, each such passage), Biton refers to a diagram that illustrates the machine in question. Mechanical components are given letter labels in the text that correspond to the diagram; the visual element is not merely decorative but is intrinsic to the description. So, while this text locates the artifacts it describes historically, no one would classify it as belonging to the genre of history. Let us tentatively place it in a box labeled "technical literature on mechanical artifacts."

Determining what else that box might contain and what divisions exist within it is a complex process that must be approached gradually. The instructions Biton delivers to his reader seem likely to be part of the reason his text fits in the box, but a "technical" passage of text need not include directives of any specific form. Vitruvius, for example, occasionally uses jussive subjunctives that direct the reader to perform construction activities, but much more often uses passive indicative verbs to show the structure of an artifact. The water organ is described in this way:

> Upon the base rods are set up on the right and left, constructed like a ladder; between these are enclosed bronze cylinders, with movable bottoms holding iron piston rods fixed in the middle, carefully crafted on a lathe. These are connected to the levers with pins, and wrapped in fleecy hides.[3]

[2] Biton, *Kataskeuai* 2.3–10. [3] Vitr. *De arch.* 10.8.1.

Vitruvius includes information about the dimensions and materials of these components of the water organ, about how they are put together and how they work. Despite its differences from Biton's instructional text, this passage also seems to fit in the "technical literature on mechanical artifacts" box.

Other descriptions of mechanical artifacts do not fit the box so well. Claudian, for example, captures the marvelous impression made by Archimedes' astronomical *sphaera* in this epigram:

> Jupiter, when he perceived the heavens in glassy miniature,
> laughed, and delivered words such as these to the gods:
> "Has the power of mortal zeal come so far?
> Is my own work now mimicked in a fragile sphere?
> The rules of heaven, the order of things, and the laws of the days –
> look, an old man of Syracuse has produced them with his art.
> Some spirit enclosed there attends the varied stars
> and spurs on the lively work with sure motions.
> The model zodiac runs through a year of its own,
> and a simulated moon returns after each new month.
> And now, this bold diligence, revolving its own world,
> rejoices, and rules the stars with human mind.
> Why do I marvel at harmless Salmoneus for his false thunder?
> One small hand has been found as nature's rival."[4]

The marvel and mystery of this intricate device are here seen through the eyes of Jupiter himself, who is puzzled by the workings of the machine, crediting its ability to model the motions of heavenly bodies to an *inclusus spiritus*. The wonder of the *sphaera*, as Jaeger observes, is connected with the conceptual jump in scale that Claudian highlights – from Jupiter to human, from the whole cosmos to the work of one hand.[5] Jaeger focuses as well on the insularity of Claudian's *sphaera*: he puts us on the outside, looking in.[6] This kind of mystifying description is very different from the approaches of Biton and Vitruvius, who lay out the machine's parts, dimensions, and functions plainly before the reader from the "inside out," showing how they merge into the completed artifact. For many reasons, including but not limited to the expectations of genre, these authors have created three very different encounters between the reader and the device being described.

[4] Claudian, *Minora* 51 (68). [5] Jaeger 2008, 125.

[6] This reading is strengthened if one reads *clauditur* for *luditur* in line 4. Jaeger notes as well that "several of Claudian's other short poems offer either a look into a closed system or a view of a figure looking into one," citing *Carmina minora* 38–9 on crystals enclosing water (2008, 125).

Some evidence does not come from separate texts at all, but is inscribed on the artifacts themselves. The most remarkable of these inscribed devices to survive is the Antikythera mechanism, whose inscriptions explain not only the cosmological phenomena the device appears to have imitated mechanically, but also the mechanics of the device itself, along with a calendar and a series of eclipse predictions.[7] Because this book focuses on textual descriptions of technological artifacts, the only material objects to be considered here are those, like the Antikythera mechanism, inscribed with text that describes the structure, construction, or use of the artifact itself.

This constraint, of course, means ignoring a tremendous amount of material evidence which could produce (and already has, in others' hands) valuable insight into ancient technologies, as well as the everyday lives, work, tools, and products of ancient craftsmen.[8] Funerary monuments, of course, provide a rich and varied store of evidence. Cuomo has made a thorough study of the funerary imagery associated with carpenters in the Roman Empire, mining the portrayals of human figures on these monuments for information about the social status of carpenters, as well as examining the depictions of the tools of their trade as a kind of insignia.[9] The tomb of C. Vedennius Moderatus[10] bears information both about the personnel involved in military mechanics (an inscription identifying the role of *arcitect(us) armament(arii) imp(eratoris)*) and the products

[7] For background on this fascinating device, see especially de Solla Price 1974; Wright 2003; Wright 2005; Freeth et al. 2006; Freeth et al. 2008; Wolkenhauer 2011, 165–8; Freeth and Jones 2012.

[8] Oleson 1986 contains a wealth of information on the surviving archaeological evidence; Oleson 2004 suggests how that archaeological evidence could be used to make claims about the literary evidence that no longer survives. Stückelberger 1994, 102, briefly notes that the manuscript drawings of Hero's water organ can be supplemented by surviving material evidence such as a mosaic depicting an organ found at Nennig. The 1931 discovery of the components of a water organ at the former Roman settlement at Aquincum is discussed in more detail by Hyde 1938. Several useful analyses of the scanty material evidence that remains of catapults are collected in Baatz 1994. Other works combine material and literary evidence to enrich their histories of specific technology types; those particularly relevant to this project include Fleury 1996; Lewis 2001; Damerow et al. 2002; Rihll 2007. Material evidence is, in short, an extremely valuable resource for studying the technological life of the ancient Mediterranean. This project, however, is concerned expressly with the literary reflection of this technological life, and how it compares with the literary reflection of other kinds of intellectual endeavors. On the material life of technological texts in the ancient world, see for example Gara 1994, 107–9.

[9] Cuomo 2007, 77–81, 92–8. Cuomo draws heavily on Zimmer's extensive collection of representations of Greek and Roman craftsmen (Zimmer 1982, particularly 161–76). More analytical reflection on evidence such as that considered by Zimmer, including questions of artistic intent, can be found in Ulrich 2008.

[10] On Vedennius, see Baatz 1994, 134, 193, 202.

of their work (a detailed engraving of a catapult of a type mentioned by Vitruvius).[11] Meißner adduces engravings of book rolls and medical instruments on the funeral stelae of doctors as evidence for the increasing importance of text to medicine.[12] Surveyors also commemorate the tools of their trade on their funerary monuments, sometimes alongside verbal inscriptions.[13] Other monuments include imagery of mechanical artifacts, such as the magnificent lifting machine depicted on the sarcophagus of the Haterii.[14] Trajan's column depicts military equipment of many kinds including ballistic machinery, notably what appears to be a *cheiroballistra*,[15] while a votive relief from Capua shows a lifting machine with a pulley system.[16]

Images like these do depict technological artifacts, and many of them also include certain kinds of inscriptions. However, they are not accompanied by any kind of verbal description, instructions, or other inscription relating to the form, construction, or use of the artifacts themselves. They rather reflect the fact observed by Cuomo, that "the 'materiality' rather than literariness of technology" was of foremost importance in the ancient world.[17] This was doubtless true, as it doubtless is today, but this project is concerned precisely with *literary* representations of artifacts, so this evidence will not come into play here. Papyri likewise survive that might be considered material evidence in their own right, like the training contracts for various technical disciplines that number among the Oxyrhynchus papyri.[18] These contracts are certainly an important part of the evidence for the practice of these technical disciplines, and indeed themselves counted among the tools of the trade, but they are not concerned with the task of describing the artifacts used in those trades.

For similar reasons this book will not focus on the archaeological remains of mechanical artifacts. These include a wide range of artifacts (or at least their components), from the metal bushings of catapults to astronomical instruments like planispheres and hemispheres.[19] Surviving

11 ILS 2034; see also Baatz 1994, 196–202; Fleury 1996, 54–5.

12 Meißner 2003, 159. Zimmer 1982 gives some examples of such monuments at 13 n. 70, 15 n. 74. On the changing significance of text-borne knowledge in ancient medicine, see Dean-Jones 2003.

13 Some such images can be found in Chouquer and Favory 2001, 288–94; more in Zimmer 1982, 196–8.

14 Zimmer 1982, 160–1.

15 Marsden 1971, 209–10.

16 Zimmer 1982, 159.

17 Cuomo 2007, 99.

18 For further discussion of these see Meißner 2003, 160–1.

19 On the former, see for example Baatz 1994; Cuomo 2007; Rihll 2007. On the latter, see for example Gibbs 1976; Drachmann 1954; Evans 1999; Taub 2002; Evans 2004.

examples of these latter instruments depict the constellations of the Greco-Roman and Egyptian astronomical systems, and the inscriptions on these artifacts were undeniably relevant to the practice of a technical discipline.[20] Unlike, for example, the funerary monuments described earlier, these artifacts were tools of their disciplines, and are in that sense perhaps more closely linked to the components of disciplinary knowledge that elsewhere might take a verbal form. However, they do not include inscriptions pertinent to the artifact itself: the planisphere does not tell the viewer what it consists of or how to use it. Such artifacts can often tell the viewer a great deal about the practice of the relevant technical disciplines, but they do not participate in the surviving literary discourse about technological artifacts.

The elaborately inscribed Antikythera mechanism is the most prominent counterexample to these "mute" artifacts. More commonplace instruments like sundials and *parapēgmata* likewise often bear such self-referential inscriptions, which are designed to help the artifact's user figure out what to do with it.[21] By contrast, the "mute" artifacts do not engage in the task of describing themselves or any other artifact for the reader. Priceless as they are on their own merits, they are not relevant to the goal here, which is to understand how authors represent technological artifacts in text.

The illustrations and diagrams that sometimes accompany textual descriptions of artifacts will be useful to this analysis even when they do not include letter or word labels of their own. The crucial difference here can be grasped immediately by comparing a bronze planisphere to a diagram of a planisphere.[22] It is the difference between a "mute" artifact and a visual representation of a comparable artifact embedded in and engaged with a verbal discourse concerning its invention, cultural significance, construction, or use. In short, a wealth of material evidence survives which provides inestimably valuable information about the manufacture and use of technological artifacts in the ancient world. However, this investigation is concerned very specifically with the *literary* reflection of those artifacts' design, construction, and use: what Jaeger calls their "associated discourses."[23]

[20] For these see Stückelberger 1994, 34–42.

[21] The most complete guides to these inscriptions are (on sundials) Gibbs 1976; see also its supplement in Catamo et al. 2000; (on *parapēgmata*) Lehoux 2007.

[22] For example, consider respectively the illustrations at Stückelberger 1994, 40, 43.

[23] Jaeger 2008, 56.

The missing evidence

The outlines of the evidence will be sharpened by a note on what is lost. Some texts are known to be missing from the traces they have left in extant authors. The Hellenistic engineer Ctesibius is credited by many authors (including Philo, Hero, and Vitruvius) with a host of mechanical innovations, but none of his work survives even in fragmentary form. We have likewise lost the treatises by the military engineers who served Philip and Alexander: Polyidus, Charias, and Diades are all mentioned by later authors, but nothing remains of their own work. Nor have the extant authors survived entire. Of Philo's *Mēchanikē syntaxis*, originally nine books, three survive in Greek and one in Latin and Arabic.[24] Archimedes is said by Proclus to have written a work (now lost) on the construction of spheres "for imitating the motions of heavenly bodies."[25] Several works by Hero, including one on water clocks and one on balances, are known only by name.[26] His *Cheiroballistra* survives only as a fragment that Schneider once identified as a piece of an engineering dictionary for the letter Κ rather than a text on a particular piece of machinery.[27]

It is possible that among the missing works were something like what we now think of as technical manuals: bare-bones descriptions of the design, assembly, and operation of machinery, written by specialists for specialists, builders, or users. No such works survive, but they were so important to the propagation of technical knowledge in later periods that there has been much speculation about their existence in the ancient world as well. Oleson argues for the likely existence of such a tradition of "subliterary" building manuals among the Romans, even though no examples survive.[28] If indeed they existed, these texts could have been purely verbal, or they could have combined verbal information

24 In Greek: the *Belopoeica*, *Parasceuastica*, and *Poliorcetica*; in Latin and Arabic, the *Pneumatica*.

25 Proclus, *In primum Euclidis elementorum librum commentarii* 41.16.

26 Tybjerg 2000, introduction.

27 Schneider 1906. It is a remarkable accident that all the parts of this hand-held catapult named in the surviving part of the text begin with the letter κ, but as Marsden shows, together they form a coherent subassembly of the device (Marsden 1971, 206).

28 Oleson 2004. Oleson chooses for his case study the woodblock pumps found at Roman sites in Britain, France, Germany, Italy, and Bulgaria. He argues that the intricate mechanisms of these devices (which were not so technologically glamorous as the bronze pumps described in works like the *De architectura* of Vitruvius, but were a more practical basic technology for household use) are too similar to be the accidental products of separate technical evolutions of the bronze pumps. On this basis he argues that the most likely explanation for the proliferation of these woodblock pumps was the circulation among Roman craftsmen of some kind of written manual for designing and constructing them.

with design drawings, graphical templates for components, or tables to help with calculations along the lines of those in the surviving mathematical papyri, or other useful tools. Traveling craftsmen would certainly have carried technical knowledge with them, perhaps without texts ever having been involved. Given the lack of surviving evidence, we simply cannot know.

In the absence of those texts, we can only make inferences from the surviving literary sources, and from material evidence such as the engravings of plans and building components that seem to have been used in ancient architecture.[29] Their form and content might have differed considerably from the literary texts that do survive; indeed they may never have existed at all and the mobile source of knowledge in Oleson's transmission model could simply have been trained workmen. While textual references like those Apollodorus gives in the preface to his *Poliorcetica* testify unmistakably to the fact that humans served as vectors of technical knowledge in the ancient Mediterranean, the knowledge they carried around with them is largely inaccessible to us.

Genres of texts

While the information transmitted from one ancient technological practitioner to another through imitation or oral instruction is not directly available to us, some of the surviving literary texts hint at the forms it might have taken. Vitruvius, for example, refers both to sources he describes as *antiqui*, which must have been written, and to *praeceptores* who might have transmitted information through any combination of oral and written methods. From this Gros concludes that the *De architectura* could not have come into being without recourse to a "littérature sous-jacente,"[30] which, though deemed by Vitruvius inappropriate to be included in a text for an imperial addressee, nevertheless provided much of the information upon which his more polished work relies. Gros labels the various textual resources that Vitruvius is likely to have drawn on as *manualetti*, *excerpta*, summaries, doxographies, bilingual lexica, and collections of models.[31] These forms are united by their summary nature, like our "handbooks" for practitioners: works densely packed with practical information but light on theory and other discursive elements. At the same time, however,

[29] On these see in particular Heisel 1993, 154–218; Gros 1996.
[30] Gros 1996, 35–6.
[31] Gros 2006, 428; see also Gros 1996, 42 for some of the controversies surrounding this material.

Vitruvius is lavish in his references to philosophical, historical, poetic, and otherwise "literary" authors, so we must not fall into the trap of thinking that ancient "technical" works did not draw quite broadly from existing literary sources.

The distinction Vitruvius draws between *antiqui* and *praeceptores*, and Gros's further subdivision of different types of texts Vitruvius might have referred to under the rubric of written sources, hints at a strategy commonly used to differentiate "technical" texts from one another. Callebat argues that in the literary culture Vitruvius worked in, texts on the *artes* tended to fall into one of two categories. On the one hand were texts written for specialists, presenting technical information without striving for aesthetic appeal (Callebat gives these the label *commentarii*).[32] On the other hand were works composed for a more general audience, aiming to propagate "moins une *compétence* qu'une *connaissance*" with the subject matter.[33] Callebat is careful to note that this *connaissance* could certainly involve practical subject matter, but that in any case these works were possessed of certain literary pretensions. They would also display an "intention théorisante," manifested in textual connections to other disciplines.

It is impossible to judge the truth of Callebat's assertions about the literary character (or lack thereof) of the *commentarii*, pertaining as they do to a mostly lost genre. More important here is the idea that works in the second category (including the *De architectura*) can be oriented toward the dissemination of practical knowledge even as they participate in the discourse of other types of knowledge. Sallmann similarly argues that Vitruvius invokes a spectrum of ancillary disciplines, not as mere ornament, but because those connections enhance the stature of architecture as a discipline and Vitruvius himself as its profiler.[34] For Vitruvius, these satellite discourses include the scientific theories he infuses into the study of architecture, the historical vignettes he uses to contextualize that work, and even the literary and rhetorical refinement that shapes the work itself. The resulting text might rely heavily on technical knowledge that could also have been propagated in some lost form of "subliterary" handbook, but the text itself takes a very different form.

We must likewise be mindful of the anachronistic associations of modern "textbooks" with institutions of technical education very

[32] Callebat 1994, 31. Compare Sallmann 1984, 18.
[33] Callebat 1994, 32. [34] Sallmann 1984, 14–15, 19–21.

different from those that seem to have been available in antiquity.[35] It is quite possible that some of the types of works now labeled "technical" were principally designed to meet a set of expectations quite different from those associated with what we call technical "manuals" or "handbooks" today. Bruun warns of Frontinus' text, for example, that even if it is in fact the kind of text we might label a technical "handbook," nevertheless it is quite possible that in antiquity "the writer of such a handbook worked differently from the way in which he would today."[36] Indeed, given the differences in technical education and propagation of texts between antiquity and today, it is difficult to imagine not finding these distinctions.

If ancient labels for types of texts are any indication of the categories recognized by authors and readers, the ancient landscape of "technical literature" was far from homogeneous. Fögen lists, for example, forms labeled *eisagōgē*, *synagōgē*, *encheiridion*, *hypomnēma*, *technē*, *pragmateia*, *epitomē*, *diēgēsis*, and *synopsis*, and alludes as well to their Latin equivalents, moreover warning that they can be difficult to differentiate.[37] Even the function these labels performed remains unclear; as van der Eijk has observed for a different subset of technical texts, "in some cases it is unclear whether a certain label is intended to serve as a title or as characterization of the form."[38] Stückelberger identifies a range of structures including disordered collections of materials rather like notes, the systematically organized "Lehrbuch," polemical treatises, single-subject monographs, accessible introductory texts and highly specialized advanced texts, works shaped for pleasant reading and daunting tables of numbers.[39] The intended audience, the principles and degree of organization, the voice the author adopts toward the reader, the stance the author takes toward other authors or practitioners, the balance of verbal and extra-verbal elements (such as numerical tables or diagrams) in the text – these are just some of the possible variations. Even in the limited territory of mechanical artifact texts, a clear-cut and unambiguous subset of "technical" texts is clearly not to be hoped for.

The spectrum of meanings associated with the term *commentarius* can give some idea of the range of texts at issue here. As mentioned previously, Callebat uses it to label works intended to communicate bare-bones technical knowledge to specialists. Saastamoinen observes

[35] Meißner 2003, 153–4, 159–67, 171–2.
[36] Bruun 1991, 15. [37] Fögen 2003, 34.
[38] van der Eijk 1997, 90. [39] Stückelberger 1998, 287.

that even though "the term *commentarius* – unlike the term *isagoga* – was frequently used by ancient writers themselves to describe their works," the genre itself was not well defined.[40] *Commentarii* could be rough or polished, include visual as well as verbal elements, and seem to have been addressed to a wide range of audiences. Nicolet has collected references to information-bearing objects, both verbal and graphical, all labeled *commentarii* by Roman authors.[41] The word can refer to memoranda or illustrated instructions, but most often denotes a text designed to accompany an object; the best known of these was written by Agrippa to accompany what seems to have been a map.

To find a way through this maze of different readings, it may help to consider some possible differences between the expectations for *commentarii* in antiquity and those for technical manuals today. Frontinus' *De aquae ductu urbis Romae* highlights the complications: if we take *commentarius* to mean anything like "technical manual," his rhetorically elaborate, multifaceted text is by and large a poor match for its modern analogues. At the same time, the work does undeniably contain a large amount of "technical" information about the water system of Rome: measurements of length and the breadth of pipes, notes on the materials used, and so forth.

This unusual combination has inspired a host of different opinions about the purpose of the text.[42] Goodyear claims that "the *De aquis* is exactly what it claims to be, a systematic account of the water-supply of Rome,"[43] and Hodge characterizes Frontinus as the technical author par excellence.[44] Bruun, on the other hand, suggests that "the mere fact that he decided to write down certain things is an expression of a moral inclination," even though it may be difficult to establish exactly what Frontinus' political intentions were.[45] Peachin claims that the work cannot be considered a technical manual on account of its incomplete treatment of the topic, and that it was instead a political pamphlet, aimed at a broad audience rather than Frontinus' successor as *curator aquarum*. This characterization is based in part on the argument that "the book provides nothing

[40] Saastamoinen 2003. As well as offering a capsule history of the controversies surrounding Frontinus' work, this chapter also contains helpful etymological and historical background to the term.

[41] Nicolet 1991, 101–14; McEwen 2003, 18–25. See also Vitr. *De arch.* 9.pr.14, 1.1.4, 7.pr.11, and 7.pr.17, as well as Pliny, *NH* 17.243.

[42] Peachin's introduction is a thorough guide to the controversy, though strongly committed to the view that the work is a political pamphlet rather than a technical work or panegyric (Peachin 2004). Other thorough and thoughtful discussions of the available viewpoints include König 2007, 177–9; Saastamoinen 2003, 25–30.

[43] Goodyear, "Technical Writing," in Kenney and Clausen 1996, vol. II, 672 n. 2.

[44] Hodge 1991, 61. [45] Bruun 1991, 15.

like a comprehensive treatment of Rome's water supply, or its management," as it omits information on important technical topics such as how to remove calcium carbonate deposits.[46]

To these arguments DeLaine counters, following Bruun, that modern expectations of completeness should not be imposed on the ancient category of the *commentarius*, and that in any case modern technical manuals are frequently lacking in completeness as well.[47] Indeed, we could argue that it is in fact *incompleteness* which would be the marker of a genuine technical manual designed strictly for the use of specialists who have obtained much of the necessary basic or tacit knowledge through other channels, such as direct imitative contact with more advanced practitioners.

It seems likely that *commentarii* as a class were not intended exclusively for the use of workmen and technicians, but were rather, as DeLaine suggests,

> a set of instructions or the text accompanying a thing to be used: that is, in essence, a written aid to those engaged in practical occupations, whether magistrates or architects, so that "administrative handbook" is a reasonable translation for any *commentarius* which serves a public office or concerns a public facility.[48]

In this view *commentarii* (or at least some of them) were written with practical ends in mind, but were principally concerned with the practicalities of the administrator rather than the technician or builder.[49] The debate over the *De aquae ductu urbis Romae* raises the possibility that *commentarii* were largely distinguished by the practical information they contained, but that this information could be about administration as well as about building water systems or catapults. Consequently, we should not be surprised to find *commentarii* packaged within other types of texts – indeed, such texts are perhaps more likely to survive, as the appealing packaging might have helped propagate the text. In short, *commentarius* seems not to have been a term that served as a fixed actors' category – not because it fails to appear as a label for texts, but because those texts seem to be so multifarious. While every ancient label was surely not as

46 Peachin 2004, 2. 47 Bruun 1991, 15; DeLaine 1996, 118–19.

48 DeLaine 1996, 118–21.

49 This connection between the *commentarius* and the texts used by Roman administrative authorities is also made in Bömer 1953, 214–15; Bruun 1991, 17–18. The so-called *Gnomon of the Idios Logos*, a kind of handbook meant to guide the Roman official in Egypt who oversaw this account, is a particularly well-studied example; see Riccobono 1950; *contra*, Lenel 1920.

problematic as the *commentarius* (the *eisagōgē*, for example, was apparently at least somewhat more tractable), we must use caution before accepting any of these ancient labels at face value.

Modern category labels have also been adduced to distinguish technical texts by their level of literary ambition. Fuhrmann defines the "Lehrbuch" as a stylistically unpretentious collection of information intended to transmit basic knowledge to a young or lay audience, and Deinlein suggests that this form matches up well with the attributes cited by ancient authors as belonging to eisagogic literature.[50] Deinlein similarly classifies three types of non-fictional literature in terms of how their level of literary style might define their intended audience: works, like the "Lehrbuch," that foreground the technical content and are aimed only at practitioners; forms designed to appeal to an educated general audience, like the didactic epistle or dialogue; and the intermediate category that Deinlein labels the "Sachbuch."[51] As we have already seen for the case of Vitruvius, Sallmann elaborates more generally on the potential of the "Sachbuch" to raise the profile of a technical discipline in the broader cultural landscape, to shape its public perception, or to enhance its author's reputation, through a style that combines "literary" and "technical" components.[52] Von Albrecht differentiates technical works from "works of fact" on the grounds that the latter "take pains to establish appealing literary form."[53]

As straightforward as this sorting scheme appears on the surface, determining what constitutes literary style presents a serious challenge. Once again Frontinus' text highlights some of the problems. DeLaine concludes that within the *De aquae ductu* "there was both a *commentarius* and a speech."[54] However, the division between these two components might have been structured counter to modern assumptions about the stylistic register of technical writing. DeLaine argues that, in fact, the long list of *fistula* dimensions for Rome's water systems belongs to the "speech" component, serving as a rhetorical performance to cement the work (and Frontinus himself) as a source of administrative authority. Saastamoinen, on the other hand, argues that such a list is devoid of literary pretensions and that the work can therefore not be considered a speech. Clearly, it is not easy to define what it is to be "stylistically unpretentious."

50 Fuhrmann 1960, 8, 34–41; Deinlein 1975, 2–4; Asper 2007, 214–314.

51 Deinlein 1975, 5–6.

52 Sallmann 1984, 13–15.

53 Albrecht and Schmeling 1997, 566; see also Saastamoinen 2003, 16–25.

54 DeLaine 1996, 134.

Furthermore, claims that an author shaped a text with a certain *literary* value in mind must be differentiated from claims about shaping a text for *cognitive* or *informational* value. Classifications of texts by their degree of literary ambition or development, like those mentioned earlier, frequently blur the distinction between writing that is designed to transmit information as clearly as possible and a maximally simple style, arguing that any elaboration is a kind of literary posturing and a departure from the "technical." This approach seems to rely on the idea that texts containing technical subject matter must be directed to one of two audiences: fellow technical practitioners who care only for a text's content and not its form, or a more generalized but learned audience that demands literary style and rhetorical polish and may not care very much about the depth of information presented. In fact, the distinction is not so straightforward: in particular, authors of otherwise "technical" texts often make use of elaborate descriptive techniques familiar from other, more straightforwardly "literary" genres of text when the goal is to put a clear picture of a technological artifact into a reader's mind.

The use of "technical vocabulary" is sometimes suggested as a feature that should confer membership among "technical texts." Some of this work comes out of modern linguistic studies of technical writing;[55] some comes from a context more closely bound to the use of specialized languages in the ancient world.[56] Fleury argues that while Vegetius should not be considered among the mechanicians, he is nonetheless a technical author. This status is intrinsically connected to his use of technical vocabulary: Vegetius is a technical author if and only if he uses precise technical terms.[57]

Defining technical vocabulary is itself, of course, a non-trivial task; Schiefsky provides a useful working definition in suggesting that "a term or phrase qualifies as a technical term if there is good reason to think that it was used in a reasonably standardized way by practitioners of a given *technē* to refer to objects, concepts, or procedures connected with that *technē*."[58] This definition has the advantage of not putting too many strictures on terminology; it acquires its status gradually in a context of use and consensus by practitioners, but a term's use by non-practitioners

[55] Notable among these studies are Sager and Dungworth 1980; Herbert 1965.

[56] Wide-ranging studies of ancient technical vocabularies include Janson 1964; Krenkel 2003; Fögen 2003; Fögen 2004; Fögen 2009, 9–25, 34–48, 61–4, 92–105. Studies focused on the technical vocabularies of particular fields include Langslow 2000; Fögen 2003; Schiefsky 2005; Conso, Gonzalès, and Guillaumin 2006; Schironi 2010; Dunsch 2012.

[57] Fleury 1996, 71.

[58] Schiefsky 2005, 253.

does nothing to disqualify it. Fögen argues that "technical texts" and the "technical language" in which they are written can to some extent be differentiated from "literary language." "Technical language" in this model is distinguished by being directed primarily toward the transmission of technical content, without any claims to add aesthetic or linguistically innovative value.[59] Yet as this definition of technical language in terms of an author's intentions (which are rarely or never clearly discernible) shows, this differentiation is far from cut and dried. Langslow describes "technical languages" as "*varieties* of a language with their own history," which may overlap with the "standard language" even though they tend toward application in a certain subject area, or among a certain group of practitioners.[60]

Some authors reflect explicitly on their use of technical vocabulary and how this might affect the reader's experience, though Fögen perhaps goes too far with the claim that practically every ancient technical text contains such "metasprachlicher Reflexionen" on the elements of language and style that characterize such texts.[61] As Vitruvius notes, the specialized vocabulary that often allows technical artifacts to be described with maximum precision may have an alienating effect on the lay reader, so the presence, absence, and marking of such vocabulary can be an indicator of the audience for which a text is intended.[62] However, the presence of technical vocabulary in a text should not be assumed to represent a conscious attempt on the author's part to render the text incomprehensible to all but a few fellow experts. This separates what Fögen calls "Fachsprachen" from "Sondersprachen": "Sondersprachen" consciously aim to limit communication to a group of experts, while "Fachsprachen" may have this effect as an externality, though they aim primarily at precision rather than exclusion.[63]

Technical texts might alternatively be defined by their inclusion of relatively cutting-edge content. Gros, for example, argues that Vitruvius' *De architectura* should not be classified amongst technical manuals aimed at practitioners, since the archaeological evidence seems to indicate that Vitruvius describes an approach to architecture that lags significantly behind contemporary practices.[64] Meißner likewise defines technical literature in the strictest sense as the transmission of novel or obscure

[59] Fögen 2003, 33. [60] Langslow 2000, 5.

[61] Fögen 2003, 31. Fögen collects a list of such self-conscious references by ancient authors to their use of technical terminology and the mechanisms of its development at 2009, 34–48.

[62] Vitr. *De arch.* 5.pr.2. [63] Fögen 2003, 34; Fögen 2009, 13–19.

[64] Gros 1996, 35.

knowledge from one specialist to another.[65] However, Meißner also notes that much of what we call "technical literature" from the ancient world probably failed to stick to cutting-edge material for a variety of reasons, including the taste of the intended audience. In the Hellenistic period, which Meißner is there referring to, the audience may have been intellectually elite, but not limited to professional specialists; in other periods authors indeed aimed for a still broader popularization.

Programmatic prefaces may seem to promise a reliable window onto an author's literary intentions. Fögen, for example, suggests that prefaces offer good guidance for coping with the "Polymorphie" of ancient technical texts.[66] It is certainly tempting to draw on prefaces that make explicit reference to a work's chosen audience, its stylistic level, and how it will set forth its subject matter. The programmatic material found in the proems to "technical" works can indeed be an invaluable guide through the tangle of different stylistic levels and degrees of technical detail these texts can involve.

However, programmatic statements are not always a reliable or complete indicator of the work's full intended readership, particularly in the case of a work addressed to a patron. The preface of Apollodorus' *Poliorcetica*, for example, repeatedly invokes an imperial addressee (probably Trajan), but the production of the machines described in the body of the text clearly requires an additional, quite different audience: whatever the level of the emperor's engagement with Apollodorus' designs, he is not a one-man production team. A patron who is explicitly signaled in the preface, and perhaps even alluded to more subtly later in the text, is at best the centerpiece of a wide-ranging possible readership with equally wide-ranging expectations. Rüpke's analysis of Caesar's *commentarii* presents a useful refinement, suggesting that *commentarii* may be identified through their inclusion of a proem that establishes the themes and structure of the work, thus setting up a "horizon of expectations" for the reader.[67] This view emphasizes the variable effects such structural elements might have on different readers, rather than imposing a rigid typology of texts through the assumption that the reader addressed in the preface determines the text's intended effects.

In the domain of mechanical texts, expectations might be delimited by authors who explicitly set out what they believe the discipline

[65] Meißner 1999, 169. [66] Fögen 2003, 36.

[67] Rüpke 1992, 210–16. The "horizon of expectations" (Erwartungshorizont) is described foundationally in Jauss 1982.

of mechanics consists of, by listing the activities or products of mechanical design and construction.[68] However, this process seems analogous to what Asper calls "inhaltsorientierte Taxonomien," which he argues are not the product of genuine consciousness of genre but rather a set of pre-theoretical "pragmatischer Produktionsregeln."[69] The central problem here is that content-oriented differentiation of texts inevitably results in what is more or less division by discipline, which in turn is merely homologous to division by groups of authors, rather than looking at markers of genre within the text itself.

Asper asserts more generally that ancient authors of "Wissenschaftstexte" developed no genuine, self-conscious theory of genre for their work.[70] In his view, the absence of such a theory is owing to the fact that these texts were not developed in an agonistic, ritualistic, or institutional environment such as state festivals or courts of law.[71] There is, to be sure, some evidence of competition in the practice of mechanics: for example, Philo's remarks praising his own innovations, the competition over medical traction machine technologies that Oribasius describes, and the competitions between engineers for state contracts recorded in Vitruvius.[72] However, these competitions typically seem to have involved judgment about the artifacts themselves rather than a gradual refinement of "rules of the game" for textual representation. In the case of the competitions for public buildings or large-scale engineering projects, documents both verbal and visual may well have been involved alongside scale models, but these do not survive.[73]

In the absence of such a competitive environment, what about a related possibility Asper raises, that of genre-formation through textual canonization, perhaps through a body like the Alexandrian scholiasts? Asper's paradigmatic example of canonization in a technical field is Euclid; he argues

[68] Vitr. *De arch.* 10.1.1–3, 10.3; Pappus, *Synagoge* 8.1024–6.

[69] Asper 2007, 50. [70] Asper 2007, 35–42, 45–53, 351–5.

[71] Asper does suggest that medicine might have been an exception to the rule for technical disciplines (2007, 48–51). Galen, for example, explicitly discusses the dominance of clarity of content over stylish form, which suggests a certain self-consciousness about genre. The medical competitions that took place in Ephesus in the second century CE were likewise supposed to have produced a set of texts containing self-reflective justifications, providing such an "institutionalisiert kompetitiven Kontext," which might have fostered something like a conscious theory of genre. It is conceivable that some analogue to these competitions existed in other technical disciplines, but no evidence for this survives.

[72] See e.g. Philo, *Belopoeica* 56.38–50, 58.43–59.3, 67.42–68.3; Oribasius, *Collectiones medicae* 49.23.1–15; Vitr. *De arch.* 10.16.3–4.

[73] On the use of such documents and models in a competitive context, see Heisel 1993, 155, 162–5. See also Herodotus 5.62.16.

that Euclid's work provoked a lineage of responses that might properly be called a "Euklid-Philologie."[74] No such text appears to have guided the formation of mechanical texts in such a standardized way. Authors of mechanical artifact texts assess their own works on the basis of others' – Hero, for example, refers in his *Peri automatopoiētikēs* to Philo's work on the same topic, and vows to make his own work easier to understand[75] – but these are too few and far between to indicate that a crucible of textual canonization was ever established as a major factor in shaping these texts as a group.

The problem of the "hermeneutic circle" becomes especially thorny when trying to define a genre or genres of technical texts, in light of the relative scarcity of evidence for the extra-textual forces that might have shaped the texts, their contemporary reception, and so on.[76] In light of these difficulties, it seems attractive to develop a scheme for differentiating these texts from one another that does not pretend to establish a clearly delineated (even if internally heterogeneous) set of "technical" texts, and does not rely on identifying a monolithic readership for any given text. Instead, it will prove useful to examine patterns in texts' fluid collections of attributes instead of imposing rigid categories, and to ask questions about what knowledge the text supplies the reader with rather than trying to reconstruct the "intended" reader's level of expertise.

Turner has suggested that the theory of literary genre might be amenable to enrichment by ideas from cognitive science about how we form categories, including

> prototype effects in genre categories, or metaphoric members of a genre category, or radial categories within our conception of a given genre, or a gradient from the categorical to the analogical in the ways literary works are connected, or (perhaps most obviously) family resemblance as a creator of genres, and so on.[77]

Turner acknowledges the vast scope and complexity of this project, and leaves its completion as an exercise for the reader. The gauntlet thus thrown down is retrieved by Sinding, who suggests prototype theory in particular as a way out of the hermeneutic circle: "we learn from examples first, build up a cognitive model, and locate new instances in relation to the model."[78] Sinding seeks to replace the traditional

[74] Asper 2007, 51–2.
[75] Hero, *Peri automatopoiētikēs* 20.1–2.
[76] For discussion of the "hermeneutic circle" in the criticism of didactic poetry, see Volk 2002, 26.
[77] Turner 1991, 150. [78] Sinding 2002, 181 n. 1.

approach to genre theory, under which categories are defined by "a set of singly necessary and jointly sufficient objectively existing properties" with a cognitive science-inspired view of categories as "elements of mental models that characterize prototypical or central subcategories and cases."[79]

Prototype theory acknowledges that humans do not typically form categories a priori with rigid and impermeable boundaries, but rather generate them gradually through a process that begins with the identification of "prototypical" category members. These prototypes are then used as a jumping-off point for experiments in category expansion, where other possible members are evaluated by comparison to the prototype. Some members will match the prototype closely, while others might possess certain non-conforming attributes but nevertheless be considered as a part of the category (as a penguin, though flightless, is considered a bird). The structure of such a category more closely resembles a cloud of members grouped around the prototype in a gradient of diminishing resemblance along multiple axes than a capsule with homogeneous contents and clear-cut standards of exclusion. This approach is quite consonant with a modified reader-response theory such as Rüpke uses to delineate the *commentarius*: author and reader become partners in the task of constructing and filling in the schema, and genre construction becomes less a matter of living up to a pre-established set of rules than an ongoing series of negotiations on the level of individual textual attributes.

Viewing the construction of literary genres from a prototype-theory point of view can be particularly useful in structuring the Protean mass of ancient technical texts. Sinding explains the construction process in terms of "schemas," which he envisions as a kind of model, a "framework" of elements and the relations between them.[80] He emphasizes the importance of the relations between elements in constructing a robust and flexible structure capable of accommodating even members that present anomalously compared to the prototypical member. Categories do not inevitably ossify, but maintain their potential for extension, whether that potential is actualized or not. Gibbs emphasizes the dynamism in applying, for each act of reading, prototypes that are "*ad hoc* and not conventional."[81]

[79] Sinding 2002, 184–5.

[80] Sinding 2002, 195. Sinding's version of the schema combines the term's use by Rosmarin, who adapts Gombrich's use of the term from the visual to the verbal domain; and in Lakoff's cognitive-psychological work: Gombrich 1960; Lakoff and Johnson 1981; Rosmarin 1985, 12–22.

[81] Gibbs 2003.

How might such an approach apply to ancient technical texts, and in particular to mechanical artifact texts? Rather than attempting to apply rigid criteria for categories like those defining Fuhrmann's "Lehrbuch," or trying to discover the "genuine" *commentarius* among the pretenders to the name, we can take a more flexible view of the frameworks that shaped the composition of these works. In working from these frameworks, authors can pick and choose features to adopt or adapt from previous texts, while making their own innovations.

Certain types of textual objects are perhaps more likely to occur in the texts we are here calling "technical": these objects include imperatival and other directive verb forms, diagrams and references to their letter-labeled components in the text itself, tables of numbers or other data, assertions claiming the rigorous standards of geometrical proof, and actual geometrical demonstrations.[82] Such objects are not, however, strictly required for admission. Vitruvius explicitly balks at including much in the way of mathematics in his text, alleging that his intended audience will not have patience for it.[83] He does provide a modest list of numerical correspondences between the size of a catapult's spring-hole, its dimensions, and the weight of the stone it can throw. Even so, the text is sparing with numbers when compared with Philo's list of those ratios, to say nothing of Frontinus' list of measurements or the elaborate data tables that characterize Ptolemy's works. A text's "technicality" is not a binary characteristic, but allows of degrees and variations.

A single "technical" text may indeed combine features associated with several different existing genres, or incorporate entirely novel textual configurations. Galen not only wrote texts intended for a number of different audiences, but also sometimes incorporated elements from several different genres into a single text. Asper construes Galen as using increasingly sophisticated literary tools to textualize increasingly complex technical knowledge: these tools include excerpts of poetry, polemical and satirical outbursts, and even "Mini-Dramen."[84] Galen incorporates such elements in his technical texts with unusual frequency, but he is not unique in using them. Eratosthenes' account of the mesolabe's use in solving the problem of two mean proportionals brings together many different genres: framed as a letter, it also incorporates elements of history, epigram, and of course mathematics.[85] Moreover, we get that account of

[82] On the density of imperatival forms in didactic prose and poetry compared with non-didactic genres, see Gibson 1998, 91–3.

[83] Vitr. *De arch.* 10.11.2. [84] Asper 2007, 336.

[85] In Eutocius, *Commentarii in libros de sphaera et cylindro* 88.3–96.27. On this work see Heath 1963, 154–5; Tybjerg 2004, 44–6; Taub 2008; Creese 2010, 196; Archimedes and Eutocius 2004, 294–8.

the mesolabe from Eutocius' commentary to Archimedes' *On the Sphere and Cylinder* as one of a string of solutions by different authors, extending the variation yet further. The surveyor's Hyginus' *Constitutio limitum* incorporates allusions to Archimedes' *Sand-Reckoner* as well as excerpts from Virgil and Lucan, while Vitruvius' *De architectura* includes a series of Greek epigrams on springs of water with unusual properties.

The forms of address an author chooses can help to differentiate among mechanical artifact texts. For example, Philo's *Belopoeica* and Hero's *Pneumatica* could both arguably be located in the working category of "technical literature on mechanics" sketched out earlier, since both texts consist almost entirely of descriptions of a single type of mechanical artifact rather than, for example, an historical event in which those artifacts were incidentally put to use. However, the two authors approach their descriptive tasks in very different ways. Philo very often addresses the reader directly; indeed, the whole treatise is framed as a kind of instructional letter.[86] Likewise, his building instructions are often framed as second-person imperatives, directly enjoining the reader to perform certain activities of building. Hero, by contrast, mostly expresses directives as third-person imperatives and impersonal constructions like δεῖ+infinitive constructions; the author–reader relationship thus takes on a very different form.

On a larger scale, the different ways entire texts are structured can also help to differentiate varieties of "technical" texts. Philo's *Belopoeica* describes just a few catapult types, in great depth, incorporating a series of knowledge bases or skill sets ranging from arithmetic to materials science. Biton's *Kataskeuai* is, like Philo's text, devoted to the description of military machinery, but structured very differently. Biton provides a sequence of discrete descriptions of engine types, each hewing quite closely to a standard pattern of elements: the engine is named, a famous historical exemplar is cited, and then the machine itself is described. While the descriptions themselves differ in certain ways, the overall structure of Biton's text resembles, more closely than any other surviving text in which mechanical artifacts are described, what Asper calls the "Rezeptstil," a form characterized by discrete sections that are internally structured and marked off from one another with formulaic language.[87] Asper observes these characteristics in mathematical and medical texts,

[86] On this form see Deinlein 1975, 5–7; Taub 2008.

[87] Asper 2007, 198.

while Biton provides an opportunity to compare those highly standardized texts with a close (but not perfect) mechanical parallel.

Some texts describe machines in terms of their dimensions, structure, and principles of operation, without any attention to outside information such as their invention or history. Consider an excerpt from Philo's detailed description of the structure of the bronze-spring catapult:

> It is necessary to forge the connecting rods that extend to both sides with a length of 5 palms plus 1½ dactyls, a width of 2 dactyls, and a depth, in the area where the spring goes, of 2½ dactyls. From the ends of the connecting rod on either side let a tooth project, however long the side support is thick, and let the tooth be slightly downturned.[88]

The catapult is in this passage more type than instantiation; the artifact is highly specified in terms of its form, materials, and operating principles, but we are not meant to associate the description with any historically specified artifact.

Other passages give the same kind of dimensional and structural detail, but link the device being described to a specific prior instantiation. Biton describes a *helepolis* siege engine in this way:

> Next after these, I submit to you the construction of the *helepolis* engine, which Posidonius of Macedonia constructed for Alexander son of Philip. The wood-working is extremely varied. For all the lengthy components and decking, use fir or silver fir or pine; for the axles and wheels, oak or ash, and the same for the beams and the supports. It is necessary for you to know in advance that you must design the sizes of *helepoleis* for attacks on walls, and that the *helepoleis* must surpass them in size ... Let our first discussion be about the axles and the beam. Let this be 60 feet in length, and let that be Λ. Within it let there be parallel axles of length 50 feet.[89]

Biton introduces this device as an exemplar with a particular historical identity, which will obviously not be maintained in building a new one. He also gives it a material and a dimensional identity created by specifying, for each component, the types of wood required and its size. The latter "identity" is also in fact remarkably fluid; Biton provides the dimensions for the components of a *helepolis* of suggested size, but also notes that those dimensions must change according to the size of the new machine's target.

[88] Philo, *Belopoeica* 64.31–65.10. It has already been noted that Philo does include some autobiographical references, but he does not give the artifacts he describes any particular historical context.

[89] Biton, *Kataskeuai* 4.1–16.

Diodorus Siculus refers to a similar device in a more typical historical narrative, his account of Demetrius' siege of Rhodes.[90] There he details the size of the machine's frame and its height, its interior structure, and the degree to which it narrowed from bottom to top. He even describes the special *antistrepta* pivots that enabled its astonishing freedom of movement. The proliferation of technical detail inside a historical narrative suggests that terms like "technical literature," which are often used as though they demarcated a clear and strict boundary, are in fact more ambiguous than they first appear. Indeed, many intriguing examples of technical ekphrasis are found precisely where these strict boundaries break down. Technological artifacts, and the knowledge that must be commanded in order to produce them, bear a cultural importance and interest that go well beyond the particular task of one expert informing another how to replicate a machine he has designed.

Consider, for example, Statius' poem on the construction of the Via Domitiana. This poem celebrates not only the convenience the road offers the traveler (who formerly had to make an unpleasant cross-country or sea journey), but also the process of constructing the road itself: cutting and reinforcing furrows in the earth, filling the bed with appropriate materials, enclosing the road with compacted blocks held together by pegs.[91] The technological achievement of building this road, the details of whose construction Statius does not shy away from representing in some detail, is obviously of interest to a broader audience than road-building professionals. Uniting the empire through improved transport technology was of tremendous importance for Rome, and for Statius himself, engaged as he was in high-profile praise of the works that were carried out under Domitian.[92] Yet Statius does not merely sing the praises of the road for its utility as a finished product. He confronts the reader with the technical details of the whole construction project, ensuring that the reader's subsequent experience of the finished road will be colored by awareness of the layers of structural complexity beneath.

There is not necessarily a great difference between the way technical treatises and other types of texts describe individual machines. A technical interlude may certainly be situated within a text otherwise generally agreed to belong to another genre, like Diodorus Siculus' description of the *helepolis*. This description is itself a detailed, self-contained episode within the account of the siege, in which the time of the narrative stands

[90] Diodorus Siculus, *Bibliotheca historica* 20.91.2–8.

[91] Statius, *Silvae* 4.3.

[92] On this see Statius 1988; Chevallier 1976; Nutton 1978.

still so that the reader can fully explore the mechanisms of the amazing engine. For example, he asks the question "how does the heavy *helepolis* change its direction of motion?," and answers very specifically that it uses the *antistrepta* pivots, which evidently work well for this purpose. As Saastamoinen notes, Fleury has suggested that the *Res gestae* of Ammianus Marcellinus should be classified as a technical text based on its inclusion of some passages describing catapults.[93] The designation of an entire text as "technical" is based on a complex balance of qualities rather than a simple binary characteristic. It is also often an oversimplification; it is much easier to categorize a single passage as "technical" than a whole work.

Intra-textual variables like the type of connection established between author and reader, or between reader and artifact, allow for more nuanced determination of genre. Does the author address the reader directly? Is the reader asked to absorb anecdotes about a machine passively, or to take on the active role of constructing a new object (or at least imagine himself involved in that construction process)? The level of engagement with the technical specifics of the artifact, or with its historical, philosophical, or aesthetic context, may vary as well. Texts generally denoted "technical" tend to cluster around the poles of this system of variables, but texts of many different kinds occasionally adopt descriptive strategies that draw from surprising places in the system: a mathematical text like Eratosthenes' may break into poetry, and a poem like Manilius' may draw on mathematics. Technical ekphrasis occurs on the level of the passage rather than the text, so the passages that make up the corpus come from many different types of "host" text. We will want to fully explore this cross-fertilization of genres rather than sweeping these variants under the rug in favor of an artificially neatened taxonomic structure. The interrelationships between attributes like those mentioned earlier create what Sinding calls "schemas"; we will see throughout the remainder of the project how they come together.

Genres of artifacts

Even if, as Asper argues, content-based categories do not define genres of technical texts in a meaningful way, they are quite commonly used by ancient authors mapping out the discipline of mechanics.[94] In fact, "genres" of artifacts are not detached from the features of the texts that describe

[93] Fleury 1996, 361. This claim is rightly characterized as "dubious" at Saastamoinen 2003, 19 n. 31.
[94] Asper 2007, 50.

them (though obviously there is not a one-to-one mapping between them). Different types of artifact lend themselves to different descriptive strategies: catapults appear more often than pneumatic devices in historical narratives; texts on harmonic and astronomical instruments can closely resemble mathematical texts; artifacts that evoke a state of wonder in the viewer may find their way into ekphrastic epigram, and so on. Complex artifacts like automata and ballistic devices are organized into variegated catalogues of multiple subtypes, but a single subtype can also occupy a full text that takes its time describing the artifact's functional principles, appearance, and construction. Such structural differences are attributes of genre, and the longer "continuous" texts (or sections thereof) appear to have been reserved for types of artifacts that possess a certain level of complexity. For example, tools like plows and the simpler instruments of surveying do not receive this treatment, whereas the surveyor's complex *dioptra* (also used in astronomical observation), catapults, and pneumatic wonders do, as do more complex agricultural devices like olive presses.

The patchy survival of ancient texts prevents access to the full spectrum of artifact genres; just as whole classes of text do not survive, there is scattered evidence for classes of artifact that are not described in the surviving texts. The techniques of designing and constructing planetary models like the *sphaera* often attributed to Archimedes are lost, as is any technical information about the flying wooden bird designed by Archytas and mentioned by Aulus Gellius.[95] The design and construction of mechanical and pneumatic wonders necessarily has an element of secrecy about it, which may have contributed to the imbalance between the numerous marvels and automata mentioned in ancient literature and the relatively few detailed descriptions that now survive.[96]

Other types of ancient artifacts are frequently mentioned as background elements in ancient literature, but are rarely the focus of a technical ekphrasis. There are, for example, surprisingly few detailed descriptions of devices like the plow and the loom relative to their importance in the ancient world. Hesiod devotes some space in the *Works and Days* to the implements of agriculture: he prescribes the time for cutting timber and the proper size for a mortar and pestle and a wagon to be cut from it, the types of wood least likely to be infested with worms, and he lists two types

[95] On the astronomical models see Schlachter 1927; Aujac 1993, 131–7, 157–71; Evans 1998, ch. 2; Evans 1999; the extraordinary Antikythera mechanism is placed in this context in Freeth and Jones 2012; the surviving material evidence for the rather different equipment used by astrologers is discussed in Evans 2004. The bird is attested in Aulus Gellius, *Noctes Atticae* 10.12.9.

[96] On these questions see particularly Long 2001; Tybjerg 2003.

of plow (the *autoguos* and the *pēktos*).[97] For all this, however, he tells the reader little about the artifacts themselves; beyond some rough dimensions and the remark that the share-beam is connected to the pole with pegs, the reader leaves this passage without a much more vividly imagined idea of a plow than he came in with. Virgil provides a description of comparable detail in the *Georgics*; there too the desirability of certain woods (elm for the beam, lime for the yoke, beech for the handle) is mentioned, and the pole is specified as being eight feet long, but again the reader is left to fill in a lot of blanks from his own prior experience.[98] Likewise, plows, scythes, and other basic tools of agriculture appear in Cato's *De agricultura*, but they appear ready-made in the text, in contexts such as lists of equipment.[99] Instructions are given as to where to purchase them, but not how to construct them.[100]

It is not because of their agricultural or domestic theaters of use that these artifacts go largely undescribed. Cato and Varro give instructions for building more complex agricultural constructions such as olive presses and aviaries (both practical and fanciful), while Pliny describes in some detail agricultural implements, including a toothed grain-grinding pestle favored in Etruria and a reaping device he says is used on the *latifundia* of Gaul; some centuries later, Palladius describes another novel Gallic harvesting machine.[101] Often, however, the *technē* that originally led to improvements in the basic agricultural technologies is embedded tacitly in the crafting of the artifact itself, and not articulated as a discussion of the challenges or advantages of its design.

Vegetius offers a parallel from military science as practiced on land and sea. He notes that *historiae*, otherwise a valuable source of information on past generations of military practice, do not describe everyday activities on the grounds that they are widely known.[102] Vitruvius makes this distinction explicit in his discussion of how technology contributes to the quality of human life. He excuses himself from including certain basic machines, saying that it is unnecessary to discuss them because they are

> at hand daily, like mills, smiths' bellows, carriages of four wheels and of two, lathes, and the other things that have common opportunities for use

[97] Hesiod, *Opera et dies* 420–5, 435, 433.
[98] Virgil, *Georgica* 1.170–5. [99] E.g. Cato, *De agricultura* 10.
[100] E.g. Cato, *De agricultura* 135.
[101] Cato, *De agricultura* 18.9; Varro, *Rust.* III.4, III.5; Pliny, *NH* 18.97–9, 18.296; Palladius, *Opus agriculturae* 7.2.2.
[102] Vegetius, *De re militari* 1.8: *de historiis ergo uel libris nobis antiqua consuetudo repetenda est. sed illi res gestas et euentas tantum scripsere bellorum, ista, quae nunc quaerimus, tamquam nota linquentes.*

in our daily lives. And so we will begin to explain about those that rarely come to hand, so they may be known.[103]

While detailed communication about these basic technologies doubtless occurred, at least among those who designed, built, and used them, they did not achieve the same prominence *in literature* as the more fully textualized communication about the types of exotic or high-status technology that lent themselves to problematization and improvement. Homer mentions spears with ash-wood hafts, but does not discuss the choice of material as part of the spear's functionality; Biton, by contrast, describes ash wood as a component of catapults superior to other, competing materials because of its strength.[104] Biton describes how the wood functions in the catapult in order to establish the artifact as part of a system of craft knowledge, and Biton himself as an arbiter of that knowledge.

The mechanical artifacts whose descriptions survive in ancient literature can be usefully grouped into a few basic genres. Military engines enjoy an unusually high number of surviving texts.[105] A second category of artifacts performs various kinds of mapping; these include graphical maps of the earth and stars, cosmological models, and more common instruments like sundials.[106] The third category comprises devices designed to carry out a single useful function, which in these texts ranges from calculating mileage to setting a bone.[107] Closely related to these are a fourth category of "wonders," which resemble the functional "gadgets" in scale and variety but are designed and described in a context of mystification and play rather than utility.[108] The final category comprises a variety of "monuments," large-scale works whose chief technological challenge lies in their construction.

Each type of artifact presents its own set of descriptive strategies and challenges. As well as imposing different types of descriptive challenges on the author, these different types of artifact and the texts that describe them also have different ways of confronting their reader with the bodies of knowledge the artifacts encode.

[103] Vitr. *De arch.* 10.1.6.

[104] Homer, *Iliad* 2.243; Biton, *Kataskeuai* 44.6.

[105] On these see for example Marsden 1969; Baatz 1994; Gara 1994, 118–22; Meißner 1999, 161–6, 185–94, 246–50; Cuomo 2007, 41–76; Rihll 2007. For claims to their utility see particularly Meißner 1999, 95–9, 161–7, 192–4, 246–55, 277–89; Cuomo 2002; Cech 2012, 204–14.

[106] de Solla Price 1974; Gibbs 1976; Nicolet 1991, 63, 71, 99–107; Heisel 1993, 184–207; Evans 1998; Lehoux 2007; Jaeger 2008, 48–68; Freeth et al. 2008.

[107] Drachmann 1963; Brumbaugh 1966; de Solla Price 1974; Aujac 1993; Taub 2008; Creese 2010.

[108] Drachmann 1963; Brumbaugh 1966; Hero 1995, introduction; Schürmann 2002; Tybjerg 2003; Berryman 2003; Berryman 2009.

As the development of military machines relied heavily on patronage from kings and emperors, descriptions of them often emphasize their utility to the state. Military devices connect the mechanical knowledge that goes into designing the device itself to contextual knowledge about physics, tactics, and in some cases the specific historical context of the deployment of these devices.

Pneumatic-mechanical wonders play with the tension between audience awareness of their complex and delicate machinery and its obfuscation by their decorative and concealing elements. The tension between the mechanical device and what it represents persists, in a more serious scientific context, in the astronomical models.

Terrestrial and celestial "maps" and timekeeping devices, as well as scientific instruments, conceal complex mathematical operations behind their projections. Descriptions of such devices thus sometimes use mathematical language and diagrams to describe the correspondence between the artifacts and the natural phenomena they model. These devices evoke the theoretical complexity of astronomy for the user through their own mechanical complexity. Other astronomical devices, such as the planisphere astrolabe, do not model cosmic phenomena directly, but rather serve as a platform for making calculations. Such devices thus make even stronger demands on the reader to understand the principles of astronomy underlying the artifact's design.

Functional devices are often praised for their utility or simplicity, particularly in comparison with competing devices. Some descriptions of the functional "gadgets" fully describe how they work, but others refuse or fail to reveal all their operational details; much the same is true for the "wonders." For the functional devices secrecy protects intellectual property from competitors, while for the wonders it maintains their air of mystery; in both cases it casts a shadow of opacity over the descriptions.

The overwhelming size and complexity of monumental constructions like the water system of Rome or the colossus of Rhodes compel the audience to contemplate the principles of design, the mechanical challenges of transporting and placing structural elements, and the work of organizing the builders. The greatest challenge of describing the monuments, like the challenge of building them, is their sheer size and complexity: even a comparatively modest building simply represents too much structural information to describe it completely. More technically oriented descriptive passages deal with this complexity by focalizing the description of the building through its construction process, as the challenge of organization and engineering is indeed what makes a building a kind of technology.[109]

[109] The most usual alternative is something along the lines of Lucian's *De domo*, in which a building is described in terms of its contents.

The categories of artifact described in the surviving texts can be compared with a few surviving ancient categorizations of mechanics. Philo's *Mēchanikē syntaxis* occupied nine books: an introduction, *Mochlica* (On Levers), *Limenopoeica* (Harbor-Making), *Belopoeica* (Artillery-Making), *Pneumatica* (Pneumatics), *Automatopoiētica* (Automaton-Making), *Parasceuastica* (Siege Preparation), *Poliorcetica* (Siegecraft), and a text on stratagems.[110] Many of these books are now lost; only the "Artillery-Making," the "Siege Preparation," and the "Siegecraft" are extant in Greek, while the "Pneumatics" is extant in Arabic. As the introduction is lost, there is no programmatic statement of why he chose these subjects for the *Syntaxis*. Particularly in the case of siegecraft and stratagems, which are only tangentially concerned with mechanics, these topics seem to have been collected together because they share a context of application rather than because they correspond to a theoretically grounded model of what mechanics should consist of. Nevertheless, many types of machines that survive in the ancient literature do appear on this list, including ballistic engines, automata, and even large structures (here represented by the treatise on harbor construction). Astronomical devices are conspicuously absent.

Vitruvius codifies the subjects that belong to mechanics differently. He initially defines a machine as "a continuous material system," which uses variations on circular motion to move weights.[111] He ultimately offers three different taxonomic systems, which are used in tandem. The first divides machines into three types: ladders (the *genus scansorium*, glossed with the Greek *acrobaticon*), pneumatic devices (the *genus spirabile*, *pneumaticon* in Greek), and traction machines (the *genus tractorium*, for which Vitruvius uses the Greek gloss *baru ison*, which seems to give more importance to mechanical equilibrium than is evident in the Latin term).[112] He casts these three categories in terms of their structure, function, and use. The first consists of uprights linked by crosspieces, and is useful for observing military engines and perhaps other things at a height; the second type is used to produce effects of sound and motion; while the third type employs mechanical advantage to raise weights to a height. Each of these three types of machine presents its own set of challenges for the designer, and

[110] See in *Der neue Pauly* Folkerts, "Philon [7]: Ph. von Byzanz"; Marsden 1971, 156. On the order of these works, see Fleury 1994, 200–1.

[111] Vitr. *De arch.* 10.1.1. On the primacy of cyclical motion to mechanics at least since the pseudo-Aristotelian *Mechanical Problems*, see Schiefsky 2007. Vitruvius later adds that rectilinear and circular motion must be combined in order to produce mechanical advantage (10.3).

[112] Vitr. *De arch.* 10.1.1–2.

garners its own kind of status. High achievement in the *genus scansorium* consists in boldness; the *spirabile* in delicate complexity; the *tractorium* in its great and varied utility. These in turn become the focal points for ekphraseis on these categories of machine.

Vitruvius' second taxonomic system differentiates machines (*machinae*) from instruments (*organa*); the former achieve their mechanical advantage by brute force and require several workmen, the latter operate by clever design and require just one careful operator.[113] Here he does not distinguish machines from instruments based on their theater of use; an engine for hurling missiles may be either a machine or an instrument. Later, however, he offers a third classificatory system, which does differentiate mechanical artifacts into categories based on their context of use.[114] These include basic technologies like the loom and yoke (which he labels "necessary inventions"), hauling machines vital to great public works, pneumatic devices, machines for amusement in peace, and those for defense in war. Vitruvius does not frame his taxonomies of machines around their practitioners; personnel are absent from his categorization schemes (and as we shall see, they are absent from his accounts of construction as well).

A few centuries later, Pappus stakes out the divisions of mechanical activity not in terms of the machines themselves, but of the "mechanicians" (μηχανικοί) who design (and perhaps build) them. He divides them into five categories: designers of pulley-based hauling machines; of military engines; of hydraulic devices; of wonders and automata (pneumatic and hydraulic); and of astronomical spheres. Pappus further subdivides the last two, whom he identifies as "wonder-workers" (θαυμασιουργοί), into those who create wonder with devices run by cords and those who use devices run by flowing water, naming Hero and Archimedes as exemplary authors on these subtopics.[115] These categories need not be taken as areas of exclusive specialization. Pappus puts Archimedes in the wonder-working category here, for example, but shortly afterward identifies him as the author of a treatise on sphere-making. Similar ground is covered by "mechanics" here as in the earlier authors who discuss the scope of the field: machines lift weights with pulleys, hurl weights with springs, use objects to move water, use water to move objects, and track the movements of time and celestial bodies.

[113] Vitr. *De arch.* 10.1.3.
[114] Vitr. *De arch.* 10.1.5, 10.2.3, 10.4–6, 10.7.4, 10.9.7, 10.10–12, 10.1.1.
[115] Pappus, *Synagoge* 8.1024–6.

Pappus also presents mechanics in terms of the mathematical or physical problems it can help solve: doubling the cube, moving a given weight with a given force, finding the angle that allows a given weight to be transferred to a given height with a given force, and so forth.[116] Pappus' text is populated by the designers of machines, but not, like Vitruvius', by the artifacts themselves. He discusses the principles under which those machines operate, and the problems those principles present or solve, but does not present machines as material artifacts in construction or use. Pappus thus offers a different approach to mechanical education from that espoused in Hero's works, where the elements of mathematical and mechanical theory are likewise present, but appear bound together with practical instruction in working with metal, wood, and so forth. Meißner, indeed, characterizes Pappus as principally interested in systematization.[117] When he does refer to construction activities, he speaks in terms of assemblies of the purely mathematical objects he portrays as the common ground for the whole discipline.

Proclus suggests yet another set of mechanical categories in his commentary to Euclid. He first says that mechanics as a whole is concerned with the sensible and material (τὰ αἰσθητὰ καὶ τὰ ἔνυλα);[118] this falls in line with his later assertion that mechanics, along with geodesy and optics, is one of the fields of knowledge that serve the practical needs of human life.[119] The study of forces, imbalances, and balances is here classified as part of wonder-working, so that the term *mēchanikē* takes on theoretical significance along with the material and practical aspects foregrounded elsewhere in his discussion. Sphere-making is, as in Pappus, listed as something separate from wonder-working. The former involves the manufacture of devices to imitate the movements of heavenly bodies (κατὰ μίμησιν τῶν οὐρανίων περιφορῶν), while the latter includes automata that imitate the movements of animate beings (τὰ δὲ διὰ νεύρων καὶ σπάρτων ἐμψύχους ὁλκὰς καὶ κινήσεις ἀπομιμουμένων).[120] While Pappus uses authors of texts on each type of mechanical activity as general proxies for their subdisciplines, Proclus lists for each an exemplary or original practitioner: Archimedes for war machines and sphere-making, Ctesibius and Hero for wonder-working.[121]

116 Pappus, *Synagoge* 8.1060–70. 117 Meißner 1999, 263–5.

118 Proclus, *In primum Euclidis elementorum librum commentarii* 41.3.

119 Proclus, *In primum Euclidis elementorum librum commentarii* 63.8.

120 Proclus, *In primum Euclidis elementorum librum commentarii* 41.19–24. On these see Berryman 2007; Berryman 2009, 27–34, 201–15.

121 On the strength of ancient categorizations like these, Fleury proposes the division of ancient mechanical texts into three major classes based on their content: theoretical mechanics, civil practical mechanics, and military practical mechanics (Fleury 1996, 46–52).

Each of these arrangements of subdisciplines indicates how the author situates mechanics in his particular intellectual environment. This is in turn reflected in the different emphases in descriptions of the machines themselves from each environment. The surviving Hellenistic texts prioritize a thorough gathering of diverse types of mechanical knowledge, incorporating both practical and theoretical elements. Likewise, Philo's *Syntaxis* promises to bring treatises on practical subjects like harbor construction together with texts like the one on levers, which seems from references in his surviving texts to have been oriented more toward a theoretical mechanics based at least partly in mathematics. Vitruvius and other Roman authors appear to strive for an even more systematic approach to mechanics, to the point that building up a system of interconnected disciplines becomes a stronger focus than innovations in the machines themselves, an interest reflected in Vitruvius' taxonomy of taxonomies of machines. In the mechanical works of late antiquity, the dialogue between past and present work takes center stage, in what Netz refers to as a "deuteronomic"[122] approach. Many of the late ancient texts emphasize commentary on previous authors, taking the same type of approach found in Pappus and Proclus, who tie each category of machine explicitly to a past author or practitioner.

Genres of text–artifact relationships

A further set of generic questions is raised by the intertwining of text and artifact that occurs when the artifact itself is inscribed. These inscriptions are experienced by a reader along with the object itself, and they often encourage the reader to become a user by interacting with the artifact. This possibility changes the inscribed text's ability to unfold the knowledge the artifact embodies. An explanatory text inscribed on an instrument such as a sundial, for example, may offer many different types of knowledge to the viewer: it can tell him how the sundial works to model cosmological phenomena, or how to use the instrument to tell time, or merely announce who set the sundial up or where its materials were brought from. For other genres of technology, such as pneumatic wonders, a designer's status comes from concealing information about a device's operational principles, so that in the rare cases when inscriptions are present, they tend rather to focus on the patronage of the artifact rather than its working principles.[123]

[122] Netz 2004, 121.

[123] E.g. the dedicatory inscription on the water organ described in Hyde 1938.

Descriptive or explanatory texts incorporated into the device itself literally bridge the gap between text and object, transforming the encounter between object and reader.[124] The spectator's experience with the inscription can be active or passive; he may be encouraged to engage physically with the artifact to perform a calculation or read the time, or simply be informed about the history of the artifact itself in a way that highlights past rather than present human engagement with the artifact. Some devices require the user to unpack some of the theoretical knowledge that went into their construction, while others need only unfold that knowledge for the designer or builder. For some devices, the display of knowledge is optional, perhaps not an absolute requirement for the device to be useful, but something that serves as a marker of intellectual status. In any of these cases, the knowledge involved may be active or passive, theoretical or practical; it may be meant to make the device's workings transparent to the reader or cloak it in secrecy.

Texts inscribed on the technological artifact they refer to are most common among astronomical devices, such as sundials and the extraordinary Antikythera mechanism. The inclusion of an explanatory text in the body of the machine is part of the context of use; these are devices whose operation depends on the user's being able to "read" the machine. Inscriptions on sundials most often consist simply of dedications to a human patron or divine entity.[125] However, some sundials contain more "technical" inscriptions, such as wind roses with the names of the winds; the hours, days, and months; or the names of the zodiac signs. One dial of this type from Delos (Gibbs 4001G) is inscribed with explanatory notes on the summer solstice, the equinox, and the winter solstice.

A more detailed functional inscription was found on a sundial from Samothrace, which is no longer extant but reported by Cyriacus of Ancona (Gibbs 8008). An inscription on this dial briefly explained that the gnomon's shadow indicated the time of day and year, highlighting the equinoctial and solstitial lines. There was evidently another inscription on

[124] Technological artifacts accompanied by inscriptions of course have some features in common with epigram. Particularly relevant here is Squire's observation that "ancient commentators were highly attuned to the visual aspects of an inscription" such as letter shape and style. He argues that in the Hellenistic period the boundary between the verbal and visual aspects of an epigram became increasingly fluid, culminating with the development of the *carmen figuratum* (Squire 2009, 148–50).

[125] The best catalogue of sundials is still Gibbs 1976. This catalogue has been slightly expanded by Catamo et al. 2000. Schaldach 1998 is a charming and practical introduction to the uses and varieties of ancient sundials in general; Evans 1998, ch. 3 is likewise an accessible yet rigorous guide. Lehoux 2007 is an excellent resource on *parapēgmata*, containing both inscriptions and extensive explanatory material.

the back, providing information about who set up the sundial as well as its design: "Odios son of Aison set me up with fifteen parts, drawing the sun, and this gnomon ten ..." The sundial's maker chose to include a variety of types of information on the dial, separated spatially according to whether the object is referenced as a constructed object or a time-telling device.

The Antikythera mechanism features a similar spatial separation of two types of inscriptions. This remarkable clockwork device's complexity is unmatched in any other surviving technological artifact until many centuries later. It served as an astronomical model and calendar, predicted eclipses, and tracked Olympiads, among other functions. The mechanism is enclosed by "doors" that open to show inscriptions; the so-called "front door" is inscribed with astronomical information, and the "back door" describes how the machine itself works, as well as discussing the cycles of phenomena like eclipses and of the Metonic calendar. This configuration constitutes a fascinating variation on the usual relationship between text and object: the machine is designed to tell the viewer how it works (to a certain precision), without relying on an external treatise.

In its current state of disrepair the device's inscriptions are largely illegible, but it does tell us about a ΧΡΥΣ[ΟΥΝ]ΣΦΑΙΡΙΟΝ right near Η[ΛΙ] ... ΑΚΤΙΝ and [Η]ΛΙΟΝ, so it looks like there was a golden ball used to model the sun.[126] Elsewhere the inscription mentions the names (theophoric or descriptive) of the planets Venus, Mars, Jupiter, Saturn, and probably Mercury, and finally ΤΟΥΚΟΣΜΟΥ.[127] The mechanism thus appears to be a device like an orrery, which shows the viewer all the motions of the cosmos as he stands outside. Still more interesting, the inscription tells us about the very nuts and bolts of the machine itself: ΣΤΗΜΑΤΙΑΔΥΟΠΕΡΙΤΥΜΠΑΝ, two trunnions around the gears, and what appears to be a reference to dragging something through a perforated hole. The inscription does not appear to contain instructions for the user; Freeth and Jones mark the apparent absence of imperative verbs, for example, and note as well that the inscriptions appear to contain specialized mechanical terminology but not specialized astronomical terminology.[128]

These inscriptions are like a treatise in terms of their function, but are situated like an epigram. The front door inscription acts like a miniature

[126] The supplementary material to Freeth et al. 2006 contains the most complete transcription of the inscriptions; Freeth and Jones 2012 includes a revised version of part of the inscription on the front door.

[127] New insights into the planets mentioned are found in Freeth and Jones 2012, sec. 2.3.2.

[128] Freeth and Jones 2012, sec. 2.3.1.

astronomical treatise, which describes the cosmic *realia* behind the model presented by the machine, while the back door inscription delves into the mechanical workings of the machine itself. Each of these doors, of course, faces part of the mechanism. The front door faces the front dial, which displays the positions of the sun and moon alongside a 365-day calendar. The back door faces the back dial, which includes a nineteen-year calendar tracking the four-year cycle of the Olympiad and the Panhellenic Games, as well as a scheme for predicting eclipses.[129] This layout mirrors the separation in focus between the two inscriptions: reading the front inscription we are driven to think about large-scale cosmic phenomena, while the back inscription ties us back down to earth, to the device itself and to human institutions like the Panhellenic Games. Even the eclipse-prediction scheme, based on a kinematic glyph-generation system, keeps the focus on the device itself by telling the reader how many hours to add to the glyph time depicted on the dial to get the predicted time of an eclipse (which is, of course, merely a particular earthbound perspective of a certain cosmic arrangement). These two inscribed plates make explicit the transition we undergo, often unconsciously, when we read a map – one side is telling us the theory behind the motions of the planets we see modeled, while the other reminds us, by making explicit how the machine works, that what we see is really just a model.

Astronomical devices are not the only examples of inscribed instruments. Eratosthenes provides an opportunity to compare how the mechanical mesolabe, used to solve the problem of doubling the cube, is presented in an inscription attached to a machine and in a separate text.[130] His text, framed as a letter to the king Ptolemy, describes an interesting composite of text and artifact: a mesolabe attached to a pillar alongside a proof of the machine's mathematical functionality, a corresponding diagram, and an epigram.[131] The letter contains as well a rather fabulous history of the problem, a detailed proof of the method for a purely geometrical solution, and a diagram corresponding to this proof. Taub describes the epistolary form as an "envelope" enclosing this mosaic of genres, and suggests that the dazzling multiplicity of forms indicates a text meant to appeal to a diverse audience, not just to mathematicians.[132]

129 Freeth et al. argue that this system was quite inaccurate (Freeth et al. 2008, 616).
130 Herrmann and Breidenbach 1952; Heath 1963, 154–5; Tybjerg 2004, 44–6; Taub 2008; Netz 2009, 160–4.
131 Ptolemy III Euergetes; see Archimedes and Eutocius 2004, 294 n. 153.
132 Taub 2008, 289. On such "mosaic" forms in Hellenistic literature, see Netz 2009, 131–2.

Eratosthenes begins his history of the problem with a story: "they say" that Minos built a tomb to Glaucos, and advised him to double it on each side to double the volume. Eratosthenes understandably writes off Minos as "mistaken," and provides a capsule history of the motivations and approaches to the problem's solution. The history, which is action-packed though brief, does give the impression of difficulty in the traditional geometrical solutions to the problem. When Eratosthenes then promises "we have conceived of a certain easy mechanical way of taking proportionals (ἐπινενόηται δέ τις ὑφ' ἡμῶν ὀργανικὴ λῆψις ῥᾳδία)," the reader is eager to hear it.

He prefaces his account of the machine itself with a few possible applications of the solution, such as enlarging various structures or calculating the volume occupied by a given number of dry or wet measures. He envisions its use in military machines as well, asserting that all the components must be increased proportionally in all dimensions if the weight is to be scaled up, an engineering problem of constant concern to Philo. Creese raises the further possibility that through the effectively unlimited number of means (μεσόγραφα μυρία) it might also be possible for the device to be useful for Eratosthenes' pursuits in the science of harmonics.[133]

After this little history and advertisement for the machine's practicality, we are plunged into the realm of geometry. Eratosthenes draws a careful picture (literally and formulaically) of all the elements in the construction, and provides a geometric proof that his method works to find the two mean proportionals between the elements ΑΕ and ΔΘ. Eratosthenes notes that the mechanical proof on the pillar is a slightly abridged version of the one he includes in the letter, so the comparison between the two must be treated carefully: the differences between them are important. The geometrical proof specific to the letter is as follows:

> For let there be given two unequal lines <namely> ΑΕ, ΔΘ between which it is required to find two mean proportionals in continuous proportion, (a) and, on a certain line, <namely> ΕΘ, let ΑΕ be set at right <angles>, (b) and let three parallelograms, <namely> ΑΖ, ΖΙ, ΙΘ, be constructed on ΕΘ, (c) and, in them, let diagonals be drawn: ΑΖ, ΛΗ, ΙΘ; (1) so they themselves will be parallel. (d) So, the middle parallelogram (ΖΙ) remaining in place, let ΑΖ be pushed above the middle <parallelogram>, <and let> Θ

133 Creese 2010, 196. Creese concludes that Eratosthenes' rejection of the Aristoxenian diatonics in favor of the ditonic diatonic found in the *Timaeus* and *Sectio canonis* was probably not based on empirical experimentation with the mesolabe because of the margins of error involved.

> <be pushed> beneath it, as in the second figure, until Α, Β, Γ, Δ come to be on a <single> line ...[134]

On the other hand, the paraphrase of the proof on the column goes as follows:

> Given two lines, to find the two mean proportionals in continuous proportion. Let ΑΕ, ΔΘ be given. (a) So I move the tables in the machine together, until the points Α, Β, Γ, Δ come to be on a <single> line ... Therefore two means have been found between the two given <lines>. And if the given <lines> will not be equal to ΑΕ, ΔΘ, then, after we make ΑΕ, ΔΘ proportional to them, we shall take the means between them <= ΑΕ, ΔΘ>, and return to those <given lines>, and we shall have the task done. And if it is demanded to find several means: we shall insert tablets in the machine, <so that their total is> always more by one than <the number of> the means to be taken; and the proof is the same.[135]

Note the presence of personnel ("I" and "we") in the proof associated with the machine, and their absence in the purely geometrical proof of the letter. The first proof is phrased, as is typical of mathematical texts, in third-person imperatives with no identification made between the author or reader and the person carrying out the geometrical construction.[136] The second proof, however, has a first-person singular "character" who handles the machine initially. This voice is quickly replaced by a first-person plural voice that accompanies the reader in extending the proof to the more complex situation arising when the lengths of the lines on the machine are not suited for the user's needs, and even adding extra tablets to the machine to find multiple proportionals. The geometrical solution thus proceeds detached from the practitioner (though it is present in the text through diagrams), while the mechanical approach to a geometrical problem conjures up hands-on, collaborative involvement with the materials of proof.

After the proof, Eratosthenes introduces a surprising new element to this technical context – a short poem praising the machine and its many practical uses (the poem is aggressive on this score, naming projects from well-digging to byre expansion). While Pfeiffer praises the epigram as "formally perfect,"[137] seeing this attention to form as itself characteristic

[134] Eutocius, *Commentarii in libros de sphaera et cylindro* 91.30–92.8. Tr. Netz; translation and commentary in Archimedes and Eutocius 2004, 294–8. Netz uses parenthesized lower-case letters to indicate steps in the sequence of constructions, and parenthesized Arabic numerals to indicate steps in the sequence of assertions.

[135] Eutocius, *Commentarii in libros de sphaera et cylindro* 94.15–96.8, tr. Netz.

[136] On the verbal forms of geometrical proofs, see Fowler and Taisbak 1999.

[137] Pfeiffer 1968, 168. See also Taub 2008, 296.

of a mathematician's precision, the poem's content is as unmathematical as can be, describing the effect of the machine only in vague poetic terms that perhaps allude to a chariot race, or to a hunt:

> ... as they run to converge, in between
> The two rulers – seize the means by their boundary-ends.[138]

In this epigram, Eratosthenes emphasizes that his solution offers the greatest ease of use, and criticizes his rivals' solutions as "impractical," though this is clearly not meant as a blanket critique of their originators, which include "divine Eudoxus." His claim for the superior simplicity of his mechanical solution is his stamp of originality, and Taub is surely correct to call the poem a *sphragis* on the letter as a whole.[139] He does not have to go into immense detail about the specifics of the machine's novel design; it is novel simply for being a machine in a mathematical world.

In this inscription, Eratosthenes brings geometrical proof (now embodied and populated by users of the machine) together with the pithy perfectionism of an inscribed epigram addressed to the passerby. In the version of the proof specific to the letter, he combines a fabulous history of the problem with an impersonal, purely geometrical description and proof of the solution. Both settings (the letter and the column) have a common element, namely the diagram. Since diagrams represent a kind of middle ground between an object and its textual representation, it is not surprising that the diagram is common to both the artifact that can be directly accessed and its representation in text.

Eratosthenes' lengthy and varied description of the mesolabe can be contrasted with another device that performs a similarly useful task of calculation, but without the poetic fanfare: Hero's odometer dial. The odometer Hero proposes is essentially a gear train housed in a box. In the configuration Hero uses as an example, the gears are connected to an axle of the carriage through a system of gear wheels and screws in such a way that the first gear advances by one tooth each eight revolutions of the carriage wheel. Inscriptions are at the very heart of this device, as Hero advises that each of the gears inside the box should be inscribed with the length of journey that corresponds to the turn of one of its teeth:

> If the 30 teeth signify 1600 cubits, then 1 tooth of the gear mentioned indicates 53⅓ cubits of traveling. Now, when the toothed gear is beginning to

[138] Eutocius, *Commentarii in libros de sphaera et cylindro* 96.14–15, tr. Netz. For the metaphor see Netz 2009, 163.

[139] Taub 2008, 295–7.

> move and is found to have moved 15 teeth, it indicates a journey of 800 cubits, that is two stadia. Therefore we will inscribe in the middle of said toothed gear "53⅓ cubits."[140]

This process is then to be repeated for each of the gears: the teeth are counted, the distance represented by each tooth's motion is calculated, and that distance is inscribed on each gear in turn. The user's experience of the odometer can be improved, says Hero, by adding another inscription, this one outside the box and designed to accompany pointers that move along a dial. Thanks to this addition, the surveyor does not need to open the box containing the odometer and manually count the progress of the revolving gears. However, ease of use comes at the cost of mechanical complexity:

> The toothed gears mentioned will be positioned so as not to touch the sides of the case; then let their axles extend upward to the outer part of the walls. Let the projecting parts be square, fastened to pointers with square holes. In this way, when the wheel is turned, the pointer will also be turned along with the axle. As its tip revolves, it will describe a circle on the other side of this wall, which we will divide into the same number as the teeth of the wheel within. Let the size of the pointer be such as will describe a fairly large circle, for the sake of the division of the teeth being at fairly large intervals. Then the circle described will have the same circumference as the wheel within, and thus on the outside we will see the length of the journey completed.[141]

The odometer is contrived for the convenience of the operator rather than the designer. The display dial system complicates the design considerably: it places limits on the sizes of the gears in the box if the axles of the pointers are to be able to move, and demands a perpendicular extension of those gears to mirror their motion on the outside of the case.[142] This is a lot of effort just to save the operator the trouble of looking inside the box and counting the teeth every day. Yet the convenience the dial provides is, like Eratosthenes' mechanical solution to the problem of doubling the cube, the attribute that differentiates this device from its competitors. Despite their common advertisement for their user-friendliness, however, the inscriptions belonging to these machines are drastically different. Hero's odometer does not praise itself for ease of use in an attached inscription, nor does it show the user the theory that

[140] Hero, *Dioptra* 34.79–89. [141] Hero, *Dioptra* 34.89–104. [142] Drachmann 1963, 164.

proves to the user the result the machine gives is the correct one (though the individual numerical inscriptions on the gears provide a quick means to check the arithmetic).

What is the difference between these types of machines that motivates the decision to include explanatory inscriptions on some but not on others? Note that the sundial and Antikythera mechanism are representational models: the inscription says something about the planetary motions wrought small in the form of the device itself. One motivation for including the text on the machine might be to explain the geometrical and astronomical modeling done by sundials or the Antikythera mechanism. This is perhaps particularly true in light of the fact that astronomical treatises generally dwell only briefly, if at all, on the mechanical equipment (such as astrolabes and armillary spheres) that supply data and assist in its interpretation. We might then call the Antikythera mechanism's back-face inscription a "treatise-replacement" usage of the inscription, since it does work comparable to a separate textual explanation of the instrument (e.g. Ptolemy's *Syntaxis mathematica*). However, these inscriptions also function to legitimize the machine as part of a world of theoretical astronomy largely dominated by those separate texts, which we could call a "treatise-adjunct" usage.

The inscription attached to Eratosthenes' mesolabe works differently, as this device is not a model in the same sense as the astronomical calculators. Of course the length being calculated has a real-world analogue in the well or whatever else is being dug or constructed, but the device is not in any way mirroring some process happening to that analogue. So the geometrical proof is included in the inscription for a reason that has nothing to do with guaranteeing the accuracy of a modeling process. Instead, it appears to brand the machine as an acceptable device for solving geometrical problems, legitimizing it in the larger world of mathematical discourse, which is a world of textual inscriptions rather than machines.

Inscriptions on technological artifacts can thus accomplish either (or both) of two apparently opposing purposes: they can serve as a capsule set of instructions allowing a non-expert to use the machine in happy ignorance of the complex theory behind it, or they can serve as a kind of marker, reminding the more experienced user of the full treatises of information that were assimilated into the original design of the object. The Antikythera mechanism manages to do both: it includes a mechanical inscription that apparently contains specialized terminology, while its astronomical inscription seems not to.

The dial on the front of the odometer is also distilling a complicated set of motions into an interesting number, but those are only the motions of gears, part of the machine itself rather than a cosmic ballet, and the user probably does not care to know about them – that is the whole point of the dial mechanism. The odometer dial is like a rule of thumb in construction, a different type of knowledge whose theoretical background is less important than the question of whether it works on a practical level. The dial must reassuringly convey to the user that its results correspond to the real distance traveled, and it does this through the bare inscription of numbers on the gears, without the kind of self-promotion and proof that Eratosthenes' mesolabe apparently requires.

The odometer is part of a world of surveyors rather than astronomers, and perhaps in this world elaborate theoretical justification is worth less in the field than a hands-on trial of the accuracy of the odometer and its dial compared with other surveying methods. Many of the surviving treatises on surveying are, of course, detailed in their explanation of the geometrical theory behind the methods of the discipline.[143] However, these explanations reside entirely within the separate treatises, and are inscribed neither on the simple-to-use odometer nor the tricky-to-use *dioptra*. In other words, while astronomical instruments attract users ranging from experts in the discipline to laymen who simply want to know the time, the data that the odometer dial produces in the field is of interest to a population of users who care more about the data itself than the theory behind the device that produces it.

Inscriptions on artifacts obviously have different kinds of work to do than inscriptions in separate texts: instead of creating a mental picture of the object in the reader's mind, they may present him with the history of an artifact, tell him how to use it, or tell him how it works. These last two are not necessarily the same thing: the first shows the device from the user's perspective, the second from the designer's. Inscriptions relating to the use of an artifact can conceal its working principles or make a proud display of them, or offer the reader a choice between full explanation and "black-box" use. Epigrammatic inscriptions provide the reader with a different experience of the artifact than those contained in separate texts, as they are concerned with a particular artifact which is concretely present at the time of reading.

143 Cantor 1968; Dilke 1971, 47–65; Guillaumin 1994; Folkerts 1992; Campbell 2000, xlv–liii; Lelgemann 2010. The surviving treatises on surveying vary greatly in the degree to which they assimilate their work to mathematics.

Conclusion

Defining a genre or genres of ancient "technical" texts proves difficult, as these texts are not subject to any systematic contemporary discussion of genre. To make matters worse, the group of texts commonly collected under the "technical" rubric is extremely diverse, including didactic poetry and dialogues alongside texts that present technical information without such literary framing. A few ancient terms define "genres" with more or less solidly defined expectations, such as the introductory *eisagōgē*, the didactic poem, and the versatile *commentarius*. These labels, however, can be deceptive, and they are certainly far from univocal. Furthermore, many of the surviving "technical" texts do not fit any of these labels very well. As Asper argues, the alternative approach of using technical subdiscipline as a substitute for genre is ultimately rather trivial and unsatisfying, and leads to many of the same taxonomic problems as attempts to use the ancient category labels.

Even the mechanical artifact texts include several subcategories. They may be dedicated to a single device or class of devices (e.g. Biton's *Kataskeuai*), or to diverse devices with a single context of use (e.g. Hero's *Dioptra*, which discusses the *dioptra*, groma, and odometer, all used by surveyors), or may alternatively contain a large number of device types (e.g. the *De architectura* of Vitruvius). Some dwell at length on the theoretical background of the device's operation (e.g. Ptolemy's *Syntaxis mathematica*), while others are oriented more toward practical, rule-of-thumb construction (e.g. the bone-setters described in Oribasius' *Collectiones medicae*). The description of the device itself may include measurements, materials, and construction notes (e.g. Philo's *Belopoeica*), or it may omit any or all of these features (as Vitruvius sometimes does). These texts may describe a device constructed in the past, or offer instructions for building a new one (Biton's *Kataskeuai*, for example, incorporates both kinds of description).

My goal here will not be to fit mechanical artifact texts into these pre-existing categories, but rather to lay out the variations in descriptive approaches as broadly and systematically as possible. I will immediately dispense with the convenient oversimplification that "technical" and "literary" texts make use of two different and readily recognizable sets of rhetorical strategies and language types. The field of source material is generically broad; passages were selected on the basis of the presence of the "technical ekphrasis" rather than by their location in a work somehow qualifying as a "technical text." Such a broad sweep yields a set of distinctions between texts that can help address some of the questions unsuccessfully approached by attempts to establish a generic scheme for "technical texts." These include the degree of expertise demanded of the text's reader, the set of rhetorical techniques an author uses for the description, and the extent to which the text builds toward related bodies of knowledge.

The final chapter in particular will pick up these questions in detail, comparing, for example, the descriptive strategies that characterize Philo's *Belopoeica* and Hero's *Pneumatica*. Philo's text creates a sustained encounter between the reader and the many different "layers" of knowledge that go into making a catapult. His varied approaches divide the text into heterogeneous sections, including an arithmetic analysis, a geometric description, a theoretical-mechanical discussion, and an analysis in terms of something like materials science. These sections differ in length and internal structure, and each makes use of different textual "technologies" including tables, formulas, diagrams, and so forth. The unifying elements are the forms of the catapults being described (quite few in number), and the figures of author and reader who appear, embodied in various ways, throughout the text. Philo's text makes challenging and explicit demands on the reader to engage intellectually with tasks, like visualizing and calculating, that represent such an important part of the work of designing and building a catapult.

Works like Hero's *Pneumatica* and his *Belopoeica* take a quite different approach, briskly stepping through a large number of different artifacts, without the persistent figure of an author, reader, or builder. Here the unifying principle is not the artifacts (each of which is described only briefly before moving on to the next) or the personnel (who come and go), but the organization of the text itself. Hero provides a series of quick vignettes of artifacts, each related to the rest through some explicitly defined principle (the evolutionary history of the *Belopoeica*, for example, or the *Pneumatica*'s progress from simple to complex). In this type of text, the emphasis is not so much on the artifacts themselves as it is on the system of knowledge into which they fit; this is borne out by the text's coherently linked structure as well as its ample historical and philosophical contextualization. This type of structure is different from that imposed by Apollodorus, for example, who does describe a series of artifacts but does not embed them in a larger connective context; he claims simply to be providing a list of useful military devices.

Such distinctions offer a pragmatic proxy for genre based on what the text does to the reader or demands that he do. Though this may not lead to a group of well-defined genres, it does offer a way of subdividing these texts in terms of how they deploy the knowledge they contain, a division that may be guided in part by the type of artifact the text focuses on. This is only a slight adjustment from Asper's dictum that textual genre is first and foremost a matter of function.[144] The approach here simply suggests more axes of variation, including the types of knowledge the text requires of a reader, the level of emphasis on systematization, the degree of textual continuity afforded by focusing steadily on a particular subset of artifacts or personnel, the logical rigor demanded of the assertions made in the text, and so forth. These variations have everything to do with the central question: how does a technical ekphrasis connect a reader to the knowledge an artifact encodes?

Rather than asking whether a text is intended for an audience of laymen or experts, or whether a reader is really intended to follow a particular set of instructions, a study that takes ekphrasis as a starting point suggests a series of cognitive and pragmatic questions. Technical ekphraseis can be differentiated from one another by the particular descriptive elements (such as appearance, texture, numerical dimensions, a viewer's reaction, and so forth) that create the vivid imaginative experience of *enargeia*. These passages may also vary in the author's choices of forms of address, which can implicate the reader more or less directly in the imagined construction of an artifact. A text might be defined by how far it carries the reader into the layered bodies of knowledge that make up a mechanical artifact, or how broadly it extends those bodies of knowledge into related knowledge systems like mathematics, harmonics, or astronomy. The technical ekphrasis that brings the artifact to life before the reader's eyes becomes the locus of a set of questions about how we engage with scientific and technical knowledge, about how we bridge the distance between words and things, and about the human-built world.

[144] Asper 2007, 19.

CHAPTER 2

Cultural Contexts

While modern technology inhabits a system of production and development that differs profoundly from its ancient analogues, the web of discourse that links technologies to their host culture was as varied in the ancient Mediterranean as it is in our own. In order to represent technology in text, ancient authors make use of a very broad range of literary techniques, and these are subject to change from one culture to another even independent of changes in the technology itself. All the texts discussed here were of course shaped by the technological, political, and economic cultures in which they were produced as well as the larger literary culture they inhabited.

The propagation of mechanical knowledge in the ancient world was built into a framework that included both literary and practical knowledge. The central importance of practical knowledge, often necessarily tacit, to technological practice must not be forgotten. Even though this book's focus will remain on textualized representations of technological artifacts, practical knowledge always underlay the design and construction of those artifacts. Most technical progress occurs through praxis, which is typically concerned with what is happening in the discipline *right now* rather than reflecting on the past or aiming at cultural positioning beyond its own utility. It does not suffer the time lag of literarization, observed by Vegetius, who connects *ars* both with the architect's everyday operations and with the theoretical principles that inform his work.[1] For Vegetius, "live" transmission of knowledge is inextricably connected both to the student's attainment of *ars* and the everyday *usus* of the trained mechanical practitioner. Knowledge gained immediately from experts through direct instruction and imitation can be up to date as information that circulates in text cannot. Practical education, transmitted by a

[1] Vegetius, *Epitoma rei militaris* IV.35; Meißner 1999, 291.

specialist who can directly address the specific needs of the student before him, offers some obvious advantages of efficiency over textually transmitted knowledge, which is in most cases not shaped for such a specific set of needs.[2]

The importance of humans as vectors for technological knowledge must not go unmentioned even though, like the "mute" artifacts, such tacit knowledge also lies outside the scope of literary discourse about technological artifacts in the ancient world. Apollodorus mentions in the preface to his *Poliorcetica* that he is providing the emperor with an assistant (ὑπουργός) whom Apollodorus has trained by working alongside him, as well as local workmen and others capable of crafting and building (ἐργάσασθαι καὶ ποιῆσαι) his constructions. Practitioners like these, trained through observation and imitation, become vectors for types of knowledge that may well have been very different from the types of knowledge propagated in the surviving literature.

Certainly modern technical and scientific practice depends heavily on practitioners' tacit knowledge. Kogge imagines a typical scene of a craftsman's transmission of tacit knowledge to a student: the master could in fact remain completely silent, using techniques like gesture, demonstration, or guiding the student's hand directly to build up his proprioceptive awareness.[3] Meißner describes a similar scene, emphasizing that written material does not necessarily have any role to play in this particular type of master–student knowledge transmission.[4] The student in such a relationship learns by observation and guided imitation, gradually acquiring the ability to do the work himself and eventually to transmit the knowledge to his own students, once again in a way that need not involve words at all.

[2] Unfortunately, little direct evidence survives for these practices in mechanics; however, other technical disciplines may provide some clues. Medical instruction, for example, may have begun as hands-on instruction transmitted within families and only later developed a literature of its own as vocational training began to transgress family borders: see Meißner 1999, 127–9; Meißner 2003, 160; for some additional ramifications of this shift, see Dean-Jones 2003. This model seems to be reflected in the Hippocratic *Oath* and is given more detail (of a speculative sort) by Galen at *De anatomicis administrationibus* 2.280–3. Meißner argues further that the expansion of literary tools in medical instruction was by the Hellenistic era paralleled in other technical disciplines. It is not clear whether this basic model, either with or without texts, was used for mechanical instruction. However, alongside textual references like Vitruvius' invocation of his *praeceptores* and Philo's mention of learning through directly observing former colleagues of his predecessor Ctesibius, it seems likely that *some* form of hands-on practical instruction was common practice in the vocational training of mechanicians.

[3] Kogge 2012, 31–2.

[4] Meißner 1999, 139. On technical instruction, see also Burford 1972, 82–103; *contra*, Nicolet 1996, 5–6; Meißner 1999, 123–46; Meißner 2003.

Cuomo argues for parallels in ancient technical practice, invoking personally transmitted knowledge that "must have required the existence of an actual, not simply a virtual, community of experts, probably hierarchically arranged, and identified by its sharing of knowledge, knowledge practice, and the values associated with them."[5] Clues about the form such organizations might have taken, like Apollodorus' reference to a system that seems to have included at least three tiers of technical professionals, are scattered here and there throughout the ancient literature on technological artifacts, and considerable evidence may also be gleaned from non-literary sources. Since this is a study of the forms taken by *literary* transmission of this information, as opposed to a broader historical or sociological study about the transmission of technological knowledge in the ancient world, it is only the first of these that will be the focus here.

Textualized forms of technical knowledge sacrifice the immediacy, depth, and nuance of hands-on instruction; in return, they enjoy opportunities for broader propagation, more reliable reproduction, and more obvious points of interconnection with bodies of theoretical knowledge. Embedding mechanical work in a textual tradition, for example, highlights its connections to mathematical texts, providing access to sophisticated geometrical techniques that may increase efficiency or scalability or (as in the case of diagrams) ease comprehension. Yet textually transmitted instruction is by no means limited to theoretical knowledge; forms of knowledge that lend themselves to textual transmission can enrich the craft knowledge most often transmitted directly between practitioners. Both Vitruvius and Pappus stipulate that the *architectus* or *mēchanikos* must master a wide-ranging combination of theoretical and technical knowledge, including theoretical mechanics, geometry, building, and drawing.[6] Hero, though his stipulations about the educational process survive only through Pappus, likewise commends the "mechanical inventor (μηχανικῶν ἔργων εὑρετής)" as a figure gifted with technical problem-solving ability as well as creativity: he should have a versatile nature (φύσιν εὐκίνητον) as well as lifelong experience with his chosen craft.[7]

As mechanical artifacts are a complex combination of many different types of theoretical and practical knowledge, so they provoke a wide range of cultural responses, including the literary responses under investigation

[5] Cuomo 2011, 161.

[6] Pappus, *Synagoge* 8 preface; Vitr. *De arch.* 1.1. The theoretical components of their prescriptions are conveniently summarized in a chart at Fleury 1994, 192, alongside much further discussion of the two authors' taxonomies of mechanical knowledge.

[7] Pappus, *Synagoge* 8.1024.6; on this title see Meißner 2003, 163.

here. Some of these responses shifted diachronically, though there was of course not any such thing as a homogeneously "Roman" way of doing mechanics or writing about mechanical artifacts, differing completely from a "Hellenistic" way. Still, it will be useful to put up some signposts to serve as a rough guide to how technical ekphraseis were shaped by the literary and technological cultures in which they were produced. A set of themes emerge, resonating more strongly in some periods than in others: these include innovation, wonder, secrecy, competition, collaboration, experience, experimentation, and craftsmanship.

Hellenistic Greece

This story begins, more or less, in Hellenistic Greece. Berryman has argued convincingly that mechanics was not a recognized category of technical activity before the fourth century BCE, and even then the development of the discipline was of course gradual; in any event no pre-Hellenistic mechanical artifact text survives.[8] The political and intellectual structures of the Hellenistic Greek world possessed a number of distinctive features that helped to foster scientific and technological development.[9] These included a rapid proliferation of cross-cultural exchanges (friendly or hostile), the association of technological prowess with military success and glory, and conscious efforts by Hellenistic rulers to gather intellectuals from all over the Greek world in concentrated working groups. Meißner locates changes in the relationship between

[8] Berryman 2009, 54–104.

[9] Such developments of course did not arise *ab nihilo*; significant technological developments had obviously taken place in the Mediterranean before the Hellenistic period, ranging from agricultural and transportation infrastructure to sophisticated projects of temple construction and shipbuilding. Moreover, many early technologies came to Greece from elsewhere in the Mediterranean. Lewis notes the incorporation of Egyptian and Babylonian technologies into the Hellenistic practice of surveying (Lewis 2001, 14–18). Heisel categorizes the architectural drawings and engravings from Ptolemaic Egypt as Egyptian rather than Greek (or Roman) on the grounds that they share many more features with pre-Ptolemaic Egyptian drawing techniques than they do with those that were in use by the Greeks prior to the Ptolemaic era; in turn, the Greek drawings adopt similar approaches to those used in Egypt (Heisel 1993, 139–44). Astronomical and time-telling technologies were also particularly strongly influenced by Babylonian and Egyptian predecessors (Neugebauer 1947; Sachs 1974; Høyrup 1997). Greek and Roman authors were of course themselves well aware of their technological indebtedness to these cultures. To name but a few examples of this sentiment in Greek and Roman literature, see Herodotus, *Histories* 2.124–5 on the project of building the pyramids; Pausanias, *Graeciae descriptio* 9.36.5, 9.38.2, 2.25.8 on the treasury of Minyas and the walls of Tiryns (though Pausanias is loath to credit human builders with this achievement); Pliny, *Natural History* 18.167–70 on the irrigation technologies used in Egypt and Babylonia. For accounts of such cross-cultural borrowings within ancient historiographies of science, see Zhmud 2006, 33–44.

Hellenistic practitioners and elites particularly in the development of military technology, where practitioners had to maintain contact with monarchs and their functionaries.[10]

Just as important were the organizational and institutional changes in the broader intellectual culture that shaped not only literary and scientific work, but technological efforts as well. Jacob focuses on Alexandria as the locus of especially intensive critical intellectual work during the Hellenistic era, where new tools of discourse were developed that could resolve scientific quandaries as well as literary ones.[11] Literary changes could likewise effect shifts in social and intellectual networks, for example, the development of a thriving epistolary culture that propagated mathematical, medical, and other technical knowledge throughout the Mediterranean, as well as the lively textual interchange of scientific knowledge among different disciplines and even with other literary genres like poetry.

Communicating the new

Military technology advanced quickly during this period; these innovations were distributed throughout the Hellenistic world, with particularly important centers of activity at Alexandria, Rhodes, and Syracuse. Rihll suggests that the center of activity created by Dionysius at Syracuse "could be interpreted as the creation of what we might now call a critical mass or a research hub."[12] This particular "working group" produced remarkable developments in engines of war, and Rihll speculates that the quinquireme developed under Dionysius at this time could likewise have been the product of "a multicultural posse of shipwrights, gathered where there was already a tradition of successfully altering existing designs."[13] While we must be cautious about imposing modern ideas about centers of technological innovation onto these arrangements, this kind of collaborative innovation may well have played an important role in technological developments in this era.

Opportunities for professional collaboration are explicitly cited by ancient authors, both near-contemporaries and later, as reasons for the technological innovations that flourished at the time. Philo explains in the preface to the *Belopoeica* how craft and scientific knowledge may be advanced beyond trial and error; he is particularly interested in the

[10] Meißner 1999, 161. [11] Jacob 1998, 24. [12] Rihll 2007, 40–1.

[13] Rihll 2007, 41. On the controversies surrounding the invention of the catapult, see Campbell 2011, 678–80.

genesis of the discovery that a catapult's firing capacity is a function of the diameter of the hole within which the spring-cord is wound.[14] He traces this discovery back to the particular social role enjoyed by craftsmen in Alexandria, where well-funded designers flourished thanks to "fame-loving and technophiliac kings."[15] Diodorus Siculus praises Dionysius of Syracuse as a shining example of a ruler who fostered technological innovation by bringing together a variety of designers and craftsmen. According to this account, Dionysius offered them not only generous pay but also additional prizes for the best-performing workmen, something like a performance bonus. Still more interesting was Dionysius' direct consultation with his workers, which in Diodorus' view created an intimacy that spurred them to further innovations.[16]

What developmental patterns might such innovations have followed? The traditional "linear" model of technological development presumes that variant forms of a type of technology arise, more or less entirely replace the older forms if they prove superior, and are in turn entirely replaced by subsequent forms. Cuomo argues that this model is flawed, whether it suggests that such linear progress is gradual or occurs by "leaps and bounds."[17] She proposes instead a "scatter" model, which allows for contemporaneous coexistence of different forms of a technology, both on the level of the complete device (e.g. the catapult) and of its components (e.g. the washers), even if those forms were originally developed at different times.[18]

Elder's analysis of families of artifacts as "copied kinds" is also useful for thinking about technological development: artifacts may diverge, but cluster around properties having to do with their shape, function, or the presence of associated technologies (Phillips-head screwdrivers and screws with appropriate heads are a modern example; specialized catapult ammunition an ancient one).[19] Hodder looks at a process related to evolution of artifacts from a slightly different perspective, focusing on shifts in the

[14] This hole is fitted with a washer, and this whole assembly is part of what Marsden calls the "hole-carrier"; on this component see Marsden 1971, 51–2; Campbell 2011, 683.

[15] Philo, *Belopoeica* 50.39–40.

[16] Diodorus Siculus, *Bibliotheca historica* 14.42.1–2. Cuomo 2007, 43, provides a dramatized account of this arrangement; see also Gille 1980, 54–8; Rihll invokes Dionysius' drive for personal glory as a motivating factor (Rihll 2007, 41).

[17] Cuomo 2007, 42, 48.

[18] Baatz 1994, 173–80, 186–91, 203–9; Cuomo 2007, 55–8.

[19] Elder 2004, 137. Thomasson offers some cautionary words about using Elder's model to define kinds of artifacts and proposes a simpler approach based more strictly on functionality or intended functionality (2009, 201–8). However, for an account of how new types of artifacts can be generated, Elder's broader model is very useful.

"entanglements" of objects over time spans larger than either of these models comfortably allows.[20]

The type of professional mechanical practice that Philo and Diodorus allude to, with its relatively high degree of mobility for practitioners and emphasis on individual achievement and reputation, is perhaps reflected in the rising popularity of the epistolary form for technical treatises of various lengths. Rosenmeyer observes a marked rise in letter exchanges in the Hellenistic period; Taub correlates this phenomenon to a specific culture of knowledge exchange between scientific practitioners, particularly mathematicians.[21] Netz likewise highlights the epistolary activities of mathematicians, particularly Archimedes.[22] The epistolary form serves to create the impression of direct connection between author and reader, and in the case of instructional treatises framed as letters, to create a kind of "instructional space."[23] Many of the surviving Hellenistic texts involving mechanical artifacts are at least framed as letters; these include the surviving components of Philo's *Mēchanikē syntaxis*, Biton's *Kataskeuai*, and Eratosthenes' instructions for the duplication of the cube by means of the mesolabe.[24] Indeed, so common is the use of the epistolary form in this period that Asper posits the form as one of just two possibilities for texts on mechanics emerging from this environment: in Asper's model these texts were either written in epistolary form, or as didactic lectures.[25]

Interdisciplinary innovations

Opportunities afforded by the political structures of the Hellenistic world for cross-cultural exchange of mechanical ideas, designs, and craft techniques were echoed by the interdisciplinary exchanges that also flourished at this time.[26] De Jong suggests that Hellenistic architectural practice was shaped by contemporary mathematical developments. For example, developments in geometry beginning around 400 BCE may have driven a change in the design principles of temples, which appear to have shifted from being laid out on an arithmetically determined grid to geometrically derived proportions.[27] Philo notes as well the application of developments

[20] Hodder 2012, 138–57.
[21] Rosenmeyer 2001, 31–5; Taub 2008, 286–7; Netz 2009, 106–7.
[22] Netz 2009, 12–15, 104–5.
[23] On the personnel of the didactic letter in particular, see Konstan 1994.
[24] Heath 1963, 154–5; Tybjerg 2004, 44–6; Taub 2008; Creese 2010, 196.
[25] Asper 2007, 75.
[26] For a brief history of some of the most prominent interdisciplinary connections, see Gille 1980, 59–68.
[27] de Jong 1989, 100.

in optical theory to architecture, as builders learned to account for perspective in the proportions of their buildings so that they would look symmetrical to a viewer on the ground.[28]

The rise of mechanical innovation in the Hellenistic world coincided with an expansion of medical and anatomical exploration and knowledge. Von Staden suggests that these phenomena may have been not only coincidental, but mutually inspired and reinforcing. He argues, on the basis of linguistic parallels between Philo's writings on catapults and the surgical devices described by Andreas of Carystus, that there was a reciprocal "cross-fertilization" between mechanical and medical thinkers in Hellenistic Alexandria.[29] Von Staden notes as well that Erasistratus' model of the heart displays several structural and functional parallels to the force pump designed by Ctesibius, and the design principles of both rely on the theory that continuous void does not exist, but interstitial void does.[30]

Hellenistic mechanical authors also drew on the developing mathematical tradition for techniques they could apply to the task of describing mechanical artifacts. Many Hellenistic descriptions of technological artifacts are "mathematical" in some respect; these are sometimes framed in terms of absolute dimensions, but more often take the form of formulaic or geometrical relationships that could be scaled up or down. Philo is particularly concerned with accurate scaling, and provides the reader with a variety of resizing methods that include looking up ratios in a table, calculation from a formula, and manual geometrical constructions.[31] Biton's treatise takes what might charitably be called a casual attitude toward mathematics, particularly the thorny problems of scaling, and certainly does not offer the cavalcade of mathematical methods presented by Philo, but even he makes frequent use of the conventions of "geometrical" prose, including letter labels and third-person imperatives.[32] Highly mathematical descriptions connect the working principles of the device being described to a developing tradition of theoretical mechanics; in doing so they provide the tools for devising new machines, rather than simple replication of the particular device in the text.

The integration of lettered diagrams into mechanical texts is stronger evidence still that Hellenistic mechanical authors adopted some of the

[28] Philo, *Belopoeica* 50.45–51.10; a structure that fails to correct these optical impressions is called ἄτεχνος (50.49).

[29] von Staden 1996, 91–5; von Staden 1998; Netz 2009, 159.

[30] von Staden 1996. On Ctesibius' pump, see in particular Oleson 1984; Oleson 2004; Oleson 2005. Asper briefly alludes to this pump in Asper 2009, 1.

[31] Philo, *Belopoeica* 51.11–55.18.

[32] On these conventions see for example Guillaumin 1994; Tybjerg 2004.

conventions of mathematical writing. While the surviving mechanical diagrams are found in manuscripts dating from the Byzantine period or later, the letter labels usually remain quite consistent from one manuscript to another, and between the diagrams and the text. This is true even for those texts that survive in Arabic; the order of letter labels tends strongly to maintain the alphabetic order of the Greek labels with the most similar sound, rather than following the order of the Arabic alphabet.[33] Integrating letter-labeled diagrams into descriptions of technological artifacts appears to be an innovation of the Hellenistic period. Architectural drawings and descriptions of buildings like the *Skeuotheca* inscription, which might respectively be the closest non-literary analogues to the visual and verbal features of mechanical artifact texts, do not feature lettered diagrams.[34]

Authors from this period self-consciously present their work against a backdrop of predecessors. Biton frames his whole text as a catalogue of military engines developed by past designers and deployed at particular battles, while Philo highlights his acquisition of information from people who had observed his predecessor Ctesibius.[35] The work of one's predecessors could be mined from texts as well as absorbed in person. Jacob focuses on the role played by institutions like the library of Alexandria as environments where different texts and versions of the same text could be compared and corrected, and where methodological models could be transferred from one discipline to another.[36] Tanner argues that embedding Hellenistic cultural knowledge in institutions like the Museum and Library of Alexandria, or the Library and Athenaeum of Pergamum, created opportunities to reflect on the history of previously collected cultural artifacts while simultaneously extracting the principles upon which new art might expand.[37] New cultural artifacts could expand upon a pre-existing tradition while self-consciously preserving the features that imparted cultural value to their predecessors. While we know little about less formalized institutions like Dionysius' coterie of engineers, it seems possible that these institutions could have played a comparable role in fostering technological innovations developed in full awareness of their predecessors' work.

[33] Philo, *Pneumatica* 68–9.

[34] IG(2)2.1054, 1068. Untermann 1983; Heisel 1993, 126–63. The findings of Marstrand's detailed architectural analysis of this inscription, published as *Arsenalet i Piraeus og Oldtidens Byggeregler* (1922), are more accessibly summarized in Hovgaard 1926.

[35] τῶν περιεργότερον τὸ τοῦ Κτησιβίου τεθεαμένων, *Belopoeica* 72.40.

[36] Jacob 1998, 22–7, 30–4. [37] Tanner 2005, 215–32.

However, Hellenistic authors' review of earlier work did not necessarily foster its repetition: the labor of exploring and revisiting past work appears as a necessary justification of one's own scientific or literary contribution, not an end in itself.[38] The "productive tension between conservatism and innovation" that Platt views as characteristic of Hellenistic religious and aesthetic life shaped contemporary technology as well: here too past work serves both as a basis for professional identity and the raw materials for change.[39] So Philo asserts that his own designs take those of Ctesibius as a starting point, but differ widely from them. He says he understood that his predecessor's design was a good one, so "we ourselves sought to bring the construction to its culmination, as a result of which the design originating with us turned out much different compared to the construction put together by him."[40]

Philo is compelled to differentiate his own technology from the designs of his predecessors; this in turn demands that he thoroughly examine the structure and function of technologies both old and new, as proof of innovation requires rigorous comparison with previous solutions to the problem. Philo's claims to innovation exemplify the particular importance in technological development of what Meißner calls "technical literati."[41] The "technical literacy" of such a specialist allows him to maintain a certain distance from contemporary praxis, which in turn encourages the creative alteration of that praxis.[42] The mechanical texts of the Hellenistic era built upon a solid basis of craft knowledge to new and innovative solutions to mechanical problems, combining the two in a portable form that facilitated reflection on the past while spurring competitive innovation.

Vitruvius brings the perspective of a Roman outsider to the Hellenistic thirst for innovation, citing the recent (*nostra vero memoria*) example of one Paconius, who was commissioned to procure a base for a large statue of Apollo situated in the temple of Diana at Ephesus.[43] Vitruvius critiques Paconius for refusing to use the proven hauling system already developed by Metagenes, instead insisting on devising his own, because he was so invested in seeking fame (Vitruvius describes him as *gloria fretus*). This new machine, however, could not be made to roll straight along the road, but instead kept veering off to one side or the other and having to be hauled backward to get it back on track.

[38] Jacob 1998, 30. [39] Platt 2011, 124. [40] Philo, *Belopoeica* 67.47–68.1.
[41] Meißner 1999, 292. [42] Meißner 1999, 170.
[43] Vitr. *De arch.* 10.2.12–14.

Ultimately Paconius was bankrupted, so wasteful was the slow progress of his novel machine. To Vitruvius' Roman eyes, this insistence on individual innovation, even in the face of successful pre-existing solutions to a problem, appears strange and impractical. To Philo, on the other hand, innovating beyond the current technological solutions to a problem is highly desirable, even necessary. Vitruvius' opinionated retelling of the story of Paconius underscores how ostensibly "value-neutral" attributes of a technological artifact like its relationship to its predecessors become loaded with values depending on how an author chooses to tell its story.[44]

"Alexandrian" ekphrasis

Cross-cultural variations in literary representations of technology obviously take a host of forms more complex than Vitruvius' disapproval of Paconius' insistent innovation. A more nuanced investigation is precisely the goal of studying technical ekphrasis in depth, and a rough description of some of its most prominent features in each of the major historical periods under investigation here is the first step. Zanker explores the artistic and literary phenomenon of "Alexandrian realism," where "Alexandrian" refers to "a literary movement rather than a historical period or a body of literature bound physically, as it were, to the city of Alexandria by the fact that it was written there."[45] Likewise Netz refers to an " 'Alexandrian' aesthetic," which is by no means firmly bound to Alexandria itself, though as Netz observes the city did of course occupy a particularly prominent role in contemporary scientific culture.[46]

The "realism" often marked as a distinguishing feature of the art and literature of this period is one facet of Stern's "perennial realism," which reflects its particular cultural and historical circumstances.[47] These included a perception of living in a highly distributed Hellenistic world, and a concomitant desire to elide the distance between cultural centers through the creation of "realistic" literary encounters between "mainstream" and "expatriate" Greek culture.[48] Zanker enumerates "realistic

[44] Retellings like these resonate with Laird's assertion that remaking a narrative changes the meaning of the story to its audience, even if the changes are to style rather than content (Laird 1999, 57).

[45] Zanker 1987, 2. [46] Netz 2009, xiii. [47] Stern 1973, 168–70.

[48] Zanker 1987, 18–27. Zanker pays lopsided attention to the perception of distance from Athens in particular, given the Hellenistic world's many centers of literary, cultural, and political power. For a more balanced set of perspectives going well beyond Athens and Alexandria, see Ager and Faber 2013. On a comparable process in later Greco-Roman culture, see Platt 2011, 215–18.

pictorialism," "realism of subject-matter," and the "appeal to learning" as principal factors in creating a sense of immediacy.[49] Briefly, these refer respectively to amassing a large amount of visual detail, the use of common sensory experiences to bridge the gap between the present day and the mythical past, and detailed reference to contemporary scientific knowledge and cult lore. Tanner describes the shift from archaic Greek sculptural "naturalism" to a form specific to classical Greece, involving a comparable process of reconciling statues' physical features to those familiar from the viewer's own lived experience.[50] The Hellenistic developments that Zanker tracks should thus be viewed as an extension of an existing tendency rather than an *ab nihilo* novelty, but they are nonetheless distinctive in focusing on differentiated *individual* forms rather than the "naturalistic" *types* Tanner cites.

Fowler also characterizes Hellenistic art, in contradistinction to its classical predecessors, as dedicated to achieving realistic effects and attentive to clever craftsmanship.[51] She notes as well the incorporation of scientific knowledge into art, for example Apollonius' allusion to the recent medical discovery of the nerves (*Argonautica* III.763), or Aratus' detailed and "realistic" references to animal behavior and its significance as a predictor of weather.[52] Asper, on the other hand, cautions that scientific knowledge in poetry should not be too closely identified with the "realistic," as this leads to the mistaken idea that poets like Callimachus attributed to science "a privileged access to reality" for which there is no strong evidence.[53] We can perhaps speak more cautiously of an increased attention to detail and verisimilitude, especially where creatures of the natural world were concerned. Nicander's parade of poisonous snakes is perhaps an extreme example, and Theocritus' botanically detailed landscapes a more typical one. Classical idealism gives way to representations, in poetry and art, of physical and physiological truths, to an interest in the way the world and its creatures

[49] Zanker 1987, 17; see also 55–102, 155–211, 113–27.

[50] Tanner 2005. He cites, for example, the differentiation of Demeter and Persephone into the figures of a mature, full-breasted woman and a more slender young woman with bound breasts, which he argues would "evoke a sensory recognition and response grounded in the body" (83). His discussion of these figures depends upon the prior (67–70) derivation of a theory of naturalism, which he describes both as "pragmatist" and "critical realist," construing cultural and natural signs as factors that can both be present in a work or style of art without necessarily opposing one another. Instead, Tanner argues that "what is distinctive about naturalism is the development of a cultural language of artistic representation which is able to put to work naturally emergent perceptual discriminations and behavioral dispositions in particular culturally organized ways for historically distinctive social and cultural purposes" (69).

[51] Fowler 1989, ch. 1. [52] Fowler 1989, 117. [53] Asper 2009, 15–16.

work, and the techniques of rendering accurate observations into artistic objects both charming and grotesque.

Platt gives a nuanced account of the "naturalism" that characterized Hellenistic sculptures of divine figures, acknowledging that some statues from this period retain "a certain hieratic formality of style" that binds them to past tradition.[54] Naturalistic elements of painted and sculpted bodies include the "soft, fleshy" bodies of Lapith women on a frieze from the temple of Apollo at Bassae, the flared nostrils that suggest Damophon's statue of Artemis is breathing, and a krater that playfully destabilizes the notion of plastic representation by depicting both Apollo and a statue of Apollo.[55] Rather than engaging in a detailed explanation of what qualifies as naturalistic, Platt argues pointedly that the approach is "contiguous to the realm of the viewer in a way that is explicitly contrasted with earlier forms."[56] In this cultural environment, even divine figures are portrayed so as to bring them into close cognitive contact with the viewer. Platt observes that a "naturalistic" depiction provokes contemplation of "the self-conscious *technē* of its creator."[57] This self-consciousness resonates very strongly with the emphasis in contemporary mechanical literature on *technē* and its importance to mechanical practice.

Zanker connects Hellenistic interest in the challenges of representation to the verbal techniques of ekphrasis and *enargeia*, both as they were elaborated in later rhetorical treatises and as they were practiced in the literature of the Hellenistic period itself. Zanker points out that *enargeia* appeared as a literary term much earlier than "ekphrasis," appearing in the "new" Posidippus papyrus and so dating at least to the third century BCE.[58] Indeed, interest in capturing a subject in minute detail reached such a pitch in the literature of the period that Lucian later sarcastically questioned the number of words that Parthenius, Euphorion, or Callimachus might require just to get the action started in the story of Tantalus or Ixion.[59]

Highly detailed works of visual and poetic art are, like the "realism" Fowler discusses, the result of privileging interest in the way things really look, in their varied idiosyncrasies, and in the mimetic technique of reproducing those features in detail. Zanker adds two more features to his model

[54] Platt 2011, 114–23. [55] Platt 2011, 115, 120, 130.

[56] Platt 2011, 117 n. 152. Squire likewise argues that "while scholars are surely right to emphasize the reflexive playfulness of Hellenistic pitchings of text against image (and vice versa), this should be understood as the culmination of an earlier and longer tradition" (Squire 2009, 161).

[57] Platt 2011, 114. On the relationship of the *technē* associated with post-classical artists to that associated with architects, see Tanner 2005, 281–3.

[58] Zanker 2007, 25. [59] Lucian, *Quomodo historia conscribenda sit* 57.

of Hellenistic art's interface with its audience: first, the frequent demand that the audience actively "supplement" the details of the artifact with additional external knowledge; second, the integration of the viewer or reader into the image or text. Supplementation requires that the audience understand and fill in the background to the pieces that are actually before his eyes, for example by knowing the mythic backstory, providing aesthetic interpretation, or recognizing allegorical features. Integration occurs when the viewer or reader is somehow involved with the object before his eyes, to the point of imagining himself a part of it as an "eyewitness" or even a participant.[60] Zanker links the integration of a viewer into sculptural scenes, themselves ranging in register from the "Boy with a Goose" to the Nike of Samothrace, with epigrammatic texts (sometimes of these selfsame statues) and mimetic hymns that address the reader directly.[61]

How might features of Hellenistic literature like its peculiar form of realism, supplementation, or viewer integration be reflected in contemporary descriptions of technological artifacts? The work that "supplementation" demands from the reader seems to correspond to a broader Hellenistic fondness for the erudite, riddling, and recondite. Hellenistic poetry is notorious for the demands it places on the reader's knowledge and his enthusiasm for puzzle-solving.[62] As Netz argues, these features pervade Hellenistic literature, including geometrical texts (which are a literary product of that culture just as the poetry of Callimachus is).[63] Hellenistic mathematicians such as Archimedes structure proofs so as to maximize surprise, delight the reader with virtuoso displays of calculation, and privilege varied mixtures of propositions and even of genres.

Some of these tendencies are also present in texts on technological artifacts, as when Philo impresses his reader with multiple calculation methods and tables of calculated numbers.[64] Such spectacles are, however, far from ubiquitous in the surviving Hellenistic mechanical texts; more often the presentation is straightforward and, while often rich in methodological and structural detail, not apparently designed to surprise the reader. On the other hand, play and surprise must have been central to the effect of some of the artifacts themselves, notably the pneumatic and astronomical

[60] Zanker 2007, 104–23, with a full suite of examples and bibliography.
[61] Zanker 2007, 109–16.
[62] While this idea is nearly ubiquitous in scholarship on Hellenistic poetry, some especially influential perspectives can be found in Fowler 1989, ch. 9; Fantuzzi and Hunter 2004, ch. 7; Tanner 2005, 261–4; Netz 2009, ch. 4. For a particularly relevant case study, see Goldhill 1994.
[63] Netz 2009, 199–210. [64] Philo, *Belopoeica* 51–4.

wonders designed by Philo or Archimedes to baffle and amaze the viewer, though the original Hellenistic texts on these do not survive. Many of these texts also require the reader to supplement textualized information with diverse kinds of outside knowledge, ranging from calculation and geometrical skill to physical intuition, as when Polybius invites the reader to think about the operational principles of Pausistratus' fire catapult.[65] Indeed, the demand for supplementation is particularly strong for texts, like Philo's, that encourage the reader to engage with the tacit knowledge that is so vital to craft work.

Zanker emphasizes that supplementation, which is fundamentally a call to the audience to become intellectually engaged with the viewing experience, naturally "merges into viewer-integration."[66] The technical texts that "reach out" on the most literal level are letters such as those sent to colleagues by Archimedes. Other texts, like Eratosthenes on the mesolabe, perhaps create a more stylized epistolary space. These letters used grammatical forms like direct address to the reader to create an intimate, collaborative space. Texts consisting largely or entirely of building instructions, which seem to flourish first during this period, likewise address the reader directly, creating the sense of a shared workspace while also introducing the element of instruction, using imperatives to deepen the immediacy of shared activity. I will here label these *kataskeuai*, using the Greek term that conventionally labels Biton's text and Hero's *Cheiroballistra*, among other construction-oriented texts.

Direct address, and imperatives that instruct the reader in envisioning an object in his mind, were also a distinctive feature of much of the epigrammatic poetry of the period. Goldhill describes these texts as involving the reader in a process of "directive reading," creating a comparable impression of shared viewing activity even over the spatial and chronological distance from author to reader.[67] Most prominent among these texts are epigrams that address or include the reader in an imagined viewing experience. For example, epigram 72 from the new Posidippus papyrus instructs the reader to "behold (θηεῖσθε)" a statue of a colt and admire it for its realism – to see how it breathes and stretches. None of this is literally visible, of course; even were the statue present, it would nonetheless be a statue, so the epigram would still demand an act of imagination, and the isolated text demands two nested acts – one to imagine a statue of a colt, and another to imagine imagining the statue alive and active.[68]

[65] Polybius, *Historiae* 21.7. [66] Zanker 2007, 75. [67] Goldhill 1994; Tueller 2008.

[68] On the practical challenges of reading Greek inscribed epigrams, see Bing 2002; *contra*, Squire 2009, 151 n. 237. On the evolving relationships between author, poetic voice, addressee, and reader

Netz observes the use of words like "imagine (νοείσθω)" in Hellenistic geometrical texts to direct the activity of the reader's mind when an element of a mathematical construction cannot be drawn or cannot be adequately understood from a diagram.[69] Hellenistic texts that tell the reader how to build mechanical artifacts frequently use "imagine" and other cognitive commands in a similar way to the mathematical texts. While texts that give instructions were certainly present in earlier cultures (for example, Egyptian and Mesopotamian medical and mathematical texts), the Hellenistic incarnation of directive texts takes a far richer set of possible forms, as the "supplemented" figure of the instructed reader as a cognizing agent develops to a degree previously unknown. The reader is asked to imaginatively manipulate and edit complex assemblies of components, estimate their physical limitations, and think about their materials, demanding a remarkable degree of imaginative sophistication that closely resembles Goldhill's well-trained Hellenistic reader of epigram.

Technical ekphraseis from the Hellenistic world of course vary in the degree to which they exhibit these features, but we can still identify a certain profile that roughly differentiates them from those written in later periods. Descriptions of mechanical artifacts from this period most often emphasize the peculiar qualities of materials and the specific structural or functional features that differentiate new devices from old. To precisely understand the differences between designs requires deep engagement with their technical details. This is facilitated by the use of textual technologies like letter labels and diagrammatic conventions adopted from mathematical texts, which were enjoying their own radical developments during this time. Mathematical texts and poetic forms like the epigram may also have served as models for approaches like direct address to the reader and imperative commands, which are a striking characteristic of Hellenistic technical ekphraseis. These texts likewise often make greater demands on the reader to bring in outside knowledge or solve problems proposed in the text. Hero will later complain in his own texts about the inconvenience of leaving such exercises for the reader, citing them as flaws in the earlier texts, but in fact they are well in line with the demands of Hellenistic mathematical texts.

from archaic and classical to Hellenistic epigram, see Tueller 2008, who argues for decisive shifts; Squire 2009, 150–68 argues for a less differentiated view. A moderate approach that focuses on the development from "naive" to "self-reflexive" viewing is presented by Tanner 2005, 233–4, 261–4. On the spatial relationships between Roman epigrams set up to be viewed alongside their objects, particularly the sense of tension or competition between their verbal and visual elements, see for example Thomas 1995; Bergmann 2007; Squire 2007; Squire 2009, 168–89, 206–9, 221–34.

[69] Netz 1999, 52–3.

The Roman world

Although the technical ekphraseis from Hellenistic literature are far from homogeneous, those in Roman literature are of another order of heterogeneity entirely, varying widely both diachronically and cross-culturally. The literature of Republican Rome includes many familiar works that are often described as *technical* texts, notably the agricultural treatises of Cato and Varro. However, the surviving Republican-era literature offers no parallels to the works of Philo and Biton – that is, no texts entirely dedicated to the description of *technological artifacts*. But of course this is not to say that the Romans were uninterested in mechanical technology; Roman literature from the Republican era features mechanical artifacts of both local and exotic origin. The former group includes the presses for olive oil and wine and aviaries with their feeding and watering mechanisms, described in different ways in agricultural works like those of Cato and Varro. The latter includes Cicero's repeated references to the astronomical *sphaera* associated with Archimedes,[70] and the fanciful aviary Varro describes that features its own miniature version of the Athenian Tower of the Winds.[71]

Mechanical artifacts, then, were to be found at least in the background of Republican-era literature, but do not seem to play a starring role. The texts themselves are likewise often structured so as to emphasize the distance between reader and artifact, quite unlike the "integrating" or "directive" work of their Hellenistic counterparts. The directive forms of Cato's instructions for building and maintaining the infrastructure of a farm, for example, appear to be addressed to the overseer rather than the owner.[72] Varro's work on agriculture is framed as an account of a dialogue, which composition is then addressed to Varro's wife. He describes the knowledge as coming from three sources: what he has observed in person, what he has read, and what he has heard from experienced people (*peritis*).[73] Both texts are thus structured to keep a kind of distance between author, addressee, and practitioner, a striking contrast to Hellenistic authors like Philo and Biton who make decisive claims to personal practice and expertise, and who (Philo particularly) suggest a path for the reader's involvement with the technical work as well.

As Rome rose to military and political dominance over the Mediterranean, many of the cultural institutions that had shaped the

[70] On the *sphaera* see Cicero, *De re publica* 1.21, *De natura deorum* 2.88, *Tusculanae disputationes* 1.63; also Jaeger 2008, 48–72.

[71] Varro, *Rust.* III.V.17. [72] Risselada 1993, 131.

[73] Varro, *Rust.* I.I.11.

practices of mechanical practitioners and authors in the Greek world changed, and those practices along with them. Rihll suggests that at least in the military domain, the various Hellenistic "innovation centers" gave way to a more centralized and homogeneous Roman system.[74] This might indeed have been plausible on the level of planned design, especially where fortifications were concerned, though certainly the surviving material evidence from other contexts of use demonstrates considerable variations in materials and craft practices. In any event, homogeneity does not of course imply stagnation; plenty of claims to improving innovations are found in literature on mechanical artifacts from the Roman world. Most of these claims, however, focus on innovations in the textual, not the mechanical, domain. Many of the surviving texts downplay detailed accounts of the author's personal experience, expertise, and mechanical innovations in favor of claims to gather the best information to create a useful text for the reader or patron. Hero, for example, claims to produce better texts than those of his predecessors (he specifically names Philo) because he makes the material easier to understand and fills in details the older authors had left the reader to figure out. Of course, the figure of the practitioner-author did not simply wink out of existence; Apollodorus, for example, prefaces his *Poliorcetica* with remarks about his own experience, and apologizes for his text's relative lack of refinement.

Meißner observes in the technical literature a trend of growing emphasis on the scientific and textual elements of technical work from the Hellenistic to the early imperial period.[75] The increased textualization of mechanical work between the Hellenistic and early imperial periods also indicates the broader propagation of a common body of theoretical mechanical knowledge, as reflected particularly by the study of the five simple machines. Schiefsky suggests that the idea of the simple machine was developed in between the pseudo-Aristotelian *Mechanical Problems* and Hero's *Mechanics*, providing a set of theoretical tools that could transform and even compete with the knowledge of mechanical practitioners.[76] Many Roman works with such a theoretical bent further integrate mechanical knowledge with related bodies of knowledge, such as philosophy or astronomy.

Technological innovations seem to remain credited largely to Greeks: important texts on technology were certainly produced in Latin by authors from Vitruvius to Frontinus and beyond, but these texts rarely

[74] On this process in the case of military technology, see Rihll 2007, 176–7, 197–209, 233–40.
[75] Meißner 1999, 140. [76] Schiefsky 2008, 18.

highlight their own innovations. On the other hand, Romans exhibited no reluctance to adopt new artifacts for their own use, patronizing technological work among other intellectual endeavors. This work, and the literary discourse associated with it, took on new forms in the Roman world, and it is not unreasonable to speak of Roman "reception" of Greek science and technology in the same way we think of the reception of Greek art and literature. Like its analogues in other disciplines, the reception of knowledge about technical subjects like mechanics involved a self-conscious translation (both geographical and linguistic) of a pre-existing body of work into a new cultural context. Schrijvers observes that Vitruvius paints the perfect architect as a Roman, possessed of the native Roman moral virtues, but at the same time learned in Greek culture.[77] Attempts to define technical knowledge in a solely Roman domain, without Greek models, were not necessarily successful, as Meißner argues was the case for agricultural texts.[78]

One of the principal challenges to the project of re-contextualizing Greek knowledge in Latin texts was the need to rework the texts composed by their Greek predecessors into a form suitable for a Roman audience, including the task of translating (or at least transliterating) technical terminology.[79] Fögen seizes on the explicit complaints of Latin technical authors about the difficulty of translating Greek technical terms as evidence for their methods of translation.[80] Indeed, "translation" may itself be a misleading term for many of these textual transformations. Fögen observes a spectrum of degrees of resemblance between a Greek source text and its reworking in Latin, ranging from more or less word-perfect renditions, to looser paraphrases, to reconfigurations involving significant abridgement, additions, or reorganization.[81] Roman authors' awareness of the range of possibilities available to them is suggested as well by the range of verbs they use to refer to their work: *vertere*, *convertere*, *transferre*, *traducere*, *interpretari*, and so forth.[82]

[77] Schrijvers 1989, 16. [78] Meißner 2003, 178.

[79] This difficulty is, of course, not peculiar to authors on the subject of mechanical technology; one thinks immediately of Lucretius' complaint (*De rerum natura* 1.832) about the "poverty of our fathers' tongue" (*patrii sermonis egestas*) where philosophical terminology was concerned; later on we find parallel complaints in Seneca (*Ep.* 58.1), the younger Pliny (*Ep.* 4.18), and Quintilian (*Institutiones oratoriae* 12.10.33–4). Quintilian shifts to a sunnier perspective at 12.10.35, suggesting how best to take advantage of the strengths of the Latin language, and minimize its disadvantages compared to Greek. Of course not all Romans felt themselves linguistically hindered at all; Cicero writes off the idea that the Latin language is *inops* as a layman's fallacy at *De finibus* 1.10.5. The complexities of this bilingual negotiation are discussed in depth throughout Fögen 2000. For a broad overview of the mechanisms of developing technical terminology, see Schironi 2010.

[80] Fögen 2003, 31. [81] Fögen 2009, 89–92.

[82] Fögen 2004, 446; Fögen 2009, 89.

Strategies for handling each specific task of this larger effort of textual transformation (such as translating individual technical terms) vary from author to author, and even from one passage to another in the same text.[83] Vitruvius, for example, uses a variety of techniques to introduce Greek terminology into the description of a catapult. Sometimes he makes an explicit translation into Greek of an existing Latin word,[84] and sometimes he simply provides the Greek terms.[85] He assumes an audience with some tolerance for untranslated Greek technical terminology, though he usually provides an explanation of the component's structure or function so as not to leave the reader completely in the dark. The relative inattention to geometry in Latin literature as compared to Greek creates a particularly noticeable terminology gap. For example, when Vitruvius makes a rare attempt at a geometrical description of the analemma on a sundial, he ultimately finds Latin terminology unequal to the task and uses Greek for technical terms such as πρὸς ὀρθάς (perpendicular), although in less intensely mathematical contexts he uses *perpendiculum* without problem.[86]

Whether Vitruvius labels features with untranslated Greek terms or provides a Latin translation, he tends to mark this act of labeling with words like *dicitur* or *vocantur*. The question still remains what exactly is intended by this translation; it is likely, as Schiefsky argues, that passages using formulae structured this way "are best interpreted not as attempts to clarify the meaning of Greek terminology in Latin, but rather as efforts to make the reference of the *Latin* phrase in question clear and unambiguous by giving the precise Greek equivalent."[87] This interpretation suggests that Vitruvius' intended reader has a certain familiarity and ease with Greek, and that Vitruvius sees himself participating in an act of repositioning Greek knowledge.

Expertise and status

A particularly striking change in Roman-era texts on mechanical artifacts, compared with their Hellenistic analogues, is an apparent shift from works imposing deep focus on a particular subject (like Philo's *Belopoeica*) to more broadly based works. These works either downplay or eliminate technical features like mathematical formulae and building instructions, or seek to work them into a larger, all-encompassing body of theoretical knowledge (like Hero's *Belopoeica*). This change might cohere

[83] Krenkel 2003, 15; Fögen 2009, 92–105.

[84] E.g. Vitr. *De arch.* 10.10.5: *posterior minor columna, quae graece dicitur antibasis.*

[85] E.g. Vitr. *De arch.* 10.10.4: *epitoxidos longitudo foraminis s–, crassitudo:–. item chelonii.*

[86] E.g. Vitr. *De arch.* 9.7.2. [87] Schiefsky 2005, 267.

with the somewhat problematic status of technical expertise in Roman society. Broadly speaking, a delicate balance had to be struck between the display of general knowledge, which was admirable, and the display of personal expertise in a particular technical area, which was not. Riggsby argues that architecture, medicine, and rhetoric were regarded in the Roman world as providing "useful, intellectual knowledge, but excessive specialization in any one of them would be inappropriate to the 'omni-competent' citizens of the political class."[88] Gros points out that the modern association between expertise and scientific or technological power and status leaves us ill-equipped to appreciate a Roman world in which this was simply not the case. We are led astray, he says, by the "crainte révérentielle" instilled in us by the very idea of specialized expertise in a wide variety of disciplines, most especially in the sciences.[89] In the Roman world, by contrast, most types of technical expertise conferred no such status in and of themselves. Some types of technical activities seem not to have been stigmatized as intrinsically sordid; so Cicero famously exempts medicine and architecture from his critique of technical trades.[90] These practices require discernment (*prudentia*) and offer utility (*utilitas*), and so are acceptable occupations for those whose class permits participation in paying work.[91] However, Cicero only allows that they are not shameful; no particular status seems to have accrued from expertise in medicine and architecture per se.

On the other hand, some level of expertise in certain types of knowledge was vital to the self-positioning of Roman elites. In describing the Roman disdain for expert knowledge Gros notes haruspicy and astrology as possible exceptions, while Horster observes the cultural cachet projected by deep knowledge of literature, particularly of poetry, rhetoric, historiography, and philosophy.[92] Some attempts to rehabilitate technical knowledge rely on building bridges to these less problematic areas of expertise. Vitruvius' *De architectura* and Frontinus' *De aquae ductu urbis Romae*, for example, seek to recontextualize mechanical technology through appeals to its connections to the *artes liberales* or its administrative value.[93]

[88] Riggsby 2007, 105.
[89] Gros 1994, 76. On the status of craftsmen see also Burford 1972, 124–52; Cuomo 2007, 77–102. On the status of *apparitores*, a group to which Vitruvius could well have belonged, see Purcell 1983; Masterson 2004, 388–92.
[90] Cicero, *De officiis* 1.150–1.
[91] On the complications of pay, see Masterson 2004, 395–8.
[92] Horster 2003, 179.
[93] See Brown 1963; Meißner 1999, 140; Meißner 2003, 162–3.

Many Romans writing on mechanical artifacts present their own authorial work as an avenue to such recontextualization. Some seem to have aimed explicitly to elevate the perceived status of certain technical occupations from *negotium* to *ars*, less a bid for material gain than for improved status.[94] Meißner observes the close connection between Vitruvius' authorial aims and his goals of securing imperial and public regard for the kind of technical work he did earlier in his career.[95] Vitruvius' technical practice appears to be long past when he takes up his pen; he no longer speaks as an active designer or engineer, principally concerned with improving the functionality or cost-effectiveness of a device. Instead, he is preoccupied with making knowledge about these devices accessible and attractive to a broad audience that includes non-practitioners. He presents his artifacts, then, at a distance appropriate not only to his audience but to his own present priorities.

How are these priorities reflected in the way he describes mechanical artifacts? Consider Vitruvius' prologue to his fifth book: "Since I observed the citizens (*civitas*) to be stretched thin with public and private business, I decided to write concisely, so that those reading in the narrow confines of their free time could quickly understand." His ability to reduce the length and complexity of his source material judiciously is presented as a service to the state, and thus a contribution to the status of his text. Such a text ought to be brief, so it must present only a well-chosen subset of the available material; it ought to present that material with the appropriate degree of complexity for a non-expert reader to understand what is going on; and it ought to be organized in a way that facilitates comprehension.

Gros argues that Vitruvius' well-known pairing of *fabrica* and *ratiocinatio* may be read as the architect–author's privileging his textual work over his building work.[96] Linking the pair *fabrica–ratiocinatio* to the duality between signified and signifier that Vitruvius discusses next, Gros suggests a conceptual link to Stoic concerns about the relationship between things and their names, and hence to an implicit argument that well-crafted discourse about an artifact can be more valuable than the thing itself.[97] Such features are hardly unique to Vitruvius' text. Athenaeus Mechanicus, who wrote his work *On machines* around the same time as

[94] Attempts to shift technical disciplines from *negotium* to *ars*, characterized as the search by "technical authors" for an audience in the swiftly changing Roman world of the middle and late Republic, are described by Krenkel 2003, 26.

[95] Meißner 1999, 179. [96] Vitr. *De arch.* 1.1.1.

[97] Vitr. *De arch.* 1.1.3; Gros 1996, 27. Cicero's use of *ratio* and *fabricata* (together with *forma* and *figura*) at *De oratore* 3.178–80 to compare human and divine art in an astronomical context (with particular reference to the *Aratea*, another text with Stoic resonance) is discussed at Gee 2000, 530–4.

Vitruvius produced his *De architectura*, echoes this sentiment about the potential of a well-written treatise to eclipse the artifacts that the treatise describes: "For when a work is well conceived, the good reputation of its architect is secured; when it is set forth in words, the arrangement will have the greatest fame in the commentaries."[98] Athenaeus lives up to his programmatic statement by distancing himself strongly from the invention or production of any of the devices he describes; he makes only one modest claim to having devised a mechanical improvement to one of them.[99]

Nor is this prioritization limited to authors writing about machines and other structures; that is, it is not specifically a rejection of the concrete and material in favor of the theoretical and intellectual, but extends as well to other spheres of technical knowledge. Aelian specifies that his work on tactics (probably completed after Vitruvius and before Apollodorus) will be a didactic improvement on those of his predecessors. He promises in particular "that the material encountered within this work will be taught with more of an eye for clarity than in my predecessors, and for showing each thing in order."[100] Aelian's celebration of his superior clarity and orderliness is, Meißner observes, part of a critique of his predecessors on the grounds that they composed their works for a readership already quite familiar with the material.[101] It also recalls Hero's claims that previous texts on artillery failed to make their material accessible to a sufficiently wide audience, that he would produce a text on automaton-making easier to understand than the one Philo had previously produced, and so forth.[102]

In fact, this is a sentiment widely expressed by technical authors of the early empire: as Arrian says explicitly and other authors of this period demonstrate implicitly in the structure and content of their treatises, they write for a group of readers who are not expected to be technical experts.[103] Rather than an elite of technical practitioners, they often represent a ruling elite, placed in charge of military affairs, the Roman water system, or other complex administrative tasks for which some degree of technical expertise is required.[104] Nor was it only the readers of technical texts who

98 Athenaeus Mechanicus, *Peri mēchanēmatōn* 15.2–4.

99 Athenaeus Mechanicus, *Peri mēchanēmatōn* 32.11.

100 Aelian, *Tactica* pr.5.1–4.

101 Meißner 1999, 189.

102 Hero, *Belopoeica* 2.1–14; *Peri automatopoiētikēs* 20.1–2.

103 Arrian, *Tactica* 1.2–3.

104 Meißner 1999, 189. This does not mean that these works could not contain up-to-date technical information. Meißner elsewhere compares Arrian's work to the imperial era literature on surveying and the *De aquae ductu urbis Romae* of Frontinus, which likewise could have served to train inexperienced administrators; here too the inclusion of up-to-the-minute innovations appears as a benefit rather than a distraction (Meißner 1999, 257).

increasingly occupied leadership roles in the Roman state, but their authors as well.[105] As the empire expanded, technical specialists were more explicitly acknowledged to have a value of their own as advisers, as when Frontinus recommends that the man placed in charge of the Roman water system should have extensive technical knowledge of it.[106] Technical expertise brought with it practical skills, which could of course be put to work in the service of administrative and legal goals, as Horster notes in the case of the surveyors.[107] Many of the texts on surveying that survive from this period seem to privilege contextualizing the discipline as useful to administrative overseers, rather than explaining its technical details.[108] The same is true for many contemporary mechanical artifact texts, so we may conjecture that the social forces that shaped the surveyors' texts operated on these authors as well, even though their background is less well understood.

Greco-Roman authors

Some patterns familiar from Hellenistic technical texts persist in works written by Greek authors under Roman rule; for example, Hero and Ptolemy use geometrical language and letter labels to define their instruments. Yet new patterns emerge too; many texts from this period build the theoretical background of their mechanical subject matter radically outwards, finding connections to other disciplines. For example, Hero finds a philosophical argument for constructing catapults, while Ptolemy uses harmonic instruments to give mathematical explanations for the structure of the cosmos and the human psyche. Another group of works omits such theoretical background in favor of an emphasis on immediate utility, as Apollodorus claims for his catalogue of military engines. These authors might be said to be using Hellenistic textual tools in the service of Roman authorial goals: for example, lettered diagrams and mathematical language are put to work describing artifacts as fitting into a systematized, interdisciplinary body of knowledge rather than showcased in deeply focused single-subject texts.

Hero and Vitruvius each describe an odometer, and the differences between the two exemplify many of the factors that tend to differentiate

[105] The socioeconomic status of technical authors in the Roman world seems to have varied widely, from the nameless *praeceptores* Vitruvius drew on to authors like Varro, Pliny, and Frontinus, all of whom were of consular or equestrian rank. On this see Purcell 1983; Schrijvers 1989, 13–15; Gros 1994, 312–16; Nicolet 1996, 15; Horster 2003.

[106] Frontinus, *De aquae ductu urbis Romae* 17.2. On this issue see also Meißner 1999, 250.

[107] Horster 2003, 191–2. [108] Meißner 1999, 255.

Greco-Roman technical ekphraseis from their Latin-language contemporaries. Vitruvius presents his odometer as part of a technological tradition handed down "from predecessors (*a maioribus*)."[109] This is not a peculiarly Roman claim; as we have seen, many Hellenistic authors make explicit reference to the place of their work in a mechanical tradition. Yet the numerical description of the odometer itself reveals a significant difference: Vitruvius fixes the wheel diameter at 4 feet, approximating each revolution of the wheel as 12.5 feet.[110] Rather than providing a mathematical formula whereby the odometer could be adapted to any desirable or available wheel size (as Hero does for his own description of an odometer[111]), Vitruvius avoids formulas and geometrical language by specifying the wheel diameter and circumference as fixed numbers. That is to say, the version of the odometer he gives his reader is presented as the *exact* device transmitted from his "predecessors," not a jumping-off point for experimentation with the *type* of device.

This is just one example of a descriptive approach very common in Latin-language technical ekphraseis, which tend to feature descriptions of artifacts already built, rather than instructions for building new machines or mathematical abstractions of functional principles. Numbers often appear in these descriptions, but these are most often the absolute dimensions of specific historical artifacts rather than the scalable proportional relationships or formulas expressed in many of the Hellenistic texts. To be sure, Vitruvius is not entirely committed to describing all kinds of artifacts in terms of absolute dimensions. He follows the tradition of Greek works on ballistic machinery by describing both an arrow-firing and a stone-throwing catapult using the *foramen* or spring-hole as the basic unit in proportion to which all the other dimensions are specified; the *foramen* is in turn defined proportionally to the size of the intended projectile.[112] The proportional sizing of the catapults is very much the exception, however.

By contrast, the odometer Hero describes in his *Dioptra* is infinitely variable; indeed, Hero does not fix even an example size like Vitruvius does (though he does give one sample calculation that shows how to relate the number of teeth on the first cogwheel and the circumference of that wheel to the amount of ground the odometer can cover).[113] Instead, after describing the construction of the odometer and its case, and acknowledging the

[109] Vitr. *De arch.* 10.9.
[110] Vitr. *De arch.* 10.9: *certum modum spatii habeat peractum pedum XII S.*
[111] Hero, *Dioptra* 34. [112] Vitr. *De arch.* 10.10–11.
[113] Tybjerg 2004, 35–9; Asper 2007, 202–4.

tricky, inexact interoperation of gears and screws, Hero simply notes that the odometer will be sized according to the spaces the surveyor has to work with at any given time.[114]

Hero's description allows mechanical flexibility as well; he suggests how to extend the number of cogs in the odometer, which can radically enhance its measuring capacity. On the other hand, he notes that it is pointless to make an odometer that measures a greater distance than its vehicle could cover in a single day, as it is easiest just to start the count over each morning. Hero's odometer is thus more grounded than Vitruvius' in practical concerns about craftsmanship and the user's changing needs, and at the same time it is described in more precise and generalizable mathematical terms. This treatise is quite reminiscent of the Hellenistic tradition of works offering instruction in a single subject, written for a relatively narrow audience (perhaps an expert or a professional in training), which make frequent use of mathematical conventions in their prose.

However, there are also many ways in which Greco-Roman texts tend to differ from their Hellenistic predecessors. The Hellenistic use of "directive reading" is largely discontinued in Hero's work and other Roman-era treatises: the populated "workspace" of the Hellenistic texts gives way to a descriptive approach, which does not reach out to implicate the reader in the act of building. These texts also diverge sharply from their Hellenistic predecessors in how they position the knowledge they disseminate. The surviving Hellenistic works generally focus deeply on the subject at hand, sometimes explicitly disdaining discussions that stray too far from the topic. By contrast, Hero and other Greco-Roman technical authors are much more likely to point out interdisciplinary connections, positioning the immediate subject within an expanding knowledge system in a way more reminiscent of Vitruvius than Philo.

Between Hellenistic and Greco-Roman texts, then, there is a shift from using technical ekphraseis to frame mechanical technology as the product of ever deeper and more specialized knowledge, to using them to build outward to neighboring disciplines. In this environment it is less imperative that a reader bring in his own supply of erudite knowledge and be prepared to fill in blanks for himself. The same distinction is reflected in Greco-Roman literature more broadly. Goldhill describes it as a shift in the types of viewer being cultivated. Whereas the Hellenistic texts are designed to create a "self-aware, oblique, brilliant" viewer, the Second Sophistic saw "a gradual and subtle shift away from these high priests of

[114] Hero, *Dioptra* 35.

the Museum" even within the thriving culture of *paideia*.[115] Goldhill notes that compared to their Hellenistic predecessors, Second Sophistic authors like Philostratus and Callistratus draw on a smaller subset of better-known models in their explications of the mythological or literary culture that underlies the paintings they describe. He stresses that this does not imply a concomitant lack of deep knowledge on the author's part; Pausanias, for example, who can hardly be accused of ignorance or inattention to detail, merely "does not challenge the reader with the playful uses of obscurity that so delight Hellenistic writers."[116] The projection of the author's (Greek) education and culture remains extremely important – but this projection is achieved less through playing with references to the obscure or challenging, and more by projecting a broad *enkuklios paideia*.[117]

Mechanical artifact texts from this period likewise emphasize accessibility to a wide audience, situating artifact-specific knowledge in a more broadly systematized scheme of knowledge disciplines. There is still a place for expertise and challenge in this literature; Ptolemy's *Harmonica*, for example, links the arithmetic ratios he discusses to a broader range of disciplines, but in order to appreciate how this is done the reader has to seriously apply himself to the complex arithmetical scheme Ptolemy proposes. The information he will need to do this is all in the text, however, and the text itself is systematically arranged, and furthermore fits together with the treatment of astronomy in Ptolemy's own magnificently systematic *Syntaxis mathematica*.

"Imperial" ekphrasis

The arrangement of information into systematically composed and interlinked texts is one characteristic of what might be called "imperial" knowledge, which König and Whitmarsh describe as depending "upon itemization, analysis, ordering, hierarchization, synthesis, synopsis."[118] Such knowledge is in large part defined by its self-conscious engagement with a store of past knowledge. The principal challenges of composing these texts seem to be comprehensive inclusivity, translation, and organization. Indeed, König and Whitmarsh emphasize that privileging catalogues and completeness should not be identified with disparaging

[115] Goldhill 2001, 157. [116] Goldhill 2001, 158.

[117] On art-historical *paideia* in the Greco-Roman world, with emphasis that this knowledge was part of an elite education, see Tanner 2005, 246–54.

[118] König and Whitmarsh 2007a, 38.

creativity, or despairing of achieving it in the face of an enormous cultural inheritance.[119]

The importance of asserting one's originality as an inventor is less pronounced in this cultural environment than in the Hellenistic world, while the importance of situating one's own text within a growing literary tradition is intensified. Solutions to the problems of a particular time and place may be reshaped upon being resurrected in a new environment.[120] So Hero can claim, with no apparent conflict, that his book on automata will reframe Philo's prior work on the subject and that it will be "quite original (καινότερον)."[121] Likewise, Vitruvius is not offering brand-new mechanical technology to his reader, nor does he pretend that he is. He is rather recontextualizing prior technologies, presenting them anew to a culture that valued technical innovation and expertise rather differently from the originating culture.[122] The artifacts may be similar or identical to those seen in earlier texts, but their literary presentation may be very different.

Descriptive strategies also seem to have drawn on certain contemporary developments in rhetoric. The refinement of the theory and practice of rhetoric throughout the Roman imperial period expanded the descriptive potential of *enargeia*, as reflected for example in the work of Quintilian, Aelius Theon, ps.-Hermogenes, Aphthonius, and Nicolaus of Myra.[123] Quintilian's *Institutiones oratoriae* is concerned, among other things, with the power of *enargeia* and the strategies an orator can use to create the effect of vividness for an audience. *Enargeia*, he claims, is a deepening of the capacity of the mind to create for itself convincing visions (*phantasiai*, *visiones*) of things that are not actually there:

> The Greeks call these *phantasiai* (while we call them *visiones*), through which images of absent things are made manifest to the mind so that we seem to perceive them with our eyes and to have them at hand.[124]

A skillful orator, says Quintilian, should be able to cause this same effect in his audience. The orator first has to summon up before his own eyes an image of what he plans to describe: in the case of an accusation of murder,

[119] König and Whitmarsh 2007a, 29. [120] Meißner 1999, 253.

[121] Hero, *Peri automatopoiētikēs* 20.1.

[122] Gros proposes a "triple originalité" on Vitruvius' part: he endeavors to create a Latin technical vocabulary; to situate Roman buildings of the late Republic (including the *theatrum latinum*, the basilica, the forum, the *domus*, etc.) in the framework of *symmetria* that characterized Greek architecture; and finally (and perhaps most importantly), to organize the massive body of knowledge he commands into a complete and coherent collection (Gros 1996, 43).

[123] Some of these authors themselves date to late antiquity, but they refer to rhetorical techniques corresponding to those known from earlier imperial periods.

[124] Quintilian, *Institutiones oratoriae* 6.2.29.

this should include the sudden intrusion of the murderer, the cries of the victim, the fatal strike, and the victim's bloody expiration. The orator generates *enargeia* from the feelings this act of imagination stirs up in him:

> *Enargeia* follows, which is called "clarity" and "vividness" by Cicero; it does not seem to say so much as show, and feelings follow no different than if we were present among the things themselves.[125]

Enargeia is what makes the image communicable to his audience, an immediate and powerful appeal to the audience's imagination that requires considerable sensitivity to the way people feel upon encountering certain spectacles. This sensitivity is exhibited, for example, by Cicero, when he imagines the effect that seeing the *sphaera* of Archimedes has upon either a Roman or a barbarian audience.[126] Accounts of a spectator's reaction to viewing a work of art, if they are themselves intended to enhance the audience's understanding of the described artifact, are a parallel to accounts of viewing technological artifacts.[127] Descriptions of artifacts that evoke their effect on a viewer rather than just their structure or function are a relatively common feature of Roman technical ekphrasis, and represent a strong connection to contemporary rhetorical thinking about how to produce *enargeia*.

Webb notes that Quintilian's description of how to generate *enargeia*, which has a specifically rhetorical context, involves several distinct "moments": the speaker summons up an image in his own mind, then encodes it in spoken words; these words are received by a listener, who constructs his own imagined picture by means of these images (*phantasiai*).[128] *On the Sublime* approaches these questions from yet another direction; while the text is rooted in the rhetorical tradition, the author describes the entire process of imagined viewing of an absent object and transferring that vision to a listener as *phantasia*.[129] In effect, *enargeia* provides the listener with access to the mind of the speaker, which is at the heart of its persuasive power. Ptolemy appeals in the *Harmonica* to the ability of a well-constructed description to bring the mind of the reader

[125] Quintilian, *Institutiones oratoriae* 6.2.31–2.

[126] Cicero, *De re publica* 1.21; *De natura deorum* 2.88; *Tusculanae disputationes* 1.63.

[127] For discussions most relevant to this context of oratorical representation of visual experiences see Platt 2011, 224–35, 299–312.

[128] Webb 2009, 93–6.

[129] Ps.-Longinus, *On the Sublime* 15.1. Webb provides a thoughtful analysis of the differences between this approach and Quintilian's at 2009, 96, and a lucid explanation of the relationship between these approaches and Stoic *phantasia* at 115. On the relationship between *phantasia* and *enargeia*, see also Elsner 1995, 26; Graf 1995; Tanner 2005, 286–7; Platt 2011, 230–4 and (in another sense) 293–332.

into "sympathy" (συνδιατίθεσθαι) with what is being described, as though it were visible.[130] He argues that this immediacy can be facilitated either through poetic imagery or diagrams, creating a seamless connection between the more traditional objects of ekphrasis and the technical objects he himself discusses. He himself uses all these strategies to enhance the ekphraseis on musical instruments in the *Harmonica*, describing both their structure and function in deep detail. Frontinus, in promising to make it as though his reader were in the presence of the features of the Roman water system, hinges the power of the technical knowledge he dispenses upon a rhetorical feat of *enargeia*.

Technical ekphraseis in Roman literature thus reflect their cultural environment in many different ways. Authors on technical topics walked a narrow line between the broader culture's distaste for too-deep technical expertise and its enthusiasm for more generalized treatments of technical topics, like Strabo's geography or Pliny's natural history. These preferences manifested in broadly systematic texts that made the most of interdisciplinary connections. Many of these texts are designed to be accessible to laymen, though others demand more from the reader while preserving the systematic approach. In systematizing texts like these, expertise in a single area is less privileged, and readers are not often exhorted to participate in the work themselves. Rather than privileging the innovation and generation of new artifacts, these approaches appear to favor collecting, organizing, and disseminating knowledge about those that already exist to a broad elite readership. Even a thorough functional description of an artifact is often juxtaposed with discussion of how it connects to other bodies of knowledge.

Late antiquity

Where the transition from the technological practices of the Hellenistic era to those of the Roman Empire may have been marked, at least in some ways, by greater centralization and homogenization of technological designs and products, the opposite dynamic prevailed during the transition to what we now tend to call "late antiquity." The consequences for mechanical engineering, however, were far removed from the once-conventional narrative of decay and stagnation.[131] The anonymous author of the *De rebus bellicis* laments evils afflicting the empire since Constantine's time, which range from debased coinage, to corruption in

[130] Ptolemy, *Harmonica* III.3.63–71. [131] Lavan 2007, xvi–xxi.

provincial governments, to barbarian invasions, and recommends combating them by means of a suite of daringly innovative military and economic technologies.[132]

This is not to say that decay was not part of the picture of technological change during this period; Lewis identifies the period around 600 CE as a particular "watershed" after which many technologies vanished.[133] Valenzani observes that in late antiquity, high-profile building projects constructed by skilled builders were increasingly paralleled by building sites exhibiting a marked decline in craftsmanship and quality of materials compared with their predecessors, like *spolia* walls, and Lewis notes a "severe setback" in civil engineering generally.[134] Yet even this process of decline might be viewed as an innovative way of reusing legacy technology, as craftsmen adapted designs to the resources available to them. The letters of Cassiodorus bear witness to an ongoing effort to preserve or restore public buildings, especially those at Rome, but elsewhere too.[135] Alongside the maintenance of existing technologies, Lavan identifies the principal patterns of technological change in late antiquity as the "recession" of more complex technologies, shifts to simpler technologies and more readily available materials, innovations made under the pressure of challenging circumstances, and the replacement of a technology from one culture by its analogue from another.[136]

Rihll identifies both longitudinal and latitudinal axes of striking geographical variation, arguing that the differences between them became more pronounced as travel became more difficult and less regular for both military and civilians.[137] Yet the silver lining Rihll finds is that "in such a climate, the natural tendency to diversity in technology (as in biology) is not stifled, and less-than-best designs are not inevitably out-competed while still young and tender."[138] Although explanatory analogies between natural and technological evolution should be applied only with great caution, the flourishing of creative technologies in the surviving late ancient texts, along with self-reflective commentary on the benefits of this creativity, cannot be denied.

132 Anonymous, *De rebus bellicis* 2–6. Baldwin tentatively dates this work to the 380s (Baldwin 1978).

133 Lewis 2007, 375.

134 Valenzani 2007, 442–6. Lewis 2007, 369.

135 See, for example, *Epistula* I.25 on the *portus Licini*, II.27 on the synagogue at Genoa, III.30 on the sewers, III.31 on aqueducts and temples.

136 Lavan 2007, xxxi–xxxvi.

137 Rihll 2007, 234.

138 Rihll 2007, 234. On the analogy between technological change and biological evolution, see Cuomo 2007, 41–59; for a more detailed consideration of the mechanisms behind such change, see Mokyr 1996.

Some branches of technical literature came to be regarded as crucial for the preservation of practical knowledge that might otherwise be in danger of being lost. Vegetius expresses frustration with the relative peace of the present day, which according to him was apt to lead to an undisciplined leisure and a neglect of military affairs.[139] Particularly alarming seems to have been the threat that knowledge of certain styles of combat, or siege warfare strategies, or how to build the arsenal of mechanical military engines, might fade away. Meißner suggests that the textual repository of knowledge (in this case of military technology) was viewed as a way of preserving knowledge even in the face of lapsed practice.[140] Vegetius' concerns about the possibility that the military knowledge of the past might simply vanish unless preserved in textual form suggest a self-consciousness of the value of textual resources such as he himself is creating. This is perhaps the other side of the coin represented by the exuberant creativity found in texts like *De rebus bellicis*; even while exploring novelties, it was still important to hang onto the proven technical knowledge that served as a jumping-off point for their invention. The idea that textual preservation of technical knowledge could be a defense against the possibility of knowledge loss may help to explain why even devices whose creativity verged on the preposterous were preserved in these texts.

Late ancient technical authors in Latin appear less concerned about the poverty of technical terminology that had plagued earlier authors like Vitruvius. Ammianus Marcellinus, for example, writing in the fourth century about various kinds of catapults, uses Greek technical terms like *cochleae* without marking them as transliterations, though the anonymous author of the *De rebus bellicis* does explain the etymology of the more exotic *tichodifrus* ("wall-chariot," a device for scaling walls).[141] At the same time, by this point there was less need to rely on Greek terminology, as Latin technical terminology had expanded. So, for example, Ammianus elaborates on the *tormentum* engine (which he etymologizes to *torquere*); Vegetius discusses the *martiobarbulus*, or loaded javelin; and the *De rebus bellicis* discusses several types of *currodrepanus*, or scythed chariot. In this later period in the development of Latin technical prose, the technical language has crystallized somewhat and the use of Greek glosses and markers of unfamiliar technical vocabulary decreases.

[139] Vegetius, *De re militari* 1.28. [140] Meißner 1999, 286–8.
[141] Ammianus Marcellinus, *Res gestae* 23.4; *De rebus bellicis* 8.1.

Innovation and preservation

Cross-cultural interchanges, differential political pressures throughout the empire, and the changing needs of the imperial apparatus itself all drove creative experimental technological developments in late antiquity. White describes a wealth of agricultural innovations such as novel types of harvesting machinery.[142] As ever, developments in technologies of civil life were accompanied by military innovations, from improved fortifications like the Theodosian Walls to new types of armor and weaponry, to say nothing of novel siege engines like the onager. Lavan notes concomitant expansions in "information technology" in late antiquity, driven in part by the rise of the codex and the increasing availability of paper from the East, synergetic with an expanding imperial bureaucracy.[143]

The anonymous author of the text known as *De rebus bellicis* makes particularly extravagant claims to innovation within the military and economic spheres. He begins his text with the claim that "everyone agrees" that nobility, wealth, political power, and eloquence provide no *utilitas*, which is solely the product of one's *ingenium*.[144] Naturally, it is the author himself who is best prepared to bequeath the relevant knowledge to his reader, knowledge that Meißner describes as newly concrete, but not derived from new principles.[145] This text innovates while self-consciously reaching back to existing stores of technical knowledge, an approach also taken by several other late ancient authors on mechanics.

It may be observed that this particular author's love of innovation sometimes gets the best of his claims to *utilitas*. This is, after all, the text that is perhaps best known for the *liburna*, a paddle-wheel boat driven by on-board oxen yoked to a capstan. The mechanism's principle of operation is murkily explained in the most unsatisfying of terms: the wheels function "by a certain marvelous achievement of craft."[146] This, it may fairly be said, is not the apex of detailed functional description of a mechanical artifact. Likewise he proposes several varieties of *currodrepanus*, a war chariot with scythes attached to its sides, on pulleys so that they can be raised or lowered at the driver's discretion. The details of their construction, and of the "automatic whips" (*verbera spontanaea*) specified for one variety of *currodrepanus*,

142 White 1979.

143 Lavan 2007, xxi–xxii. Cassiodorus celebrates the power of paper to drive bureaucracy at *Variae* XI.38.

144 *De rebus bellicis*, pr.6.

145 Meißner 1999, 279.

146 *miro quodam artis effectu operantur*; Anonymous, *De rebus bellicis* 17.2. The practicality of such a vessel is discussed at Hassall 1979, 84–7.

are left almost entirely to the reader's unguided imagination, an omission of detail particularly striking in contrast to the detailed description of a javelin's construction that precedes them.[147] Though we might yearn for additional details, however, this text does vividly illustrate the zest for creative problem-solving that characterized many technical texts in late antiquity.

Palladius Rutilius Taurus Aemilianus likely wrote his *Opus agriculturae* in the late fourth or early fifth century CE. Despite the claims Palladius makes in his preface to present the reader with a no-frills practical work suitable for *rustici*, Diederich has noted his widespread use of rhetorical techniques like *variatio* and personification; he also closes the work with a book in poetry, the *De insitione*.[148] Much of Palladius' technical information overlaps with Columella, and much of the *Opus agriculturae* might fairly be regarded as a reworking of that text. However, he includes some novel material as well, notably an excursus on the materials from which structures like cisterns can be built, and a description of an innovative harvesting machine.[149]

This machine (*compendium*), says Palladius, is used in places like Gaul where the ground is level and the fields large; it also discards the straw, so he specifies that it is useful only where straw is not itself a valuable commodity. He provides considerable detail on the construction and use of the device. First, a two-wheeled cart is built, on which small recurved teeth are lined up, with as much space between them as the height of the ears of grain. An ox of special docility is yoked with his head directed at the cart; as he drives the cart through the field the ears of grain are all captured by the teeth and the straw stripped away.[150] While the description of the device is unique in the surviving literature, White finds archaeological evidence to group it with five "identifiable representations of machines belonging to the same genus as Palladius's harvester, but to a different species, the *vallus*."[151] A harvester of one type among many variants, whose use is specialized for a particular topographical and economic environment: while this sort of variation and specialization of course did not appear out of nowhere in late antiquity, it is emphasized very strongly by authors like Palladius and the author of the *De rebus bellicis*.

"Deuteronomic" ekphrasis

Late ancient literature more broadly considered underwent a similar kind of creative, ramifying expansion. Roberts characterizes late ancient

147 Anonymous, *De rebus bellicis* 12–14; contrast the missiles described at 10–11.
148 Diederich 2007, 258–64.
149 White 1979, 39. On innovation in Palladius, see also Diederich 2007, 73.
150 Palladius, *Opus agriculturae* 7.2.2.
151 White 1979, 42.

poetry as favoring a "jeweled style" generated through detailed piecewise descriptions and a "taste for the densely textured play of repetition and variation."[152] Roberts notes in particular that a special connection between *variatio* and ekphrasis appears at this time; exhaustive enumerations are privileged in a style of ekphrastic description that Aquila Romanus calls *leptologia* and Demetrius *akribologia*.[153] The overall ekphrastic effect is "as though texts were put under a microscope, magnifying the constituent parts at the expense of the whole."[154]

The body of available literature was considerably expanded through commentaries and other types of text that more or less explicitly referenced earlier works. Through the processes of exegesis and commentary, texts on many topics, including technical subjects, became ever more convoluted. Commentaries on religious texts and the philosophical works of the (so-called) Neoplatonists, along with commentaries to literary texts like Servius' commentary on the *Aeneid*, made serious work of linking up the texts of the archive. Other, more playful, forms of intertextual ransacking flourished in this period too, such as the *cento*, the patchwork poems whose rules of composition were formalized by Ausonius among others. The *carmina figurata* written by poets like Optatian represented another form of poetic play with familiar objects; Optatian's *Carmen* 20 is of particular relevance here as it both describes and takes the form of a water organ. The rhetorical advances in the art of description that had occurred earlier in the empire's history also continued to be a focus of improvement during late antiquity. Menander Rhetor's work, for example, codifies the strategies developed by earlier rhetorical theoreticians into instructions about the proper way to describe and praise all kinds of things, from cities to harbors.

Netz situates late ancient geometrical writing within a "deuteronomic" commentary tradition: "a large part of intellectual activity in Late Antiquity was involved not with writing about things, but in writing about books."[155] Cuomo also addresses this approach, noting that deuteronomic textual culture creates opportunities for later authors to recast primary-text material in more complete or accessible forms than the original authors. Among the "manipulations" Cuomo identifies in these texts are the supplementation of steps in a proof skipped by the original author, reworkings of terminology, variations of diagrams, and expansions of sub-cases and generalizations of a particular problem.[156]

[152] Roberts 1989, 38, 40–1.
[153] Roberts 1989, 47. Aquila Romanus, *De figuris sententiarum et elocutionis* 2: 23.16–17 Hahn; Demetrius, *Eloc.* 4.209.
[154] Roberts 1989, 55. [155] Netz 2004, 121. [156] Cuomo 2000, 256.

Commentary of this kind, far from being mindless repetition of dusty old material, is a creative engagement with the intellectual territory staked out by a predecessor. This engagement may simplify or complicate; difficulties may be explained away or new ones identified. By striving to represent all the variations of a problem, a deuteronomic author is performing a task different from that of the encyclopedist. Rather than producing a survey of all the relevant intellectual territory, the explorer of sub-cases exhaustively combs a particular area of that territory for everything it can yield up.

Oribasius, for example, describes in his fourth-century *Collectiones medicae* a whole lineage of variants on the *trispaston* (a triple pulley system, used in a medical context for setting bones).[157] These variants thoroughly cover the possibilities for the *trispaston*'s design, and contextualize those possibilities as a competitive or collaborative technological evolution. Oribasius provides a history of the *trispaston* that focuses primarily on what each new player had to say about previous designs, and the arguments they made for the superiority of their own. That is, it is not so much a history of a technology as a history of commentary on that technology. Likewise, Pappus and Proclus both commented at length on Ptolemy's descriptions of astronomical equipment, citing occasions when Ptolemy did not include all the necessary information and expanding upon the material he does provide.[158] They fill in the missing information from their experiences of a similar device, providing much more material detail than Ptolemy himself does. Each new text in such a chain really is new, and represents an improvement over old technological solutions or textual representations. While the nature of the commentary is to look backward to the original, it also looks forward to the needs of a new crop of readers, and a good commentary strives to innovate and to improve understanding.

Intense consciousness of one's predecessors in invention, and the idea that new approaches to old problems can foster real improvements, can be a strong driver of competition. "Along with the rhetoric of 'past is good,' we also find a notion that 'new is better,'" Cuomo remarks on the perception among late ancient mathematicians that improvements in their work meant real benefits to their communities.[159] Innovation can also create a desire for secrecy: although communication is critical for defining and

[157] Oribasius, *Collectiones medicae* 49.23.1–15.

[158] Pappus, *Commentaria in Ptolemaei syntaxin mathematicam 5–6*, p. 3; Proclus, *Hypotyposis astronomicarum positionum* 6.2–9.

[159] Cuomo 2000, 261.

developing technology, there are times when it is deliberately choked off as a way to control that technology.

The interplay of innovation and secrecy forms the core of Oribasius' story of the *trispaston*'s competitive evolution. The *trispaston* emerged as a large-scale ship-hauling pulley device credited to Archimedes, which was later scaled down into a device suitable for setting bones and other medical tasks. For his account of the later development of the device, Oribasius draws on the story of a multigenerational family of *trispaston* designers, who seem to have lived in the Hellenistic period. Aristion, the first designer in the family, apparently designed his *trispaston* as an improvement to an existing model; his version added an axle, so that there would be a middle axle that always remained fast. Oribasius then goes into some detail about a new version of the machine designed by Aristion's son Pasicrates, a description that is as interesting for its obscurities as its clear details. After providing a list of the machine's outer dimensions, we get a tentative glimpse of the machine's interior:

> As Pasicrates said, in the middle of the sides, which are dug out into shallow cuts, there are two hidden axles (ἄξονες κρυπτοί), holding wheels near the sides. Each wheel is bored with a hole that goes all the way through. These holes are made for tying the ropes.[160]

The word *kruptos* will prove to be important to the way Oribasius handles the description. We are told how the cables enter the holes in the two (front and rear) axles and are stopped with knots, and pulled back through the holes in the case (the front, back, and connectors all have holes bored in them). When the machine is to be put into operation,

> when [the cables] are pulled by hand, using the cable in the middle of the back, and the hidden cables are wound against one another, as was stated, the external ends of the cables are drawn forth, and by these ends the straps from the stretched bodies are stretched out.[161]

Once again we see (or rather don't see) the mysterious hidden cables. This description is like a spectator's account rather than a builder's, coming from the outside of the obscuring case rather than from the perspective of a designer who understands the machine throughout. Drachmann points out that the description is so very external that the inner workings of the box are missing, and that no possible textual emendation would change this. Hence he suggests that the aim of disguising the device's workings, so

[160] Oribasius, *Collectiones medicae* 49.23.2. [161] Oribasius, *Collectiones medicae* 49.23.14.

important in this competitive environment, was so thoroughly achieved that Pasicrates did not actually know how it worked.[162] Whether or not this was actually true of Pasicrates, the successful concealment of technological information is central to Oribasius' own representation of the family dynamics of knowledge exchange.

Pasicrates' criticism of the original design (as he understood it) follows: he argues that since the only applied force is from the rear cable being pulled by hand, not enough force would be available to pull a large joint into alignment. He settles on the solution of increasing the size differential between the front and rear axles.[163] It is hardly surprising that Pasicrates thinks poorly of the original machine, given that (at least as he is presented by Oribasius) all the information he has about the instrument is what can be conjectured from outside the box. There is no component in his description of it that would give the user any kind of mechanical advantage, and so he produces his own new and improved design, which allows the back axle to be turned via the handles, which presumably serve as a kind of crank.[164]

The story of the *trispaston* design does not end here. Pasicrates himself has a son, also named Aristion, who disagrees with his reasons for altering the machine:

> But Aristion, the son of Pasicrates, objected to his father as being ignorant of the original design of the machine; he says that in the original apparatus (ὀργανοποιία) the axles did not move in hollows in the sides, but around pivots nailed to the sides, with helical bearings; it is possible to see something of this kind in water-lifting machines, which are turned easily because of the design.[165]

Aristion criticizes his father's (literal) lack of insight into the machine, saying that hidden bearings would allow a machine with the external appearance described by Pasicrates to work. At last Oribasius himself weighs in on the issue, supporting Pasicrates' claim that though the hand alone is not strong enough to exert bone-setting force, a properly constructed *trispaston* does not rely on the hand for its force. Finally, having considered all these possible variations on the form and the different reactions to each of them, Oribasius proposes his own design, which adds a third external axle behind the one that had been in the rear.

This story exemplifies several ways the flow of knowledge about technological artifacts might be controlled. The casing of the *trispaston*

[162] Drachmann 1963, 180. [163] Oribasius, *Collectiones medicae* 49.23.15.
[164] Drachmann 1973. [165] Oribasius, *Collectiones medicae* 49.23.19.

has a lid which serves specifically to hide the workings from prying eyes, whether they belong to patients who are more likely to be impressed by a bone-setting "black box", or to fellow doctors who would like to get their hands on novel technology to help them attract patients. Furthermore, this particular branch of the *trispaston*'s development is happening within the relatively closed community of a single family. While there was not a broadly standardized toolbox for controlling access to knowledge about an artifact along the lines of the modern patent, practitioners and authors appear to have been quite cognizant of the value of technological knowledge and the people who possess it.

The *Collectiones medicae* covers a broad range of medical topics, and often seeks in particular to present these subjects in the context of their intellectual history as problems and debates. Oribasius reaches back as far as he can in this history, to the near-legendary Archimedean origins of the *trispaston*, and then narrows his focus to the beginning of its use in a medical context. He is deeply interested in the progress made in the design of the machine, and particularly in the flow and evolution of knowledge about it: what did Pasicrates know about the original design, and where did his knowledge fail? In what sense is his own design a product of his imperfect knowledge? How did the critical, competitive approach of Pasicrates' son (whose knowledge was flawed in its own way) to the previous iterations of the design shape his own new design?

This is the close-knit, competitive design atmosphere we have seen in Hellenistic writing on the development of mechanical artifacts, framed for the reader in a new way. For the most part, Oribasius holds this competition at arm's length. He focuses not so much on providing the reader with knowledge about the *trispaston*, but rather with knowledge about knowledge about the *trispaston*. Finally, however, he does engage himself in the story by providing his own design: the "deuteronomic" approach to describing technical problems and their solutions entails the resurrection of and re-engagement with those problems, with a particular interest in how the knowledge involved in those problems is transmitted and manipulated. Secrecy about knowledge (whether religious, alchemical, or technical) was of particular concern in late antiquity.[166] The fact

166 Long 2001, ch. 2, particularly 64–71 on the secret-keeping methods of dyers. While Long says that there is "little evidence for craft secrecy within the ordinary crafts of the ancient and late ancient worlds" (71), this would seem to be an exception. The differentiating factor may be that the *trispaston* represents a kind of "high technology," a world somewhat different from that in which the "ordinary crafts" were designed and created.

that Oribasius frames this technological evolution as a kind of contest of deceptive secrecy may speak specifically to his late ancient context.

Textually mediated evolution of a problem or a technology demands a group of texts whose authors are aware of and interested in what their predecessors had to say about the issue, and furthermore in flagging their references as such for the reader. A reference to past problem-solving may tell about the whole chain of development (as Oribasius does), or it may just point the reader one textual step back along the way. Though the problem under scrutiny may remain the same from one text to another, each author in the series will handle it in his own way. This may result in progress toward completeness, abstraction, or specificity, depending on the problem and how it is used by successive authors.

Conclusion

This chapter has sought to adumbrate a few of the most prominent intersections between the features of the surviving literature in which mechanical artifacts are described and the larger intellectual cultures, scientific and otherwise, in which those texts were produced. Of course it is only possible here to attempt this task at a very coarse granularity, and even then the findings here are tentative, being tendencies rather than immutable laws.

Strong features of the Hellenistic practice of mechanics include an approach to engineering that values innovation, continuity of development, and exposure of the theoretical underpinnings of the mechanical work at hand. These values make their presence known in texts through praise of ingenuity or novelty, praise or blame of predecessors, and frequently through presentations of machines in a style very close to that of contemporary mathematical writing. In order to vivify their descriptions and draw the reader in, Hellenistic authors on technical subjects may use the same techniques of "directive reading" as their contemporaries in other genres such as epigram.

Many technological artifacts, as well as much of the literature that transmitted and decoded the knowledge they embody, were incorporated into new intellectual cultures that arose with the expansion of Roman power. These strategies remain active in the Greek-speaking cultures of the Roman Empire, while the somewhat different cultural priorities of Roman elites reshaped these technologies as they were integrated into Latin technical literature. The development of rhetorical theory throughout the Roman Empire, with its explicit emphasis on the persuasive power of *enargeia*, provided a wealth of descriptive strategies to authors trying to bring objects of all kinds before their readers' eyes.

In late antiquity, literary forms like the *cento* reflected a delight in the play of exploratory recombination, reuse, and evolution of existing materials, a culture that managed to look back toward tradition and forward toward innovation at the same time. Explicit reference to the connection between current technology and its predecessors is common in this "deuteronomic" culture, as it was in Hellenistic culture. Late ancient authors on technical topics may (more often than their Hellenistic counterparts, more like Roman authors) focus centrally on the textual tradition of those connections. At the same time, of course, innovations and improvements continued to be enacted on the artifacts themselves, work that could be carried out in the persistent extra-textual domain of tacit knowledge exchange and might never find its way into these chains of commentary texts at all.

CHAPTER 3

The Ekphrastic Complex

Introduction

Technical ekphraseis weave together descriptions of the artifacts themselves with the stories of invention, acquisition, and so forth that often accompany the artifact into the text. Rhetorical authors, indeed, explicitly recognize the skillful capture of such attendant circumstances as an important part of the effect an ekphrasis has on its reader. They do not give this component of the description a separate name, but for our purposes here it will sometimes be useful to refer to it separately, so let us call it the "parekphrasis." The whole assembly may best be referred to as an "ekphrastic complex" as a reminder of its functional unity, despite the slight artificiality of the separation imposed for the analysis here. The ekphrasis is dedicated specifically to the task of describing the object to the reader through attributes like its form, materials, mechanical function, or construction process. The "parekphrasis" contextualizes the artifact described in the ekphrasis: it may include stories of a device's invention or use, or convey the subjective experience of people who come in contact with the artifact as spectators or users.

Both ekphrasis and parekphrasis are products of the interplay between some attributes that change from one culture or genre to another, and others that remain relatively stable. The verbal methods potentially available to an author for evoking a vivid imagined picture of an object do not normally change radically from culture to culture. Indeed, although there have been vast improvements in our ability to reproduce complex diagrams easily, someone attempting to describe a machine in words today has approximately the same toolbox available as our ancient authors. This stability reflects what Zanker calls "the eternal realism," and Stern "perennial realism," the recurrent literary mode that aims at "fidelity or representation, rendering the precise details of the real thing or scene."[1] In Zanker's

[1] Zanker 1987, 3–8; Stern 1973, 168–70.

view, the effect of realism is achieved by literature that is highly detailed, makes an effort to relate to the reader's everyday life, and puts a premium on plausibility.

However, caution should be exercised in defining what is meant here by "realism." In some (and only some) cases of technical ekphrasis, there was indeed, at some point, a particular historical artifact in the world that the author seeks to describe. In other cases, however, it is a *type* rather than a specific *instantiation* of artifact that is under consideration, and moreover it is even possible to describe an artifact that never existed at all. Even in the first case, however, two authors might provide completely different accounts of the same artifact, as the *helepolis* of Demetrius is described quite differently by Diodorus Siculus and Vitruvius. Each author's choice of descriptive tools is shaped by contemporary cultural attitudes toward technology and by the demands the text's form places on its content.[2]

Any artifact (indeed, any material object at all) can be viewed as composed of "layers" of attributes, including the parts and materials it is made of, its dimensions, the kinds of motion it can undergo, and in the case of an historical artifact the physical events it actually did undergo. All this represents a large amount of information, which is not best communicated to the reader by simply listing everything there is to know about the artifact. Instead, representational power lies in the skillful choice of some subset of all these "layers" of the artifact, deployed by an author in such a way as to bring the artifact to life through the text. The particular subset an author chooses to represent in the ekphrastic complex varies from one genre or culture to another, the choice motivated by the context of the work and the demands of varied readerships.

The "cultural particularity" of ekphrasis is thus divided: the goal of enabling a state of *enargeia* in the reader remains, while the tools engaged to produce it and the literary form of the text in which it appears change over time. Such fragmentation inheres in the term "ekphrasis" itself, even though it is often treated as having a single meaning. Scholz refracts the term into its many usages: it is regarded variously as a rhetorical figure with a particular effect on the reader, as a rhetorical exercise with a canon of appropriate objects, as an element of a text whose deployment can be

[2] This tension is akin to Tanner's definition of "agency" as "the capacity to transpose inherited schemata from one context to another and to transform and elaborate the inherited repertoire of cultural schemata … or schemata of social interaction" (Tanner 2005, 145). There is of course some circularity here, as a text's form and intended audience is itself shaped by cultural priorities and preferences; on these problems in the context of didactic poetry, and particularly on the "hermeneutic circle," see Volk 2002, 25–43.

analyzed in a functionalist way, as a literary genre, as an intertext, or as a broadly defined literary mode that contrasts with (for example) dialogue.[3]

Choosing any one of these (and the choice is unavoidable, whether or not it is explicit) will impose a certain structure on how the text is written and read, governing both literary factors (genre, style, intertextual relationships) and cognitive goals (realism, clarity, specificity). The former are much more subject to diachronic change, while the factors that lend vividness to a verbal description of technology in Hellenistic Greece and to one in late ancient Rome (and to one in modern-day New York) will have much in common. Hence Heffernan's *Museum of Words* begins with a claim that "this book is two things: a continuing meditation on the poetics of ekphrasis as a synchronic literary mode, and a selectively diachronic study of its development from ancient times to our own."[4] Heffernan and Scholz both distinguish between the definition of ekphrasis as a rhetorical element with the constant goal of producing a clear picture in the reader's mind, and as a chunk of descriptive text whose style, placement, and significance will vary from one literary culture to another.

The ekphrasis

Ekphrasis can be viewed as mapping an object in the world (or in the author's imagination) to an image of that object in the reader's mind. However, authors make stylistic and structural decisions that complicate that mapping and draw attention from the object to the medium of the text itself. Some authors have a designer's eye for the mechanical or structural features of a device, and will describe it in terms of its construction process, the physical properties of its materials, or the mechanics of its operation. Other texts approach the device from an outsider's perspective, dwelling more on its appearance or historical achievements than on the mechanical principles of its operation. Descriptions of an object's dimensions and the materials it is made from feature in both kinds of texts, and many also lay out the spatial arrangement of a device's components. These descriptive approaches can make use of visual information, or they can invoke the other senses to bring the reader in closer contact with the object of description.

This is not to say that the ekphrasis gives no indication of the object's cultural context; indeed, the features of a machine that an author chooses to bring out in the ekphrasis are a product of the particular cultural

[3] Scholz 1998. [4] Heffernan 1993, 1.

situation the text affords the artifact. Technology may be positioned as something to be understood from a close technical perspective and so constantly improved, or as something to be understood from an overseer's perspective and so watched over and controlled. Technological development may be represented as the work of individual innovators or as a collaborative effort of gradual improvements.

Numbers

Using numbers in a description has a peculiar power to cement an object as a concrete artifact in the reader's mind, giving a description a sense of realism and accuracy whether or not it is deserved. For example, Diodorus Siculus, like Plutarch, supplies quite a lot of information about the *helepolis* Demetrius Poliorcetes used to attack Rhodes, including its dimensions:

> Having prepared a lot of material of every kind he constructed a machine called the *helepolis*, which greatly surpassed in size those that had come before it. Its base was square, each of its sides nearly 50 cubits, constructed from squared beams bound together with iron. The middle area above he divided with beams about a cubit apart from one another, so that there would be room for those who were going to push the machine forward. The whole mass was wheeled, supported on eight large, thick wheels, 2 cubits wide at their rims, armored with strong plates. For sideways motion *antistrepta* pivots were constructed, by which the whole machine easily took on motion in any direction. From the corners came columns equal in length, a little less than 100 cubits long, so connected to one another that when the whole nine-story construction was complete, the first story was 43 *akainoi*, and the highest was 9.[5]

This description gives the spatially oriented reader enough information to construct a fairly detailed mental image of the *helepolis*: its trapezoidal shape, breadth, depth, and height. The numbers Diodorus gives are not obviously rounded; the first story is 43 *akainoi* (a measure of 100 square feet) and not 40, and the last is nine and not ten. (Only by comparing the passage to the description in Plutarch, who gives the length of the side of the base as 48 cubits rather than "nearly 50," do we begin to suspect rounding.) Numbers like these give a reassuring sense of reality to a description; they seem to suggest that the author is describing a specific instantiation of a machine, not necessarily implying that he has seen the object himself, but certainly that he is making a scrupulous record of the details.

[5] Diodorus Siculus, *Bibliotheca historica* 20.91.2–8.

It is a bit of a surprise, then, when Vitruvius also presents the reader with a sizable array of numbers in his description of the same machine, but they differ from Diodorus'. Vitruvius' version of the *helepolis* is 135 feet high and 60 feet wide, and he specifies further that it could withstand a blow from a 360-pound stone.[6] The machine Diodorus describes is about 23 meters wide, whereas the one described by Vitruvius is something like 18 meters, and Plutarch's description earlier makes it around 21 meters. The heights of the "three machines" are respectively about 46, 41, and 29 meters. The machine is the same in the historical sense that it was used by Demetrius to besiege Rhodes, but the machine that appears in our mind as we read these two descriptions is different, because the numbers are different. This is one of the pitfalls of assuming that a "realistic" ekphrasis corresponds to an object in the world with the same attributes.

Where did these numbers come from? After the siege of Rhodes, Demetrius was supposedly forced to leave his siege engines behind (according to Pliny, they were then sold off by the Rhodians to pay for the Colossus).[7] However, according to Vitruvius the story does not end there. The *helepolis* was taken as a special prize by the Rhodian engineer Diognetus, and set up in the city with an inscription identifying it as a public offering (*munus*, in Vitruvius' Latin paraphrase) from Diognetus to the people. If indeed this was true, this machine, set up as a public monument, might have been witnessed and its details propagated; eyewitness measurements could be estimates, which would differ slightly from witness to witness, and after all the two sets of dimensions do not differ by orders of magnitude. What is quite clear, however, is that they were not copied from some common text describing the size of the machine, nor measured by anyone with an eye on precision. The numbers in every text conceal stories such as this; behind every number is the question of its provenance, which can only rarely be answered.[8] Paradoxically, while those numbers reify the machine in our imagination, there is no guarantee that the reality we imagine corresponds to the historical reality of a given machine. Once the machine and text exist as separate entities, the text takes on a life of its own, and in the course of its being transmitted and referenced by other texts, the numbers may well change.

The power of quantitative measurements to lend a specific historical identity to a technological artifact is illustrated even more vividly by Varro. In the third book of his *De re rustica*, he describes two different

[6] Vitr. *De arch.* 10.16. [7] Pliny, *NH* 34.41.
[8] On stylization in monetary figures in Greek and Roman texts, see Scheidel 1996.

aviaries. The first is the standard type of aviary, designed for profit (*fructus causa*).[9] It consists of "a dome or peristyle covered with tiles or netting" (*testudo aut peristylum tectum tegulis aut rete*), equipped with a "low, narrow door," ideally of the spiraling shape used to allow bulls into the theater for fights. Its windows should be "few, and such that trees and birds outside cannot be seen," so the birds inside do not get depressed as they reflect on their own captivity. The description gives an idea of the general shape and layout of the aviary, but its precise form and size are undetermined, dependent on the location of the notional structure.

Compare this description with the second type of aviary, which was built for amusement (*animi causa*).[10] This aviary is a specific object with a defined location (near the narrator's villa near Casinum), based on another specific object with a defined location (at Marcus Laenius Strabo's house at Brundisium) rather than a generalized description of an object type like the first aviary. It has not only a location but also a set of external and internal dimensions:

> Along the banks of this stream there runs an uncovered walk 10 feet broad … between these [walls] lies the site of the aviary, shaped in the form of a writing tablet with a top-piece, the quadrangular part being 48 feet in width and 72 feet in length, while at the rounded top-piece it is 27 feet … Below the base of the columns is stonework rising 1 foot 9 inches above the platform; the platform itself rises about two feet above a pond, and is about five feet wide, so that the "guests" can walk around among the benches and the small columns.[11]

This marvelous aviary receives a very different type of descriptive attention than the first type. Rough details of form are given for both, but only the second is pinned down to its exact dimensions, making of it a particular *thing* rather than a type. It is, indeed, a very particular thing; a spectacle in miniature, providing the birds with a formal Roman-style dining room and even a miniature Tower of the Winds. In this text, this difference of dimensional definition is strengthened by a difference in the verb forms; while the first type is described in terms of what kind of thing it ought to be, through a string of constructions with *oportet*, the second is described in terms of what it is.

Dimensions are not always specified in such absolute terms; particularly when the goal is to explicate a general mechanical principle rather than describe a particular instantiation of an artifact, relative magnitudes

[9] Varro, *Rust.* III.v.1–3. [10] Varro, *Rust.* III.v.8–16. [11] Varro, *Rust.* III.v.10–14.

may be a better guide than absolute dimensions. So Vitruvius compares the mechanical function of a lever to that of a pulley:

> In this way too, when an iron lever is applied to a weight that a multitude of hands cannot move (*quod manuum multitudo non potest movere*), when a fulcrum is placed underneath as a center, which the Greeks call *hypomochlion*, and if the tongue of the lever is slid under the weight, then when pressed by the strength of one man it lifts the weight (*caput eius unius hominis viribus pressum id onus extollit*).[12]

This is a standard problem of ancient mechanics (as it is in statics today), but Vitruvius handles the problem differently from other authors on the subject. Hero, for example, provides a precise formula for the conditions of equilibrium between forces applied at the lever's two ends.[13] Philo's *Mēchanikē syntaxis* is supposed to have included a treatise on levers, which in light of his use of exact quantities and mathematical formulas in the surviving parts of the *Syntaxis* is likely to have been similarly numerically exact. Vitruvius, by contrast, does not deal in specific numbers, but rather with vague quantities like the *manuum multitudo* above. He continues in the same vein as the discussion goes on, referring to "a few hands" (*paucis manibus*), "a little uneven weight" (*paulo et impari pondere*), or "a great weight" (*amplissimam pensionem*).

Vitruvius approaches the idea of an exact ratio most closely in the discussion of balances that follows the explanation of the lever. Here he notes that the mass of the balancing weight should diminish proportionally as it is moved further from the center.[14] Even here, however, he avoids numerical examples and mathematical formulations of the problem. Vitruvius' description of his odometer, which has the wheel diameter fixed at four feet, contrasts in one way with the odometer for whose dimensions Hero gave his reader quite precise instructions for varying alongside his single numerical example. His description of the lever contrasts with Hero's rather differently: this time Vitruvius' description is numerically under-specified compared to Hero's. Vitruvius occupies two different territories – the numerically invariable instantiation and the vaguely quantified rule of thumb – neither of which encourages the reader to perform his own mental calculations and variations. Hero more often occupies a

[12] Vitr. *De arch.* 10.3.2. Here I follow Callebat and Fleury, who approve Schneider's omission of the hard-to-justify *citro* and interpolated *ultro* in describing the fulcrum's position (Vitruvius 1969, vol. x, 117–18).

[13] Hero, *Mechanica* 2.8–9, ed. Carra de Vaux. On the different ways Vitruvius and Hero handle the problem of the lever, see Fleury 1993, 64–9.

[14] Vitr. *De arch.* 10.3.4.

third territory, where quantitative relationships are more precisely determined, but often in formulaic terms suitable for resizing as needed.

Questions about the provenance of numbers that make it into a text, and the sense of precise historical identity they lend to the artifact described, are thus mirrored by questions about what the reader of a particular text is supposed to do with them. These issues play a role in the ongoing debate over the purpose of Frontinus' *De aquae ductu urbis Romae*. DeLaine argues that the numbers in this text serve as a marker that the reader has entered the second of two parts of the text, having moved from the historical–geographical section into the more "technical" section. The numbers tell the reader that he is now in "the world of mathematics, or at least that part of it which can be applied in everyday life and is therefore useful for the practical man to understand if not actually use himself."[15] Frontinus goes to some pains to ease the reader's entry into this world, as he includes definitions and origin stories for the units of measurement.

This section, then, begins in a relatively accessible style, with eisagogic characteristics such as definitions of terms. In DeLaine's view, however, the long string of numbers Frontinus presents may represent a stumbling block for the reader amid these introductory techniques. This is particularly likely, she argues, given that the *De aquae ductu urbis Romae* was probably intended to be heard rather than seen, and that under these circumstances the list of fistula sizes would have been particularly "somnolent."[16] She suggests that the list would not have been merely boring but also probably incomprehensible, asking, "Were the Romans any better than we are at understanding multiple fractions or of retaining more than two or three complex numbers out of a series?"[17] DeLaine is perhaps a bit too quick to judge how good "we" are at these tasks, and the interest such lists of numbers might hold. Individuals' natural number-handling capacities vary considerably, and practice improves them in any case (as attention to a season of baseball or a political race may attest, depending on one's proclivities). It is not unreasonable to think that the Roman elites charged with overseeing the water supply (if they were indeed among the intended audience for the text) would have dealt with such data frequently, and gotten good at understanding and keeping track of what they were hearing.

On the other hand, DeLaine is surely correct in her assessment of the relationship between these numbers and anything that could properly

[15] DeLaine 1996, 126. [16] DeLaine 1996, 128. [17] DeLaine 1996, 128.

be called mathematics. She asserts that the numbers are "statistics rather than calculations," and suggests the eisagogic tone coupld be compared to that of Columella's excursus on land measurement in the fifth book of his *De re rustica*.[18] Columella's treatment is indeed thorough; he presents an extravaganza of units going up to the *centuria* and down to the half-*scripulum* of the *iugerum*, taking time along the way to note regional variations of terminology.[19] At the same time, Columella emphasizes that taking measurements is not properly part of the farmer's job, but rather belongs to the surveyor. Here he compares his stance to that of architects who think it unworthy of their time to take the absolute measure (*modus*) of their buildings when it is only the proportion (*ratio*) that is their proper concern. So he frames the excursus on measurement as a generous extension of his subject matter for Marcus Trebellius, a friend in need: for him he stoops to provide something closer to measurement than mathematics.

The numbers that feature in a technical ekphrasis may thus serve different purposes. Precision (which is not to say accuracy) can be a powerful tool for convincing the reader of the truth of the description. Accumulating a large set of numbers certainly lends an impression of descriptive completeness, as well as suggesting mastery over a system. Alice König delightfully describes the numbers in Frontinus' text as "a veritable flood of aqueduct-related data," and situates them in the larger context of Frontinus' mastery over the "minutiae" of the water system, which allowed him to detect and combat the fraud being perpetrated.[20] While quantitative control does not always produce results as dramatic as these, other number-crammed technical ekphraseis may likewise be intended to give a comparable impression of mastery.

Imagining materials

Identifying the specific materials a machine is constructed from can help to bring it another kind of imagined solidity, one which does not necessarily raise the same problems of provenance that the numbers do. For example, both Diodorus and Vitruvius tell us that Demetrius' *helepolis* was covered in mats of hair and hide to protect it against the impact of oncoming missiles and fire, but this information does not compel the question "how did they know?" as in the case of the numbers. Building materials, though they serve the same function of bringing a specific machine to life in the viewer's imagination, are a very different matter at

[18] DeLaine 1996, 127. [19] Columella, *Rust.* v.1.4–13. [20] König 2007, 186.

the authorial end, because they can be suggested generically by the type of machine. Even if he had never seen the specific machine in question, even if the reports he received of it did not contain that detail, Diodorus could still tell us about a *helepolis* with covering mats because they are a common feature of machines of that type.

It is a relatively easy task to "cloak" a mental image in a skin of an arbitrary material, even if the resulting composite is something we have never seen before. If I ask you, for example, to picture an elephant made of alabaster, or oak, or wholewheat bread, you will have no trouble doing so, even though you may never have seen such a thing. Incorporating materials with familiar physical properties into a description can bring even a machine the reader has never seen into vivid detail; they provide a bridge between the machine in the text and the reader's own experience.

Pliny's description of the materials used in the foundation of the vast temple of Diana at Ephesus brings even the hidden parts of this wonder of the world before the reader's eyes in such a way that he can easily picture their function in the building:

> They built it in marshy ground, so that it would not feel earthquakes or fear their fissures; on the other hand, so that the foundations of such a great mass would not be located in a slippery and unstable place, they strewed them underneath with crushed charcoal, and then with fleeces of wool.[21]

This passage answers two technical questions that the engineers of the building must have asked themselves: How to protect the building from earthquakes? Then, how to keep the thing standing in marshy ground? The enormous temple was a building feat whose structural demands were novel problems for its architects, and still more unfamiliar to Pliny's casual reader.[22] Yet the description of the familiar materials in the foundation makes the engineering principles intuitively clear. I have never tried to sink foundations for any sort of building in any sort of ground, and yet I can easily imagine how weight will settle on the base layer of crunching charcoal. I am familiar with the properties of damp fleeces, and can imagine their absorbing, stabilizing action on the heavy foundations they support. By appealing to the reader's intuitive sense of the behavior of known materials, Pliny provides a clearer picture of the weight and

[21] Pliny, *NH* 36.95.

[22] The temple was supposedly built around 560 BCE with the aid of Croesus, destroyed by Herostratus in 356 BCE, and reconstructed in the Hellenistic period (see in *Der neue Pauly*: Scherrer, "Ephesus: Artemisium"). So it is conceivable that a contemporary of Pliny could have been familiar with the reconstruction, but of course the details of the foundation are in any case invisible to a spectator.

structure of the gigantic building's foundations than a raft of numerical values alone ever could.

Description of the materials used in the construction of a machine may also suggest (whether factually true or false) that an author has a craftsmanlike familiarity with the devices he describes. Philo describes the development of "Celtic" and "Spanish" swords made of a particularly springy kind of steel, and describes how the material is developed and tested.[23] Likewise, he says he himself observed at work craftsmen who once associated with Ctesibius, absorbing their material knowledge. Philo warns that the strength of the bronze plates of a catapult depends on employing the whitest, purest cast metal. Oribasius advises that all the medical traction devices should be made of the springiest wood (ἐκ τῆς εὐτονωτάτης ὕλης), namely oak, and in second place (ἐν δευτέρᾳ δὲ τάξει) ash wood.[24]

Hero likewise cautions the reader that the springs and bowstrings of a catapult should be made of certain kinds of sinew:

> It is necessary to use sinews, either from the shoulders or the back, of all animals except pigs – for theirs is unsuitable (ἀποίητα). It is necessary to keep in mind why the back and shoulder sinew of other animals is so useful. It has been discovered that the most exercised sinews of the animal turn out to be the most elastic: from the deer it is those from the feet, of the bull those on the neck. Keep this in mind for other cases.[25]

Hero's knowledge of the material behavior of all these types of sinew suggests a practical expertise. Descriptions like these (particularly in the *Belopoeica*, which is dedicated to a historical review of ballistic device designs rather than the proposal of new innovations) do not necessarily imply that Hero came up with these guidelines himself; εὕρηται suggests that he is repeating material previously discovered. Hero connects himself to a line of prior practitioners, discovering the subtle variations in the properties of the materials around them through trial and error and passing that knowledge on to their successors. Hero's parting imperative to "keep this in mind" (διανοοῦ) in future endeavors involves the reader in this succession. He is to keep these guidelines in mind, remembering that a catapult is not a geometrical skeleton guided predictably by the abstract principles of mechanics, but a physical structure of wood, metal, and sinew whose performance depends on the variable material properties of its components. The complex and variable properties of building

[23] Philo, *Belopoeica* 71.24–72.5.
[24] Philo, *Belopoeica* 70.17–20; Oribasius, *Collectiones medicae* 49.3.
[25] Hero, *Belopoeica* 29.1–16.

materials are still better learned through experience (one's own or handed down) than through *ab initio* attempts to understand the physical principles of those properties.

Beyond the visual: tactile and auditory engagement

Enargeia, as its etymology suggests, is canonically a visual phenomenon; there are many references in ancient rhetorical works to goals like "placing before the eyes" (*sub oculos subiectio*), and the bulk of ekphrastic description appeals to the sense of vision.[26] The materials an object is made of are more often described as experienced by sight (that is, in terms of their color, brightness, etc.) than by other senses (that is, in terms of their texture, smell, etc.). However, while the preference for visual data is certainly strong, it is far from a hard and fast rule that an ekphrasis must bring the object to mind solely through appeal to its visible qualities.

Cicero defines *illustris oratio* as the type of oratory that simulates having the subject of discussion before one's very eyes, "for this sense is touched most of all – but the others as well, and the mind itself can most of all be moved."[27] Hermogenes identifies hearing as a particular source of ekphrastic pleasure (γλυκύτης), giving many examples from poetry.[28] Forms of ekphrasis that target action as well as form, like the ekphrasis of the *tropos* and ekphraseis of *pragmata*, practically require that the author be able to appeal to the rest of the senses in order to replicate the experience of witnessing the original event. For example, Apollonius' blow-by-blow account of the boxing match between Amycus and Polydeuces is full of visual detail, but also tactile and auditory information: Polydeuces' hands are "smooth-rolling" (ἐυτρόχαλοι), the gauntlets are "parched" and "dry" (ἀζαλέους, ἐσκληῶτες), while the blows fall with a variety of thuds and crunches (δοῦπος, βρυχή) and the fighters incur tactile events such that the reader cannot help but feel in sympathy, like chattering teeth and snapping bones.[29]

The tactile sense in particular is concerned with bodily realities; Zanker attributes this focus particularly to Hellenistic poetry, but these non-visual senses are addressed in later poetry as well. For example, Rees, drawing on Jenkyns, notes that the ekphrasis in Catullus' poem 64 appeals to a variety

[26] See, for example, Dionysus of Halicarnassus, *Lysias* 7.1; *Rhet. ad Herennium* 4.55.68; Quintilian, *Institutiones oratoriae* 9.2.40.

[27] Cicero, *De partitione oratoria* 20.

[28] Hermogenes, *Peri ideōn logou* 2.4.

[29] Apollonius, *Argonautica* 2.67–97.

of senses.[30] The tapestry the ekphrasis describes is, of course, rich with visual spectacle. But the ekphrasis has an auditory component as well; it concludes with Bacchic revelry and a rich cacophony of sound:

> plangebant aliae proceris tympana palmis,
> aut tereti tenues tinnitus aere ciebant;
> multis raucisonos efflabant cornua bombos
> barbaraque horribili stridebat tibia cantu.
>
> Others struck their drums with palms upraised,
> or stirred piercing jangles from rounded bronze;
> horns blew repeated raucous buzzings,
> and the barbarous flute howled with hair-raising song.[31]

From the cry of *euhoe* with which the band appears, to the thump of palms striking drums and the crashing, buzzing instruments they carry, the conclusion of the ekphrasis is richly auditory. Elsewhere in the ekphrasis, we find novel sound-words like "tide-resounding" (*fluentisonus*), and the sensory experiment is carried even further with the wonderfully synaesthetic "softened by the delightful scent, the palace laughed" (*quo permulsa domus iucundo risit odore*).

Machines in motion are noisy, creating a world of possibilities for ekphrastic representation of their auditory component. Pappus, in his discussion of the mechanical effects of the simple machines, vividly describes the "clashings" (ψόφους) of the wedge's splitting action.[32] Vitruvius' odometer features a bronze vessel into which stones fall to mark the passing miles.[33] The verb *sonare*, which describes the noise of the stone falling into the bronze vessel, can describe a variety of phenomena, ranging from drums to the crashing of wind and waves at sea to the groans of animals and people. These sounds have in common high volume and hollow resonance, which will be produced by the hollow bronze vessel as well. The addition of the auditory detail of these punctuated clangs makes it easy for the reader to imagine the work of the odometer as the carriage rolls along.

Other machines are specifically designed to make noise; this is a job for wonders such as those described in Hero's *Pneumatica* and his *Peri automatopoiētikēs*. Specific and detailed description of the audible features of these machines will of course allow the reader to imagine how the device sounds when it is working properly, and so to appreciate it from the external perspective of a spectator. Pneumatic wonders are, of course, designed to be a mystifying spectacle; at the same time, the text is designed to take

[30] Rees 1994. [31] Catullus, *Carmina* 64.261–4. [32] Pappus, *Synagoge* 8.1122.
[33] Vitr. *De arch.* 10.9.

the reader inside the machine, introducing the hydrodynamic principles of its operation. Hero uses a description of the sounds produced by the device to portray both the "insider's" and "outsider's" views of it.

For a device designed to use flowing water to make a group of artificial birds alternate between singing and silence, Hero begins by matching the device's default sound to the song of a specific bird: it produces the sound of the blackcap (μελαγκορύφου).[34] Hero describes other noisemaking wonders as imitating the call of this bird; for all these machines the reference gives the reader a specific bird call to hear in his mind. Letter labels are used to identify the various pipes and their parts, specifying that ΗΘΚ should be "suited to make noise." What the noise sounds like depends on the level of water in the device:

> When the water pours in through funnel ΕΖ, it will happen that the air in the base gets squeezed out through pipe ΗΘΚ and gives off a sound. If, however, the end of the pipe (ΗΘΚ) bends into the water, the sound will be perceived burbling out like the voice of the blackcap. If the water is not near it, there will only be a hissing. For the noises come through the pipes; differences in the sounds happen when the pipes are made more slender or wider, or longer or shorter, and when the part submerged in the water is either greater or lesser, so that in such a way the different songs of many birds are produced.[35]

The "many birds" are an intrinsic part of the design Hero describes; these figures are supposed to be arranged near watery scenery like a fountain or cave, and the figure of an owl placed beside them. The owl is designed to turn toward the birds, which fall silent while it "watches" them and resume their song when it turns away. This is the machine as the spectator sees and hears it, the "outsider" perspective.

Meanwhile, inside the machine, varying the pipe diameters, lengths, and degree of immersion in the water will produce a variety of sounds. Even a reader unfamiliar with the principles of fluid dynamics can imagine the hissing burble of a pipe partially submerged in water; this imagined scene in turn makes clear the position of the pipes' exits and their varied relationships to the surrounding water. It is relatively easy to imagine the effects of even the more difficult pneumatic principles involved (for example, that pipes of different lengths and radii will produce different notes when air is blown through them) when their results are described

[34] *Sylvia atricapilla*, whose song is a high, sliding babble and so particularly suitable to a device like this, which produces a variety of warbling sounds (Pijanowski et al. 2011, 207).

[35] Hero, *Pneumatica* 1.15.

qualitatively as, for example, sounding like the blackcap. This is true despite the lack of detailed, numerical explanation about the dynamical principles involved. Never in the work does Hero use dimensions or even proportions to describe the devices.

It happens that, for both the "outsider" and "insider" spectator of the pneumatic device Hero describes, the physical principles of its operation are most easily comprehensible in auditory rather than visual terms, and so these are the terms Hero uses to describe the device. Other types of operation, such as the springiness of a cable or the smoothness of a piston, are most readily approached through the sense of touch. Hero's *Belopoeica*, for example, contains several passages describing construction methods that contain distinctly tactile information. One such moment occurs in a discussion of how the spring (for the *palintone* model of catapult, or the frame for the *euthytone* model) is tightened:

> When the spring has been threaded through, whenever the holes receiving it admit it only with difficulty because of being filled, it is necessary to push bolts that are rounded, smooth, and narrow at the ends through the aforementioned holes, and knock them in with a hammer. When it seems that adequate space exists for receiving the spring-cord, then it is necessary to insert it. But whenever it receives [the spring] thus with difficulty, then it is necessary to take an iron needle and draw the beginning of the spring-cord through the hole in it, and then, pushing the needle itself through, to drag the spring-cord after it.[36]

Verbs of motion like "push," "knock," and "drag" (διωθεῖν, ἐνκρούειν, ἐπισπᾶσθαι), and textural descriptors like "smooth" (λείας), give a strong tactile impression of the work being undertaken, of handling components that slide easily against one another and components that stick in place. A description of an object's tactile attributes is especially effective when directed at a reader who can imagine himself touching the object, and so those actions are often (as here) delivered through direct imperatives and other personal verbs.

Narrativized description

Modern approaches to ekphrasis have sometimes attempted to draw a hard line between description and narration, assigning each a separate set of functions in the text and generally casting description as subordinate to narration. Lessing takes the stance in *Laocoon* that signs

[36] Hero, *Belopoeica* 28.46–65.

(Zeichen) juxtaposed in space can only express objects (Gegenstände) that are spatially juxtaposed (entirely or piecewise), while consecutive signs can likewise only express objects appearing in series.[37] That is to say, space-occupying objects can only be represented in a visual medium like painting, while time-occupying actions can only be represented by signs encountered consecutively, as they are in text. Opposing objects to actions in this way requires an unrealistic degree of separation between the verbal and visual, which Bal terms the "dichotomistic fallacy"[38] and Squire rejects as an artifact of the "Lutheran segregation of sign from referent."[39]

Lukacs makes the more polemical claim that "objects come to life poetically only to the extent that they are related to men's life; that is why the real epic poet does not describe objects."[40] This claim can be problematized by epic poetry itself (for example, Homer's description of the shield of Achilles, which may portray scenes of human life but is not itself a feature of those scenes), and by reflecting on the degree to which human life consists of interactions with the objects amongst which we live.[41] Genette's famous formulation of description as the *ancilla narrationis* is often interpreted reductively as discounting its importance.[42] These arguments all make a special virtue of the actions that cannot be replicated by (stationary) visual media, and ignore the possibility of describing an object undergoing a process that admits of narration.

Indeed, Webb suggests that ancient students of rhetoric would naturally have classified the most common types of subject matter for ekphrasis (persons (*prosōpa*), places (*topoi*), times (*kairoi* or *chronoi*), and events (*pragmata*)) among Theon's "'elements of narration' (*diēgēsis*)."[43] Webb clarifies this relationship when she notes that Quintilian's differentiation between a bare statement of facts (*narratio*) and a "narration with *enargeia*" is equivalent to the distinction Nicolaus makes between *diēgēsis* and ekphrasis.[44] In this tradition, then, there is no opposition between description and narration. However, it should be noted that Theon and

[37] Lessing 1965, 157. [38] Bal 1991, 27.

[39] Squire 2009, 94–5. In this chapter Squire identifies the numerous problems posed by Lessing's privileging of the verbal over the visual, particularly for the reader who wants to use *Laocoon* to understand the relationship between art and text in antiquity.

[40] Lukács 1971, 137. Noted in Fowler 1991, 26 n. 6.

[41] On the place of the shield of Achilles in contemporary literary criticism on ekphrasis, see Mitchell 1994, 176–81. Lessing claims that Homer cannot be said to be describing an object precisely because he shows the shield in the process of construction (Lessing 1965, 177.) This claim is called into question for ancient literature by Theon's category of the "ekphrasis of the *tropos*."

[42] Genette 1982, 134. Genette clarifies that "ancillary" must not be interpreted as "superfluous" at Genette 1988, 48 n. 16.

[43] Webb 2009, 63. [44] Webb 2009, 88.

ps.-Hermogenes both recommend fitting an ekphrasis to the narrative context or setting in which it is embedded, suggesting that it might be subordinated somewhat to narrative just as Lukacs and Genette claim.[45] However, ekphrastic treatments of *pragmata* and the ekphrasis of the *tropos* in particular seem from the rhetorical manuals to be exempt from subordination to narrative *diēgēsis*.[46]

The claim that description is irreconcilable with narration leads some critics to the further claim that ekphraseis must themselves be free of narrated action. Heffernan selects a handful of more recent critical stances that have developed from these:[47] ekphrasis has been delimited as "a way of freezing time in space,"[48] the verbal equivalent of the "pregnant moment" in art,[49] and the depiction of objects or people in stasis (involving suspension of time and subordination to narrative).[50] Heffernan himself, casting ekphrasis in the role of narrative-delivering midwife to the "pregnant moment," is willing to grant ekphrasis a certain "dynamism."[51] However, since he defines ekphrasis as a verbal representation of a visual representation,[52] its dynamism is limited to imagining the events surrounding the moment of the event depicted in the visual representation, never actually allowing the events to unfold in the ekphrasis itself.

This differentiation is artificial and unnecessary. If we divest the word "description" from adjectives like "mere" that tend to be appended to it and recognize its heritage in the Latin *descriptio* or *explicatio* (rhetorically speaking, the Latin siblings of the Greek *ekphrasis*[53]), we come to recognize its power to re-present an object in text, to bring it vividly before the reader. This representation may use the means we most often associate with the English word "description," namely catalogues of features of the object that are potentially available to the senses, or it may reach beyond these means to verbalize the object in other ways. Such approaches frequently break the stillness ("stasis," "freezing," "pregnant moment") that some critics set up as a defining criterion of description or ekphrasis, and enter the territory of narrative.

Already Genette discounted the claim that a descriptive interlude should be thought of as halting the narrative (unless the narrator

45 Aelius Theon, *Progymnasmata* 119.22–7; ps.-Hermogenes, *Progymnasmata* 10.13–22; see also Webb 2009, 66.

46 Webb 2009, 67–75.

47 Heffernan 1993, 5.

48 Krieger 1992, appendix.

49 Steiner 1988, 13.

50 Genette 1982, 134.

51 Heffernan 1993, 113.

52 Heffernan 1993, 3.

53 Most probably, as Zanker argues, *descriptio* originated as the Latin equivalent of the Greek ὑποτύπωσις or διατύπωσις, earlier rhetorical terms functionally equivalent to the relative latecomer *ekphrasis* (Zanker 1981, 305 n. 33).

himself explicitly engages in a contemplative pause).[54] Yacobi yet more fully embraces the possibilities of "narrativized description," a hybrid term whose very awkwardness calls welcome attention to the artificial partitions so often bludgeoned into the inclusive ancient rhetorical conception of ekphrasis. She locates the roots of narrativized description as far back as Homer, who chooses to reveal objects such as Hebe's chariot and the shield of Achilles (which no one invalidates as ekphraseis) through narratives of their construction or assembly. In scenes like these, says Yacobi, "the elements of the 'body' are projected into the flow of story time, rather than enumerated in spatial coexistence."[55] Nor is there anything exclusive about these scenes' ability to communicate the flow of time through ekphrasis; Yacobi argues that this technique is always available to an author.

Where Lessing saw the shield as the breaking point of description and the starting point of narration, Yacobi sees it as description blossoming into its full potential. Homer can make us see the shield or the chariot by describing the finished form piece by piece, once each piece is in place, or by describing the pieces as they come together. Both approaches verbalize the artifact, encoding it in text so that it can come to life again before the reader's eyes. There is no reason to impose an artificial barrier between the two techniques the poet uses, because ekphrasis is defined by the result it achieves, not by the semantic category of its object or the duration of the object's lifetime that the author describes. This duration could be one "pregnant moment" of finished perfection, or the gradual perfecting of assembly in time; it is at the author's discretion to decide which approach brings the object into the reader's mind most clearly. Hebe's chariot and the shield of Achilles are each in their way functional objects as well as beautiful ones, and this perhaps explains the decision to describe them through their process of construction.

Seeing a device composed of multiple parts come together is a particularly compelling way of getting a sense of its functionality; a description of the construction process is a kind of verbal equivalent of the modern exploded diagram. Its effect can be appreciated by comparing two descriptions of the colossus of Rhodes, an artifact famous both for its immensity and for collapsing in an earthquake. A typical gazetteer-style description of the statue is given by Pliny:

> But the colossus of Rhodes exceeded all in wonder; Chares of Lindos, a student of the Lysippus mentioned earlier, made it. This statue was 70 cubits in height; after sixty-six years it was felled by an earthquake, but even in

[54] Genette 1980, 99–100. [55] Yacobi 1995, 612.

> ruins it is a marvel. Only a few can embrace its thumb, and its fingers were greater than many whole statues. Vast caves gape in its broken-down limbs; rocks of great size are seen within, whose weight stabilized it when it was standing. They say it was created in twelve years for 300 talents, which they had obtained from the war apparatus of King Demetrius, abandoned with his siege of Rhodes out of frustration and tedium. In this same city there are other colossi, numbering one hundred, smaller than this one, but they stand alone all over the place and so make it famous.[56]

Pliny is interested, as is so often the case with him, more in the anecdotal than the technical details of the statue. He describes the basic dimensions and expense of the statue, and notes that it was stabilized with stones, but that is where his interest in the statue as a structure ends. He is more concerned with elements similar to his other engagements with monumental constructions: the name of the builder, the story of its creation out of the spoils of war, and the fact that it fell down in an earthquake. Pliny's description of the vain attempts by visitors to put their arms around its thumb makes vividly clear both its great size and the fact that it is defunct; in zooming in to the hand alone Pliny makes it occupy the reader's whole field of vision, suggesting the staggering size of the whole thing. The statue's immensity is highlighted as well by contrast to the smaller colossi that are left standing, available for viewing around the city. Pliny makes of the colossus a spectacle rather than a technological achievement – that is, we see it from a tourist's perspective rather than an engineer's.

Pliny's approach is very different from that taken by Philo Byzantius "Paradoxographus" in his *De septem orbis spectaculis*.[57] Where Pliny shows us the statue at the end of its monumental lifetime, Philo takes us back to its very beginning, emphasizing the technology of its construction.

> The architect reinforced it within using iron flanges and square stones, the crossbars of which show signs of having been rounded with a hammer, and the hidden part of the labor is greater than that of the parts that are seen. An admirer from among the onlookers is baffled at what kind of fire tongs were used, how great the bases of the anvils were, and what kind of strength the workmen had, in order to forge bars of such huge size.[58]

Philo proceeds to counteract the viewer's bafflement by explaining how the colossus was erected: first the architect built its feet to scale for a statue

[56] Pliny, *NH* 34.41.

[57] The authorship of this work and its date are uncertain; it likely dates to late antiquity, though its author was long identified with the Hellenistic Philo of Byzantium (see in *Der neue Pauly*: Brodersen, "Philon [17]"; Philo Paradoxographus 1992, 14–15).

[58] Philo Paradoxographus, *De septem orbis spectaculis* 28.

70 cubits high, upon a marble base. Since the base was already as high as a statue of normal size, the colossus could not be built upright, but had to be made piece by piece from the ground up. So large was the statue that even its limbs each had to be made of several pieces:

> And after this he had the same design in mind for the work, for it was not possible to move the limbs from the mines. When the casting was done on the parts that were already made, the divisions of the bars and the weld of the flange were made carefully, and when the rocks were placed inside he secured the counterweight. So that he could watch over the design of the work in safety, they piled earth in a huge heap around the limbs of the colossus as they were constructed, and he fashioned molds by burying the already constructed limbs under earth.

Philo starts with the perspective available to a contemporary spectator of the now fallen statue: the appearance of the inside of the statue and the intricate internal structure that once supported it. We see the rocks Pliny talks about, used as a counterbalance; in this account we see them as the statue is rising up in construction, while Pliny only shows them lurking in the *specus* of the fallen statue.

Just as an exploded diagram depicts a complete piece of machinery through its separated components, often with arrows showing how they will merge together, this description tells us about the construction process and encourages us to envision the colossus being assembled and raised. Of course, this account is itself based on details visible, after construction, only once the thing has collapsed and split apart. Pliny offers a kind of sightseer's description of a remarkable but decidedly defunct artifact, where the onlooker tries to put his arms around the thumb like a modern tourist trying to scale the lions in Trafalgar Square. Philo, on the other hand, makes his sightseer boggle at the construction process that lies, latent yet impossible to ignore, within every piece of the statue. As readers, we stand amazed alongside his hypothetical *thaumastēs*, trying to envision the incredible smithy that created this enormous thing – and then we realize that what we are looking at is only one of the many crossbars within the statue. Pliny's description concentrates on the spectacle of the fallen colossus, which can be visited as a point of interest in a world of Greek technological culture now subordinated to Roman imperial power. Philo's description, on the other hand, brings that world back to life by narrating the creation process of the colossus, with all the engineering challenges that made it so wondrous.

Besides stepping through the temporal stages of its construction, an author can also gradually reveal an object through steps in the spatial domain, by walking the reader through it. This technique is, of course, particularly effective for buildings, since this is the way we encounter them in person, but the same spatial indicators work for other complex structures that we read through piece by piece. The largest surviving "walkthrough" is the *De aquae ductu urbis Romae* of Frontinus, who, as noted earlier, gives his text the multivalent label *commentarius*. Indeed, Evans argues that Frontinus' text appears to be modeled on the *commentarii* of Agrippa, as (among other indicators) the descriptions of the individual waterways seem to verbalize maps of them.[59] Although he is selective about the information he includes, Frontinus effectively takes his reader on a walk through the complex Roman water system. Like Vitruvius, Frontinus integrates the task of thoroughly describing technology with the interests of Roman elites, who are more concerned with administration than with the technology per se.[60] The technique of the walkthrough, guiding the reader through the extensive system in great detail, naming points of interest and numbering the distances between them and the widths of the connecting pipes, produces an impression of synoptic comprehension and control.

DeLaine describes the work as a combination of two basic parts: first, a "potted" history of aqueducts combined with geographical and legal information; second, a description of each Roman waterline.[61] These different parts add up to a kind of advertisement for Frontinus himself, "a man of encyclopedic learning but also of practical experience, in fact the complete Roman senator of the austere school, plain in speech but well-educated, a later Cato or Agrippa."[62] The aqueducts are to be celebrated as a crucial part of Rome's imperial identity. DeLaine then goes on to claim that "the natural corollary to this is that the technical details in the *De aquis* have little importance *for themselves*."[63] This is related to her claim that the measurements he gives are not important in themselves, but only as a shower of data meant to overwhelm the reader or listener with Frontinus' comprehensive administrative knowledge of the water system.

In point of fact, however, Frontinus indicates that a comprehensive descriptive approach is absolutely essential to a practical understanding of the water system. The passage that opened this book is worth a second look at this juncture:

> It did not seem irrelevant to me to include the lengths of the rivers and the conduit of each in the appearance of the works. For since the greatest part

[59] Evans 1994, 56–63. [60] Bruun 1991, 17–19, 66–87, 140–200.
[61] DeLaine 1996, 121, 124. [62] DeLaine 1996, 129. [63] DeLaine 1996, 130.

> of this office is the safekeeping of these, it is only right that their overseer should know which require the greater expenditures. Indeed, it was not enough for my concern to put them individually before the eyes, but I also took care to write about the forms of their conduit, so that from these it would become clear where the valleys were and how large, where rivers were crossed, where hollows in the sides of mountains demand assiduous care in protecting and fortifying the stream. From this that utility appears, that we can have the thing as though it were in view, and consider it as though standing right beside it (*ut rem statim veluti in conspectu habere possimus et deliberare tamquam adsistentes*).[64]

Frontinus' stated goal here is precisely what contemporary works on rhetoric advise: the completeness of the description is designed to make something absent virtually visible and present. One of the challenges of this task is ordering such a complete catalogue of information comprehensibly, since an excess of descriptive detail can make an ekphrasis less clear, not more.

In order to help the reader understand this complex system, Frontinus takes him on a "walk" through its various waterways. The walk is essentially a very long ekphrasis, impossible to capture here, but we can at least look at a few passages to see how Frontinus handles it. He ties the location of each waterway to familiar above-ground landmarks of the Roman road system, meanwhile integrating the water system with the road system's imperial heritage:

> Marcia begins at the Valerian road, near the 36th mile marker, three miles along a side road, to the right for those coming from Rome. On the Sublacensian road, which was first laid under Nero's reign, it is around the 38th mile marker, to the left, within a distance of 200 paces, on the left. The water of its springs stands nearly still beneath stony arches, quite green in color. Its conduit has a length from source to city of 61,710½ paces; in a subterranean stream 54,247½; on above-ground works 7,463. Out of the latter, at some distance from the city in many places across valleys 463 paces are on arched works; closer to the city, from the 7th milestone 528 paces are in substructures, the remaining 6,472 paces on arched works.[65]

Frontinus maps out an imagined journey to the beginning of the Marcia aqueduct, and then guides the reader from this starting point along its course to Rome, numbering the distance with paces like a surveyor. The way he directs his reader to the aqueducts' sources is worth noting: the source can be found off the road "to the right for those coming from Rome." Frontinus thus provides the reader with a subjectively oriented

[64] Frontinus, *De aquae ductu urbis Romae* 1.17.
[65] Frontinus, *De aquae ductu urbis Romae* 1.6–8.

perspective of motion along a path, creating the effect Langacker refers to as "fictive motion."[66] The reader is implicated in the walkthrough through Frontinus' orienting maneuver.

In passages such as this, Frontinus gives his reader enough information to create a detailed mental map of the water system. The precision (once again, this is not the same as accuracy) of the numbers, combined with the geographical landmarks, gives an impression of comprehensive control. It is, of course, not a perfect map. For example, in passages like the one mentioned earlier, Frontinus does not always say exactly where the water travels above ground and where below, so it is difficult to imagine looking at the system from outside. Instead, he shows the system from the water's perspective; how long it needs to travel from its spring, through tunnels and aqueducts, where it will finally emerge. Frontinus talks in particular detail about how the water is handled once it comes into the city:

> From these [aqueducts] six are drawn from covered pools on this side of the 7th milestone on the Latin road, where they deposit sediment while, as it were, taking a breath from the rush of their streams. Here too their supply is measured. Julia, Marcia, and Tepula appear at the same point. As we demonstrated above, Tepula had formerly been added to Julia's stream, but is now cut off, taking its supply from the pool of that same Julia, and arrives in its own channel under its own name. The three of them are brought from the pools on the same arches. The highest of them is Julia, Tepula is lower, then Marcia. At the level of the Viminal hill they are enclosed, and come down all the way to the Viminal gate, where they emerge again.[67]

The personification of the waters, with their individual *nomina*, as "taking a breath" (*respirantes*) from their "rushing" (*cursus*), lends a charming liveliness to the system. Of course there is a component of administrative truth to the first set of descriptors, and of hydrodynamic truth to the second, but that just makes the account more vivid. Passages such as this deepen the "insider's perspective" on the system, the narration of the watercourse. The water system's many paths unfold gradually from springs to city, and the waters travel in a temporally extended process (*nunc, deinde, rursus*) that would be impossible to describe comprehensibly without this narrative evolution.

The "walkthrough" technique is not limited to structures that a human being would literally be able to walk through; Vitruvius frequently uses this technique to structure descriptions of complex machines. In his tenth book, for example, he describes a complicated traction device, whose

[66] Langacker 2008, 523–30.

[67] Frontinus, *De aquae ductu urbis Romae* 1.19.

construction begins with a support pole held up by cables, to which are attached sockets with blocks and pulleys. The blocks are threaded with guide ropes, which Vitruvius uses to organize the remainder of the description of the machine:

> Then they are brought to the lowest block and drawn through from the inner part across its highest pulleys. Then they are brought to the highest block and drawn through from the outer part into the inner across its lowest pulleys. When they descend to the bottom, they are brought across from the inner part and across the second set of pulleys to the outside, and brought back up to the second pulleys; having been drawn through they return to the bottom and from the bottom are brought back to the top; drawn through across the top [pulleys] they return to the bottom of the machine. At the base of the machine is a third block; the Greeks call it *epagonta*, and we call it *artemon*. This block is connected to the base of the block with the three pulleys, across which ropes are drawn through and given to men to pull. And so three groups of men, without a capstan, quickly draw a weight to the top.[68]

The cables in this pulley system are our guidelines through the blizzard of components involved, leading the reader's eye from the cables that attach the pole to the ground, to the blocks and pulleys suspended from it, through that complex arrangement, until the ropes terminate in the hands of the machine's operators. Though Vitruvius is explaining a complicated "contrivance" (*machinatio*), he makes its description straightforward by taking his reader along the ropes, along the path of force that makes the system work. A directed walkthrough like this makes a complicated device's operational principles much more easily comprehensible than a naive spatial description like "there is a pole, and three pulleys above and six below, and ropes connecting them." Like Frontinus' walk through the water system, this rope's-eye view of the *trispaston*, from the supporting pole to the experienced (*periti*) workmen operating it, gives an impression of physical comprehension even to a reader with no specialized knowledge of mechanics. Such an approach makes technical discussions accessible to an amateur readership, which is a vital technique for an author hoping to reach an audience of Roman elites.

Ekphraseis can thus be extended narratively in both the spatial and the temporal domains: the first is like a tour through a machine, and the second shows the machine in the making. Temporal extension, achieved through narrative descriptions of the construction process or the operation

[68] Vitr. *De arch.* 10.2.8–9.

of a movable device, is found in Hellenistic Greek authors like Philo, Greco-Roman authors like Hero, and later Greek authors like Oribasius and Eutocius, as well as in the "technical interludes" of historical authors like Diodorus Siculus and Ammianus Marcellinus. Meanwhile, the spatially extensive "tour" is a common feature of less technical texts where the reader approaches the artifact not from the perspective of the designer or builder, but as a tour guide or overseer who requires a comprehensive but not necessarily a technical understanding of it. Such passages may mark an understanding of technology as cultural accessory or administrative tool rather than as something primarily of mechanical interest.

The parekphrasis

The importance of "parekphrastic" material for inducing *enargeia* was well known to ancient rhetorical authors. Dionysius of Halicarnassus, writing on the extraordinary vividness of Lysias' prose, credits this effect to his ability to capture attendant circumstances (τῆς τῶν παρακολουθούντων λήψεως). Likewise, the pseudo-Demetrian *De elocutione* connects the creation of *enargeia* to the faithful representation of accompanying circumstances (τὰ παρεπόμενα τοῖς πράγμασι).[69] The illuminating effect of attendant detail may be related to Maximus of Tyre's observation that memory of one thing can be stimulated by the sight of related objects, as when lovers find pleasure in seeing things they associate with memories of the beloved, like a lyre or a racetrack.[70]

In most passages the parekphrasis is ancillary to the ekphrasis, clarifying the object for the reader using techniques other than the pure description of the ekphrasis itself. Clarification may include extra detail such as a story about the invention or deployment of the device, or it may effectively reduce the amount of information the reader has to deal with by signposting the most important features of the ekphrastic description. The parekphrasis can also supplement the description in the ekphrasis by reporting the subjective response of spectators to the object. Where the ekphrasis tells how something works or how it is made, the parekphrasis

[69] Dionysius of Halicarnassus, *Lysias* 7.1; ps.-Demetrius, *Eloc.* 217.1. On this topic see Zanker 1981, 297–8. Comparable passages not mentioned by Zanker include ps.-Hermogenes, *Progymnasmata* 10.14 (Ἐπιχειρήσομεν δὲ τὰ μὲν πράγματα ἐκφράζοντες ἀπὸ τῶν προγεγονότων καὶ ἐν αὐτοῖς γινομένων καὶ ἐπισυμβαινόντων ...), and Nicolaus, *Progymnasmata* 68.6–70.5 on the descriptions of external factors that separate ekphrasis from plain narrative.

[70] Maximus of Tyre, *Dissertationes* 2.10.10–14; on this passage see Platt 2011, 228–30.

tells what it is like to use it and who made it; where the ekphrasis tells what something looks like, the parekphrasis tells what it is like to see it.

Invention

Every new invention is connected to its predecessors by a tense balance of analogy and innovation. Analogy, the structural and functional resemblance between novel devices and their predecessors, serves to guide the development of new inventions, as designers use proven techniques to winnow out promising solutions to mechanical problems from the baffling array of potential approaches. Popular accounts of invention often downplay this continuity. There is something rather attractive about the idea that inventors produce their work through spontaneous, discontinuous inspiration: the light bulb over the head. However, a closer look at the real, gradual and competitive, process of invention indicates that continuity and analogy are essential for developing successful technology (light bulbs included). On the other hand, the differentiating power of innovation is essential to motivate the functional improvements, repurposing, appropriation of technologies into new theaters of use, and all the other shifts that keep technology evolving.[71]

Both of these forces, the differentiating dynamism of innovation and the stabilizing guidance of analogy, provide features that can be enlisted by an author in representing a technological artifact for the reader. The principles of analogy allow an author to describe a novel device, which the reader might never have seen, using descriptors belonging to a more familiar or simpler one. Meanwhile, the differentiating force of innovation allows an author to single out and indicate the novelties of a new device, thus focusing the ekphrastic description precisely on what is new and interesting.

In antiquity as today, a typical approach to explaining a complex device or set of physical principles to a novice reader is to preface it with a more or less fictionalized account of its invention.[72] Creese observes that Nicomachus of Gerasa uses stories of discovery to make the early stages of his instructional program more comprehensible, as "they lead their readers and their protagonist together from ignorance to knowledge."[73]

[71] On the modern analogue of this interplay, see Hindle 1981.

[72] Healy persists in an oddly uncritical reading of White's assertion that "no Greek or Roman ever tells us, either in words, or in iconography, what he, or his society, wanted from technology, or why he wanted it" (White 1980, 236–7; Healy 1999, 347–8).

[73] Creese 2010, 88.

By describing the process by which the inventor is said to have acquired the idea for the device, an author makes that device more comprehensible to the reader by moving stepwise into its description. Stories about how James Watt invented the steam engine after being inspired by the sight of a steaming teakettle, or how Isaac Newton formulated his theory of gravity after the apple incident, are modern representatives of this category.[74] Stories like this are easily absorbed into popular culture, and turn up in all kinds of literature.

The same can be said for classical literature. Daedalus, the ancient world's inventor par excellence, figures often in invention stories. Ovid's *Metamorphoses* provides an especially compelling example: the poem refers to two inventions, and explains the second by analogy to the first.[75] In book 8, Daedalus devises the flying machine with which he hopes to escape from Crete. Ovid describes this supremely unfamiliar device to the reader in terms of a more common device: the panpipes, whose invention was already described in book 1. There Argus, who is supposed to be guarding Io, asks Mercury to tell the invention story of the panpipes, a device familiar to us and to Ovid's readers, but novel and unfamiliar for Argus.[76] Mercury accedes, telling him the sad tale of Pan and Syrinx, and the events that finally moved Pan to bind graduated lengths of reeds together with beeswax to make the *fistula* (1.711–12).

So, when the reader reaches the flying machine in book 8, he has already been given an image of a structurally analogous device. Ovid prompts recollection of this device even as he emphasizes the novelty of Daedalus' creation:

> [Daedalus] spoke, and aimed his thought at unknown arts,
> renovating nature. He put feathers in order,
> beginning from the smallest, following a little longer,
> as you'd think they would grow on a slope: so once the rustic
> panpipes arose, bit by bit, from unequal reeds.
> Then he bound them, at the middle and the ends, with thread and wax,
> and bent them once assembled with a slight curve,
> so that they resembled real birds.[77]

The flying machine is an invention of extraordinary creativity, so much so that Daedalus alters nature itself (*naturamque novat*) through its invention. At the same time, like any other invention, this machine does not

[74] Basalla 1988 is a good guide to the sort of fictionalized punctuality that gets superimposed on the more or less gradual process of technological development.

[75] Thanks to Alessandro Barchiesi for pointing out this comparison.

[76] *quaerit quoque (namque reperta fistula nuper erat), qua sit ratione reperta* (1.687–8).

[77] Ovid, *Metamorphoses* 8.188–95.

come out of nowhere. It is connected to its predecessors through structural and functional analogy; by pedigree it is half bird, half panpipes. Ovid does not so much describe the appearance of the machine as highlight these connections. Like the wings of the birds it imitates, the flying machine is constructed of feathers of various sizes, bound in a flexing curve. The feathers are positioned so that they appear to have grown on a slope (*ut clivo crevisse putes*); perhaps this means that their shafts are oriented toward the wing's curved leading edge. Like the panpipes, its technological ancestor, the device is constructed from hollow shafts, arranged in parallel from largest to smallest[78] and bound together with beeswax, with the addition here of linen thread.

Thanks to these analogies, Ovid scarcely has to provide any direct description of the flying machine for the reader to be able to visualize it quite clearly; we can easily merge what we know about panpipes with what we know about bird's wings to get an idea of its structure and appearance. The minimal need for description leaves Ovid free to emphasize the process of creative innovation. Daedalus' mechanical invention, like Ovid's own poetic invention, is the result of a process of creative imitation, a productive tension between the canonically successful and the brilliantly new. Ovid lives quite consciously in a populated literary world, where every creation builds off what has come before, while at the same time making it new. So Ovid's Daedalus, who likewise inhabits a world already populated by the first inventions, takes one of those primordial devices and adapts its structural principles to make something fantastically new – from a rustic instrument to a flying machine. Ovid himself meanwhile adapts an old description into a new one, calling attention to the adaptation to highlight how invention works. In drawing out the resemblance between the flying machine and the panpipes, Ovid makes a cultural claim about the role of the old in inspiring the new, by highlighting the similarities between a simple and familiar device and one that is more complex and less familiar.

Ekphraseis often appeal to images already stored in the memory, whether the text has itself earlier installed them or makes use of images from elsewhere stored in the reader's mind. *Enargeia*'s reliance on the

[78] The line (*a minima coeptas, longam breviore sequenti*) presents obvious topological problems, but Anderson argues that the manuscripts do not admit of emendation. The machine is of course a *pair* of wings, tapering to either end, which could arguably be imitated iconically in the two hemistichs. Indeed, *technopaegnia* including both an extant *Pteryges* by Simias, and a *Syrinx* ascribed to Theocritus, are noted at Fantuzzi and Hunter 2004, 40. Compare Optatian's fourth-century *Carmen* 20, a *carmen figuratum* where lines of differing length reflect the different lengths of the pipes in the organ. In any case, the point here is about the tapering shape of the panpipes, which of course looks more like one wing than two, but Daedalus was no doubt capable of extrapolating, and so are we.

audience's pre-existing supply of memories is mentioned in ancient rhetorical literature. Quintilian, for example, offers as part of his advice on achieving *enargeia* that "minds most easily admit what they recognize."[79] Augustine addresses the problem that this recognition raises: left to their own devices, minds may conjure up images that are generically plausible, but not specifically accurate.[80] Augustine had, of course, seen the city of Carthage, and hence he has an internal image of it, which he can use when he describes the city to others.[81] This *phantasia* is the product of sensory experience, and this (and not the word "Carthage") is stored in the memory as the referent or "word." On the other hand, says Augustine, he has never seen Alexandria, but if called upon to describe it he uses some cobbled-together internal *phantasia*: "I formed an image of it in my mind as I was able; and this is with me its word when I wish to speak of it." However, he says, if he could somehow make those who have seen Alexandria see his image of it, it would be very surprising to find a match between his image and their own experienced image of it. Here is one of the basic problems with ekphrastic "realism": every listener must delve into his own internal library of memories to provide the raw materials for imagining sights unseen. This is an uncontrollable element of ekphrasis, but its impact can be lessened by providing specific details about the object, such as the use of mechanical analogy to give clues to the object's appearance.

The power of mechanical analogy to reveal the process of invention is apparent in more technology-centered texts as well, where it also functions as a focusing descriptive tool. For example, Vitruvius prefaces his complex description of Ctesibius' water clock with what is billed as the story of its inspiration by a simpler, structurally analogous device he invented as a boy in his father's barber shop. Vitruvius promises that the account of this device is useful for didactic purposes,[82] and it is worth reviewing in full:

> Ctesibius was born at Alexandria, the son of a barber. He is said to have excelled beyond the rest in cleverness and great industry, and to have

[79] *facillime enim recipient animi quod agnoscunt*; Quintilian, *Institutiones oratoriae* 8.3.71. On this point, see Webb 2009, 109–11.

[80] The passage is discussed at Webb 2009, 122.

[81] *apud me ipsum inuenio phantasiam Carthaginis*; Augustine, *De trinitate* 8.6.9. Compare Cicero, *De oratore* 8–9 on the creative process behind Pheidias' statue of Zeus: he modeled it not on any specific person, but on the basis of an internal image of beauty that lay in his mind (*ipsius in mente insidebat species pulchritudinis eximia quaedam*), and in contemplating this image he shaped the statue. On this passage see also Platt 2011, 227–8. Compare also Philostratus' *Vita Apollonii* 6.19 on the ability of *phantasia* to render images of the unseen, whereas mimesis can only reflect what has been seen; this passage is discussed at Tanner 2005, 285; on rational *phantasia* more generally, 259–61.

[82] Vitr. *De arch.* 9.8.2: *uti fuerint ea exquisita, dignum studiosis agnoscere.*

> enjoyed technical (*artificiosi*) things. And so, given that he wanted to hang a mirror in his father's shop in such a way that a hidden string would lift the weight when it was raised and lowered back down, he arranged a mechanism in the following way. He fixed a wooden channel beneath a beam, and there placed pulleys; he ran the string through the channel up to a corner and there he constructed small tubes; inside these he carefully placed a lead ball to be let down by the string. And so, when the weight compressed the air's density as it ran into the narrow tube, and in its rapid exit through its opening, because of the compression of the air's density it squeezed out a compressed mass into the open air with its outpouring, as a clear sound.[83]

To Ctesibius' mechanical eye, the device is more than a noisemaking toy; he observes its behavior and deduces that the air forced through the tubes was pressurized, and extracts this principle for application in hydraulic machines, automata, and the water clocks that Vitruvius is about to explain. The little device in the barbershop provides an ideal testing ground for the development of Ctesibius' physical intuition (*his principiis usus hydraulicas machinas primus instituit*). If we are to believe the story, the mechanical "sandbox" in the barbershop taught him the basic principles that allowed him to develop the complex machines he became known for in adulthood.

Regardless of whether we give this charming little story any credence, it certainly uses a stripped-down device to teach the *reader* the principles necessary to understand the more complex description of the water clock with which Vitruvius follows the story. We have just been walked through a process of imagining how a fluid flows through a system of pipes, so that the principles of both the basic filling of the water clock (which is necessary for telling time) and the extraneous flourish of toy trumpets (which is not) are already familiar even to the non-technical reader. Vitruvius meanwhile introduces a new kind of mechanical activity (the geared rotation of drums that marks off longer periods of time), and new technical terminology (*terebrata*, *phellos*, *tympanum*), which the reader comes to understand more easily because he is not simultaneously trying to understand the hydraulic action. Just as Ovid makes the flying machine more comprehensible by comparing it to a simpler device with an analogous structure, so Vitruvius makes the water clock more comprehensible by beginning with a device that puts the same operating principles to work in a stripped-down, easily accessible context. In both cases the process of invention is used in the parekphrasis as a window onto this analogical development from simple to complex, from familiar to novel.

[83] Vitr. *De arch.* 9.8.2–4.

This technique of reducing a complicated machine down to just the most critical elements is particularly useful when the reader must compare multiple machines in some detail; this task could quickly become overwhelming if the field of interesting features were not somehow reduced. Philo uses such a focusing technique to describe a different invention of the prolific Ctesibius: the bronze-spring catapult. Philo recalls his discovery that Ctesibius had invented this device but had not transmitted information about how to construct it. When Philo realized the value of the machine, he reports that he developed a plan to investigate and perfect it, but he stresses that based on what he has learned from people who observed Ctesibius at work, his own machine is quite different. Here is a tension between analogy and innovation: Philo plans to create a machine with the same functionality as a previous invention, but makes a strong claim that the device itself will be quite different.

Philo plans to tell the reader what is known about Ctesibius' machines, and then to compare it to his own design, showing that the latter is superior. His description focuses quite precisely on the novel features of the bronze-spring catapult:

> Since we intend to explain about a peculiar property of the springs, we supposed it was good to investigate the old-fashioned machines in this regard first and to infer the cause of their being able to send a missile far …
>
> Now it is necessary to make a complete inquiry about the whole issue, as we intend to present a design (θεώρημα) that is completely unique in arrangement and a radical departure from its predecessors. The force of the arms is entirely responsible for the missile being carried far, and it will be increased by the springs. Therefore it must be investigated what the existing arrangement is as far as the springs go.[84]

Philo intends here to compare his machine's performance with that of Ctesibius' original, which leaves him with the nontrivial task of describing both machines in enough detail to give an accurate picture of the functional characteristics that would make one superior over the other. Handled less skillfully, such a description would require an overwhelming amount of information and would be easy to get lost in.

However, Philo uses the focusing power of the parekphrasis to refine his description. Before he begins the detailed ekphrastic description of the way the two machines work, he "zooms in" on the specific changes he has made to the older design. This refinement makes it possible for Philo not only to describe the springs as they work in the machine, but also to

[84] Philo, *Belopoeica* 68.8–25.

provide an abstract digression on the mechanical principles involved (relying, unfortunately, on the attractive but incorrect assertion that a heavier weight drops faster than a lighter one), and on the material properties of the specially forged bronze he recommends for the task. Finally he "zooms out" again to discuss the performance of the springs in the context of the bronze-spring catapult taken as a whole machine.[85] He does all this without overwhelming the reader, because he precisely established in the parekphrasis the novel features that make his machine a radical improvement over Ctesibius' design.

Contexts of use: praise and blame

The parekphrasis can help to introduce an object to the reader in other ways as well. Techniques that effectively reduce the amount of information the reader has to keep at his fingertips are especially useful for presenting an artifact in detail as an innovative, functional piece of technology. On the other hand, particularly in situations where the object is already familiar or simple enough that the ekphrasis itself is minimal, an author may use the parekphrasis to *increase* the amount of information by giving the object a background and context of use. Besides deepening the reader's understanding of a specific device, contextual information also illuminates the role technology plays in its culture of use: whether it is valued for innovation and specific functionality, or symbolically as a reflection of imperial and administrative power, and so forth.

Pliny's *Natural History* mentions several technological artifacts, all presented mostly through stories of their provenance or invention rather than through technical descriptions of the way they work. This is not to say that Pliny never treats these objects as functional devices; however, he devotes minimal detail to their working principles and effectiveness compared to information about the people and places associated with their creation, destruction, travel, or other such events.[86] Their functionality becomes a part of their historical background, and is hence part of the context of the parekphrasis rather than the vivid description of the ekphrasis.

For example, Pliny's account of Augustus' transformation of an obelisk into the sundial known as the *horologium Augusti* combines an invention story with a story of blame. First, the sundial-obelisk is compared to

[85] Philo, *Belopoeica* 72.6–73.32.

[86] On Pliny's accounts of "time and the measurement of time," including the changing length of days in the year, sundials, and calendars, see Healy 1999, 360–70.

another one which Augustus set up in the Circus Maximus; that one was apparently nine feet taller. Pliny calls attention to the Egyptian origin of the obelisks; the production of the one used in the *horologium Augusti* is credited to one "Sesothis," and, like its counterpart in the Circus Maximus, it supposedly includes Egyptian inscriptions about natural philosophy.[87] Upon its removal to Rome, the obelisk was set up in the Campus Martius, and placed upon a stone surface sized to match the length of the obelisk's shadow at noon on the winter solstice.[88] The shadow's changing length throughout the year was matched to inlaid metal strips in the stone,

> a thing worth knowing about, stemming from the talent of the mathematician Facundus Novius. He added a golden ball to the apex, in whose vertex the shadow would be collected into itself, since otherwise the apex's shadow would be cast enormously; they say that this approach was inspired by the human head.[89]

Unfortunately, the sundial has told the wrong time for thirty years, for which Pliny proceeds to give several possible explanations: perhaps the sun's path or the earth's position has changed, or perhaps local floods and earthquakes have wrenched the gnomon out of its true position.

The description of the obelisk's invention and its function as a sundial is presented from the perspective of an observer, who sees how the shadows compare with the inset metal markings from day to day, rather than the perspective of an astronomer who knows the theoretical principles behind its operation. It is not that Pliny does not describe the sundial at all; he reports how tall it is (and how its height compares with another obelisk), and that it is surmounted with a golden ball, and even that it includes markings determined by the shadow cast by the sun on the winter solstice (a very technical astronomical note by Pliny's standards). But he describes the sundial in just enough detail that the reader can picture it, and then moves on to the subject of real interest to him: the stories that attach to it and the people those stories involve. These include its provenance in Egypt, its erection in Rome by Augustus, its improvement as a sundial by Facundus Novius, and a whole raft of speculation about what caused its malfunction. But the technical details of its failures as a sundial

[87] On the obelisk's quarrying and manufacture in Egypt and its transfer to Rome, and for a more reliable account than Pliny's of its hieroglyphic inscription, see Swetnam-Burland 2010.

[88] A very detailed examination of this monument, including considerable guidance on the technical details of ancient sundials, may be found in Buchner 1982. Buchner argues that the sundial was set up so its shadow would trace a path along the Ara Pacis Augustae; *contra*, see Schütz 1990.

[89] Pliny, *NH* 36.71–3. On the use of the golden ball to prevent the shadow of the obelisk's cusp being "swallowed" at certain times of day, see McEwen 2003, 248.

are omitted, and the theories about why the problems might have arisen are not at all motivated by any specific technical detail of its inaccuracy.[90]

This approach, in which the appearance and functionality of the sundial take a distant second place to its historical context, is typical of Pliny and in many ways typical of the presentation of technological artifacts in Roman literature.[91] Though credit for engraving the markings on this sundial is given to the apparently Roman Facundus Novius, the astronomical and geometrical knowledge required to make such an object is more often represented as the province of educated Greeks. Vitruvius, for example, rejects the possibility of improving on the Greek legacy of sundial varieties.[92] Pliny is concerned mostly with where they were brought from and the patrons who set them up in Rome; the position of this passage in the *Natural History* among other obelisks Augustus set up in Rome testifies to this. He contextualizes these artifacts first in terms of their provenance and second in terms of their performance, which is typically poor because changes in latitude affect the function of a sundial. Yet the sundials remained on display: their primary cultural role at Rome was evidently to embody the imperial power that collected them rather than the intellectual power that devised them as accurate time-telling devices in the first place. Such an approach favors parekphrastic context heavily over ekphrastic description of the device itself.

Plutarch's description of the glory Demetrius Poliorcetes achieved by his collection of siege technology initially has much in common with Pliny's approach, though in this case the story is of praise rather than

[90] Pliny discusses another malfunctioning sundial at *NH* 7.214, a story that he attributes to Varro. There too, the precise reason for the sundial's problems goes undisclosed, and once again an entertaining story is the main event for Pliny, as he reports that they continued to use the sundial anyway for ninety-nine years. Pliny there says his reticence on the details of the sundial itself is owing to Varro's own failure to record them (*sed neque facti horologii rationem vel artificem significat nec unde translatum sit aut apud quem scriptum id invenerit*), but Pliny on no occasion reports on the *horologii ratio* in much detail. Hannah argues that the problem is the change in latitude from the sundial's origin in Catania to Rome, which caused either the hours of the day or the seasons of the year (either of which may be indicated by the word *horas*) to miss their intended markings significantly (Hannah 2008, 750). No doubt this is the case, but Pliny gives no indication of any of this information.

[91] Pliny's relative lack of concern for technical detail contrasts, for example, with Ptolemy's much more detailed criticism in the *Syntaxis mathematica* of the equinoctial rings that were set up in a public place in Alexandria (identified by Hipparchus as the "Square Stoa"). As Taub describes it, "Ptolemy implies that the rings were fairly useless, at least in the state he found them, since they did not give accurate readings and sometimes even indicated two equinoxes during the same day" (Taub 2002, 134). Ptolemy's response to this unacceptable performance is dense with mathematical detail, an approach very different from Pliny's style. Another perspective on the representation of transplanted Greek technology at Rome (the *sphaera* of Archimedes) may be found in Jaeger 2008, 48–68.

[92] Vitr. *De arch.* 9.7.7.

blame.[93] Plutarch begins by describing the impression of power generated by his ships and engines:

> For with its size it stopped his friends short, with its beauty it even delighted his enemies; this is said with more truth than skill. And his enemies, brought to a standstill, marveled at his sixteen-banked and fifteen-banked ships as they sailed past their land, while his *helepoleis* were a wonder to the besieged.[94]

So impressive was this weaponry that Demetrius' great enemy Lysimachus, who had previously fought against him at Soli, retreated in admiration upon viewing it. So far, Plutarch's description is quite like Pliny's; he is concerned mostly with the political and military power embodied by the devices, while the description of the technology itself is limited to vague allusions to the types of machinery involved. But when Plutarch focuses on the *helepolis* used to besiege Rhodes, he not only gives a precise and detailed account of its structure, but even delves into the problem of keeping such an enormous machine upright while moving:

> He went to war against the Rhodians, since they were allies of Ptolemy, and he moved his greatest *helepolis* against their walls; its base was square, having each leg of the under-frame 48 cubits; it had a height of 66 cubits, converging with beams at the top narrower than those of the base. Inside it was divided into several stories and spaces; the face pointed toward the enemy was opened up with doors at each story, and through these all kinds of missiles plummeted; for it was full of fighting men of every kind. And it did not shake nor lean while moving, but, advancing unshaking and balanced, with a rush and a great force, it offered wonder to the soul and joy to the eye of those who saw it.[95]

He does not continue the narrative of the *helepolis* any further than this. Vitruvius reports that a disastrous fate awaited the machine, mired in sludge on account of its great size.[96] Plutarch, however, leaves it at this moment of triumph. This description is not at the height of technicality (this is, after all, Plutarch, who writes off the engines of Archimedes as distracting playthings[97]), and yet it foregrounds the machinery in a way that Pliny generally does not. Like Pliny, Plutarch is interested in the history they were part of, and he is particularly focused on the role of

[93] On the perceived relationship in antiquity between ekphrasis and encomium, see Webb 2009, 78–81. The relationship was only made explicit in rhetorical manuals that likely postdate Plutarch considerably, but the phenomena the manuals discuss had in most cases been in practical use long before the composition of the manuals themselves.

[94] Plutarch, *Demetrius* 20. [95] Plutarch, *Demetrius* 21. [96] Vitr. *De arch.* 10.16.4.

[97] Plutarch, *Marcellus* 14.4.

Demetrius as patron (and, if we are to take him at his word, engineer). Yet despite this commonality of purpose, Plutarch and Pliny have very different expectations for the cultural value of technology; where the technical performance of the sundials was of small importance for Pliny, it is the technical success of Demetrius' machinery that takes center stage in Plutarch's narrative.

Parekphrasis and subjective experience

The parekphrasis can also complement the descriptive work of the ekphrasis proper by providing a different perspective, revealing the subjective response of a spectator instead of that of the author or designer. The astronomical *sphaera* often attributed to Archimedes, a mechanical model of the cosmos, inspired several examples of this technique, in a wide variety of genres, from philosophical prose to poetry. Many descriptions of spectators' reactions to it survive, but little information on the device's actual appearance or design. The surviving descriptions of the *sphaera* do not dwell on the technical details of its operation, but rather on its ability to transport the viewer out of his familiar earthbound perspective, affording him a sweeping view of all the planets and stars working together. It is no wonder the *sphaera* appears in Cicero's *De re publica*, prefiguring the *somnium Scipionis* at the end, where Scipio Africanus is amazed by the exact same kind of transportation the ordinary viewer enjoys when he looks at the model.[98]

Philus describes what it is like to view the *sphaera*, but first he provides a short history of the device. He begins by clarifying that there are two kinds of *sphaera*: one a solid map of the stars and one a kind of clockwork device.[99] One of each was apparently looted from Syracuse by Marcellus' grandfather, whom Philus commends for not taking anything else. The solid sphere, says Philus, was dedicated by Marcellus at the temple of Valor, where it became well known to the Roman viewing public and much admired for its beauty. Spheres of this type were not a novelty; Philus reports that Gallus told him Thales of Miletus was the inventor of this type of sphere, and that the designs of the stars were added by Eudoxus and later put into verbal form by Aratus.

The mechanical sphere was kept at the house of Marcus Marcellus, where it was seen by C. Sulpicius Gallus, who in turn evidently explained

98 Gallagher 2001, 514–15.

99 On the different cultural resonance of the two types of spheres, see Jaeger 2008, 53–68; on the mechanisms of the latter type, see Freeth and Jones 2012, sec. 2.2.

its operation to Philus in such a way that he came to admire it even more than the famously beautiful solid *sphaera*:

> But this type of sphere, in which the movements of the sun and moon were present and those of the five planets that are called "rovers" from their wandering, could not have been marked out on that solid sphere, and in this one the invention of Archimedes was to be admired, that he had thought through how a single turn could preserve the unequal and varied courses in unlike motions. When Gallus moved this sphere, he made it so the moon followed the sun in the same number of rotations in that bronze form as days in the sky itself, so that in the sphere that same diminishing of the sun occurred, and the moon reached the point where the earth's shadow was at the moment when the sun ... from the region ...[100]

If the part of the text that precedes the lacuna is indicative of the text as it originally stood, this is not a technical exposition, not a description of how the device works in the mechanical and geometrical terms its designer might have used. That such a detailed mechanical description is also lacking in the other texts where Cicero mentions a *sphaera* suggests that the absence here is not only due to the lacuna. His is an outsider's perspective, but the distinction of the two kinds of *sphaera* does contextualize the device within a genre of artifacts.

Philus is in no position to evaluate the sphere from a technical standpoint, although he does give some sense of the aesthetic considerations that give a device status in the eyes of casual viewers; the solid, engraved sphere is more popularly admired (*nobilior in vulgus*) because it is lovelier (*venustior*). This characterization has been taken by Büchner to indicate a distancing mechanism between an intellectual elite (in this model practically coterminous with the Roman social elite) and the uneducated masses.[101] This *sphaera* is appealing, as Jaeger puts it, as "a choice piece of booty from the city that exemplified rich spoils."[102]

The mechanical *sphaera*, by contrast, is more than a lovely bauble; what earns Philus' admiration is not exactly the device itself, but rather the intellectual process it encodes. He is impressed by Archimedes' ability to visualize a system that in one rotation carries out exactly the varied motions of all the planetary subsystems.[103] It is Archimedes' synoptic

[100] Cicero, *De re publica* 1.21. Unfortunately, a long lacuna in the text follows.

[101] Cicero 1984, 103. [102] Jaeger 2008, 57.

[103] Archimedes is very frequently associated with this type of device, whether or not he had anything to do with their original development or the construction of any given known *sphaera*, possibilities that are respectively quite dubious and almost certainly untrue. For a cautious analysis of the likely date of such mechanisms, with excellent bibliography, see Keyser 1998. For a more recent

vision, more than his technical skill, that makes Philus describe him as more than humanly brilliant (*plus in illo Siculo ingenii quam videretur natura humana ferre potuisse*). The brilliance of the *sphaera* is tied up with its inexplicability; Philus cannot say how to build one, and probably neither can Gallus, because it is a product of a nearly divine intellect. They can, however, comprehend what it represents, because the unique brilliance of the designer is (somewhat paradoxically) to make the astronomical phenomena represented immediately obvious to anyone who sees the machine, be they ever so barbaric. Cicero asserts in *De natura deorum* that a comparable device supposedly designed by Posidonius could even be appreciated by a Scythian or a Briton; as

> each of its rotations cause in the sun, the moon, and the five wandering stars what occurs in the heavens with each day and night, who in that barbarian country would doubt that this sphere was the product of perfect reason (*quis in illa barbaria dubitet quin ea sphaera sit perfecta ratione*)?[104]

For Cicero, the *sphaera* is a kind of specialized container for the genius of its creator: it is a portable device, which functions to astonish the viewer by its brilliant design, rather than functioning primarily as an orrery in the sense of predicting astronomical events like eclipses. This explains the lack of physical description of the machine: it resonates here because it marks a transfer of intellectual capital from Greece to Rome, rather than because of any one specific mechanical detail.[105] Jaeger writes that Gallus "finally takes over himself and describes its workings in detail," thus enriching the *sphaera* with the "layers of significant associations" that complete the image.[106] In point of fact, however, this is not exactly what Gallus does here: he tells what the machine *does*, from the perspective of an outside observer, even what it *is*, in the sense of what it represents, but not how it *works*, as a mechanical artifact constructed from carefully sized gears, trunnions, and so forth. The *sphaera*'s inner workings are not described (at least in the text that survives); it remains a "black box."

In a letter written on behalf of the king Theoderic, Cassiodorus predicts that two different types of celestial model, a water clock (*horologium*) and a sundial (*quod solis immensi comprehensa illuminatione distinguitur*), will have a similar mystifying effect on Gundobad, king of Burgundy. The letter stresses that the delight and enjoyment of viewing such a device are

look at the mechanisms that might have been involved with the *sphaera* that Cicero describes, see Freeth and Jones 2012.

[104] Cicero, *De natura deorum* 2.88. [105] Jaeger 2008, 55–7.

[106] Jaeger 2008, 56.

everyday experiences for the likes of Theoderic but will indeed appear miraculous to the Burgundians (*quod nobis cottidianum, illis videatur esse miraculum*).[107] At the end of the letter, the kind of wonder Cicero envisions being sparked by the *sphaera* receives a practical, political application: "when they have recovered from their stupefaction, they will not dare to say they are equal to us, among whom they know wise men conceived of such things." The devices are to be dispatched along with technicians who will set them up properly, no doubt further enhancing their mystery.[108]

Writing in Theoderic's name to Gundobad about the celestial models coming his way, Cassiodorus takes a similar line, though he strategically excludes Gundobad himself from the ignorance of his people. Gundobad is rather the channel through which the Burgundians will receive Roman culture, and the time-telling devices are the tool by which they will learn to "examine the most subtly wrought things and praise the ancients' discoveries."[109] As the Syracusan *sphaera* brought Greek intellectual culture to Rome, so will the water clock and sundial bring it, now wholly identified with Roman culture, to Burgundy.

Both Cicero and Cassiodorus envision the celestial model from the perspective of the ultimate layman. As Roman laymen, they themselves are one step removed from the science and craft that made the original model possible, but a much wider gulf separates the device from these hypothetical barbarians. From outside the cultural tradition of the educated Roman, the *sphaera* must be completely mystifying – yet its mechanical intricacy and ability to model the heavens, they suppose, would still not be obscured. In neither case do the mechanical operations or technical details of the devices take center stage – indeed, neither author lays claim to particular expertise about them, as Cassiodorus emphasizes the uniqueness of Boethius' knowledge, and Cicero of Archimedes'.

Sextus Empiricus has a comparable reaction to a *sphaera* that, like Cicero's, is credited to Archimedes (though once again he probably did not have anything to do with it).[110] Sextus, like Cicero, devotes more attention to the genius intellect the machine encodes than to how it works mechanically. The context of the passage is another indicator of his priorities, as it

[107] Cassiodorus, *Variae* 1.45.

[108] Compare the *ballista fulminalis* described in the anonymous *De rebus bellicis* (18.4–9): the imperial addressee is counseled that the one-man operation of that device is vital to its mystifying effect (*uidelicet ne si hominum turba huius ministerio inseruiret, minueretur artis inuentio … quanto minus ad armandam se uirorum ministerium quaerit, tanto est ingenii uirtute praestantior*).

[109] Cassiodorus, *Variae* 1.46.

[110] Sextus Empiricus, *Adversus mathematicos* 9.115.

appears in a philosophical discussion about the orderliness of the universe. Cicero invokes similar ideas in *Tusculan Disputations*: the *sphaera* shows that there is an intelligent demiurge running the universe in an orderly fashion, as we can see from the amazing precision of Archimedes' system, crafted as it was by a mere man, genius or no.[111] Ovid likewise removes the *sphaera* from a technical astronomical role to a kind of wonder:

> This is a modest space, which supports the atrium of Vesta …
> a globe stands, suspended in trapped air by Syracusan art,
> a little version of the immense heavens,
> and the earth is as far from the top as it is from the bottom,
> which it causes to happen by its round shape.
> So it is with the temple exterior; no angle juts out on it,
> and a rotunda protects it from the rain.[112]

In this case the *sphaera* is probably either the solid type or an armillary sphere.[113] Here the celestial model is described in a context where the astronomical knowledge it embodies is of still less interest, though the allusion to Archimedes stands. In this poem the *sphaera* is eclipsed by another famously round construction: the foundation of the temple of Vesta, as Ovid lowers his gaze down to the earth itself, its roundness and spatial balance.

None of this is a "misunderstanding" of a device that was really intended to be a serious scientific tool. The significance of each type of *sphaera*, in these literary contexts, turns out to be how it makes us feel when we look at it, the way it makes us confront the limitations of our everyday spatial perspective. It functions to take us out of ourselves, out of the world, and show us how things look from somewhere else, as well as functioning as a reassuring symbol of the orderliness of the universe. All this extra-scientific significance means that the texts describing this extraordinary machine say very little about the device itself, but quite a lot about the experience of seeing it.

Nor can this kind of revelation only be produced by a complex device like the *sphaera*. Vitruvius describes the realization that a sundial's gnomon sparks in the educated viewer: that it is not just a stick that casts a shadow, but a model of a vast cosmic system that integrates the whole world into something larger:

> These things are arranged by the divine mind and hold great admiration for those who contemplate them, that the shadow of the gnomon at the

[111] Cicero, *Tusculanae disputationes* 1.63.
[112] Ovid, *Fasti* 6.263–83. [113] Evans 1998, 82.

> equinox is of one length at Athens, another at Alexandria, another at Rome, not the same at Piacenza, and the other parts of the world.[114]

The variable length of the gnomon's shadow is keyed to the earth's shape and position within the cosmos, too large to be perceived directly. As with the *sphaera*, recognition of what the gnomon is modeling expands the viewer's perspective, reminding him that he stands within an enormous cosmic system governed by a divine intelligence (*divina mente*). This reminder has the emotional effect of provoking great wonder (*admirationem magnam*), which moves the thoughtful viewer.

The appearance of certain types of artifact is sometimes cited as important to their very functionality as well as the aesthetic impression they create for the viewer; Cuomo cites this as evidence that the priorities of ancient military engineering were not necessarily the same as they are today.[115] Philo critiques the unrefined appearance of the frame of a catapult in its unfinished state:

> The frame constructed and prepared in this way is ready to shoot, but it is a little unsuitable to the eye – for it appears headless, as in fact it is. So that it may appear beautiful to the eye, and the spring be completely covered up, a covering is constructed on either side in the manner that we are going to describe.[116]

In this case, the solution to the catapult's aesthetic shortcomings also has the practical effect as well of giving the device's spring some extra protection. At other times, the appearance seems to be emphasized more for its own sake, as when he advises that if you wish to add a decorative border of boxwood because you are aiming at good looks, then you should go ahead and do so.[117]

The catapult's appearance indeed seems to have been crucial for establishing its very identity as a machine.[118] Philo tells us that Ctesibius took care over the external appearance, "so that it would look machine-like" (ὅπως ὀργανικὴ φαίνηται).[119] As in the case of the medical traction devices described by Oribasius, and the descriptions of the colossus of Rhodes written by Pliny and Philo Paradoxographus, a technical ekphrasis may make a distinction between the "insider's" and "outsider's" perspective. Philo portrays Ctesibius imagining the catapult from the perspective of

114 Vitr. *De arch.* 9.1.1.
115 Cuomo 2007, 54–5. "Shock and awe" campaigns do, however, seem to court a similar effect.
116 Philo, *Belopoeica* 66.24–30. 117 Philo, *Belopoeica* 66.50–67.2.
118 Meißner 1999, 92; see 91–4 on these issues more generally.
119 Philo, *Belopoeica* 78.17.

an outsider: making its outward appearance correspond to certain standards is a way of proving to a viewer who was not involved in its design or construction that it is in fact "the real thing." Philo is by no means the only author to take note of the role played by appearance in reinforcing the effectiveness of a military onslaught. Onasander's *Strategicus* advises the would-be general to keep his army's weapons and armor bright and shining, on the grounds that such a force of soldiers will be more terrifying to the enemy.[120] Cuomo mentions the copper-clad catapult found at Cremona, and invites her reader to "imagine the sun flashing ominously" on it.[121]

On a more mechanical note, Ammianus Marcellinus is likewise interested in the psychological as well as physical effects of weaponry on the enemy. He concludes his description of the ballista by describing the devastating effect of the projectile on its target:

> When the very tip [of the arrow's] point has come to the extremity of the cords, driven by an internal thrust from the ballista it rapidly flies out of sight, meanwhile sending off sparks from its excess heat, and very often before the projectile can be perceived, pain senses the deadly wound.[122]

The reader's point of view is instantaneously transferred from the site of the ballista, where we have just been watching an engineer (*artifex*) adjust the arrow and strong youths (*validi iuvenes*) ratchet up the tension, to the stricken enemy who falls victim to the missile. Ammianus includes in nearly all of his descriptions of siege engines such a report from the enemy's perspective. Fleury argues that he is striving specifically to astonish his reader by presenting him with "mécanismes extraordinaires," which Fleury identifies as one of the primary purposes of the *commentarius*.[123] The beauty or surprise of technological wonders can also have dramatic effects on the viewer, beyond sowing admiration or distress in the mind of one's enemy. Fleury compares Ammianus' engines to Varro's elaborate aviary, a passage that Meißner cites as an example of how a technological artifact's aesthetic appeal could be closely linked to its functionality.[124]

Hero's *Peri automatopoiētikēs* describes the design and manufacture of automatic puppet theaters, designed either so that figurines inside a theater that moves from place to place perform acts like kindling fires or squirting milk or wine from a thyrsus, or so that puppets perform a whole play within a stationary theater. As might be expected,

[120] Onasander, *Strategicus* 28.1–7. [121] Cuomo 2007, 54.
[122] Ammianus Marcellinus, *Res gestae* 23.4.
[123] Fleury 1996, 61. [124] Meißner 1999, 94.

Hero is quite concerned in this work with what a spectator would see, and the text contains several descriptions of viewers' possible reactions. Some are more or less purely matters of aesthetics or style, as when he invokes the need for a "modern"-looking device crowned by an architrave "finished" with mouldings, featuring "plausible"-looking background figures who "appear to be moving in accordance with the established subject" and act "just as in real life," as well as accessories such as "a garland on four sides, with whatever kind of braiding one wants, harmonious to the eye."[125] Hero shapes the view of the artifact from the perspective of the reader (who may be intending to design and build his own automata) by imposing the expectations and experience of a hypothetical viewer. This viewer apparently demands a device that combines novel storytelling with reliable performance, up-to-date design, and aesthetically appealing components.

He may also bring a healthy skepticism to the performance, as Hero warns that the measurements he gives were chosen so that the viewer will not suspect that a person is hidden inside working the machinery.[126] This warning is reminiscent of Philo's caveat that a catapult should have a "machine-like" appearance, in that Hero too suggests the possibility that if a device is not built to match certain previous expectations on the part of the viewer, it will be suspected of not being properly "mechanical." While the automaton is designed for novelty, its designer must nevertheless remain mindful that the viewer will bring his past experience into the viewing process, and anticipate his suspicions and preferences accordingly.

Describing the automata in terms of what the viewer might see requires many quick imaginative transitions from "inside" to "outside," as Hero describes components to the reader, and then notes that those components must be hidden from view. The ability to reveal the artifact's interior as well as its exterior to the reader is a particular strength of these descriptions: the ekphrastic impression of seeing that the author hopes to engender in the reader has as its target more than the outward physical characteristics of the object. Here we may think of Lessing's claim for the power of poetry to see more than these outward features: "For the poet a garment is no garment; it covers nothing; our imagination sees everything

[125] Hero, *Peri automatopoiētikēs* 2.12, 25.7, 24.1, 21.1, 22.4, 15.2.

[126] Hero, *Peri automatopoiētikēs* 4.4: μειζόνων γὰρ γενηθέντων ὑπόνοιαν ἕξει τὸ ὅραμα ὡς ἐντός τινος ταῦτα δημιουργοῦντος. This problem is addressed as well at Tybjerg 2003, 451. The viewer's suspicion was quite possibly justified by real events; in a later period of fascination with automata, the example of the chess-playing "Turk" of course springs to mind (on this subject, see Standage 2002).

within."[127] Unlike visual representation, verbal representation gives the author the power to reveal not just the surface characteristics of the artifact, but its entire physical form, as well as attributes that lie beyond the merely physical.[128]

There is a tension in the design of wonders, to create a device whose actions surprise and mystify the viewer, and yet provoke speculation into how they were achieved. This response was observed already by Aristotle, who remarks that wonder is the beginning of philosophy, both for the individual and mankind as a whole.[129] Automata, says Meißner, are rightly identified by Aristotle as a paradigm of this type of wonder, as they bring about an experience of surprise for most observers, but for a small group of experts they also bring the impulse to overcome that surprise by unraveling the mystery.[130] Viewers with different levels of expertise will approach the automata with somewhat different mindsets: some will remain content with their initial bafflement and surprise, while others will try to think their way past the outside coverings that hide the mechanisms from view.

Tybjerg observes that Hero advises the reader to hide the working mechanisms not only for the automata, but also in his works on catoptrics and pneumatics.[131] This ensures that the audience is blind to the mechanical causes of the actions they see, so that Hero establishes the designer as "epistemically superior" to the spectator. Tybjerg exemplifies this epistemic distance with a passage in Hero's *Pneumatica* describing a figure of Hercules who fires an arrow toward a snake. This passage, she argues, is divided in two parts: first the spectator's experience of the outside of the device, second the mechanisms present inside, known only to the expert. Hero's textual representation of the automata, the bulk of which shows us

[127] Lessing 1965, 99.

[128] Galen compares Polyclitus unfavorably to Nature in that the sculptor produced only skin-deep symmetry, and even then was merely imitating Nature's craftsmanship (*De usu partium*, II.441 Helmreich). As Tanner notes, the topos of the artist's comparatively superficial knowledge of the human body compared with the doctor's is found as well in the Hippocratic *On the Art* 12 and *Regimen* 1.21, though the comparison in the Hippocratic texts is not as highly developed as Galen's (Tanner 2005, 203 n. 259). In his own works, particularly *De usu partium* and *De anatomicis administrationibus*, Galen is of course highlighting his ability to give *verbal* access to the whole creature, even as he offers dissecting instructions that would if followed give the reader *visual* access.

[129] Aristotle, *Metaphysica* 982b12–983a21. The objects of wonder listed by Aristotle begin with the immediate problems faced by early man, who then expanded his speculation to greater things like the reasons behind the movements of celestial objects; Aristotle later goes on to name automata, the solstices, and the incommensurability of the square's diagonal with the length of its side as suitably provocative topics.

[130] Meißner 1999, 58–9. [131] Tybjerg 2003, 445.

an "insider's" perspective, enriches his design recommendations by intermittently guiding the reader to imagine what the viewer will see.

Describing the viewer's experience can enrich the ekphrasis by prescribing the appearance of the artifact based on an imagined conception of what the viewer sees. The viewer's experience may be directly connected to the device's proper functioning, as when the reader is guided to create a properly terrifying siege engine or a properly mystifying automaton. On some occasions, notably in texts that purport to provide the reader with instructions for building such a device, this "outsider's" view is complemented by an "inside" view of the inner workings in the description. In other texts, like Oribasius' description of several generations of competitive traction-machine development, elements missing from the description put the reader into the same "outsider's" perspective as the observers described in the text.[132]

This representation of the impression an object makes on a spectator is a common adjunct to ekphraseis of artistic as well as technological artifacts. For example, Homer prefaces the famous ekphrasis on the shield of Achilles with a claim about the effect it will have on anyone who sees it: "Such beautiful arms will be prepared for him, that anyone among the multitude of men will marvel when he sees them."[133] Gutzwiller observes that references to the wonder felt by a spectator are so common in Hellenistic epigram as to constitute a trope.[134] The tapestry on display at the marriage of Thetis and Peleus in Catullus 64 is entrancing; it sparks a desire to keep on looking at it in the minds of the (human) spectators until they are "sated with greedy gazing."[135] Virgil's Aeneas, awestruck by the depiction of the Trojans' story on the walls of Juno's temple at Carthage, experiences a variety of emotions: his soul is nourished (1.464), he sighs and releases a flood of tears (1.465), he is reduced to stunned stillness (1.495).

These canonical examples are far from the only passages in classical literature where an ekphrasis describing the appearance of an object is supplemented by an account of how spectators react to it; rather, such accounts are an intrinsic part of the task of description. That such viewer responses were deeply connected with the techniques of *enargeia* is confirmed by

[132] Oribasius, *Collectiones medicae* 49.23.

[133] Homer, *Iliad* 18.466–7.

[134] Gutzwiller 2002, 96–7; she cites *Anthologia Graeca* 16.96, where the reader is invited to marvel at a gilded sculpture of Heracles subduing the Ceryneian hind.

[135] Catullus 64.265–8. Fordyce notes a parallel to the formulation *cupide spectando* at *Odyssey* 4.47, used to describe the amazement of the youths traveling with Telemachus upon the sight of Menelaus' palace (Catullus 1990).

Quintilian, who emphasizes them in his definition of the term. Here he describes the cognitive effect of Cicero's vivid description of the dissolute Verres:

> Is there anyone so far removed from imagining appearances that when he reads in the Verrines: "the praetor of the Roman people stood on the shore, sandal-clad, wrapped in a purple *pallium* and a tunic down to his heels, leaning on this little tramp," he does not seem only to gaze on those figures themselves, the place, their demeanor, but even constructs for himself from these the things that have not been said? I myself seem to see his face as well, his eyes, and the loathsome caresses between the two of them, and the silent aversion and timid shame of those who were present.[136]

Note how central the perspective of the spectators is to the mental impression Quintilian gets from Cicero's description of Verres: their horrified reaction becomes Quintilian's, showing how *enargeia* can turn readers into spectators. Scholz describes the reaction of a reader who experiences the same emotions when he reads Cicero's description that the crowd did when it saw Verres as "a case of 'seeing' seeing itself."[137] Cicero's original description included a vividly pictorial description of Verres, loaded with terms (like *muliercula*) designed to provoke an emotional reaction in the reader mirroring that of the Roman spectators to Verres, though their response is left undescribed in the original passage. Despite this lack of literal description, it is just this emotional effect that Quintilian takes away as his central impression of the scene. And why should this be surprising, when for an actual spectator the reactions of the crowd around him are likewise an immediate and sometimes overwhelming part of the experience of viewing? Whether the spectators' response is actually described in the text or is part of the reader's effort of "supplementation," it is an important part of the immersive effect of *enargeia*.

The appearance of a person or object, and its effect on a spectator, are deeply intertwined aspects of ekphrastic description; each illuminates the other. Even though we, in reading a passage like this, are just spectators of spectators, sharing their mindset is an effective way of gaining knowledge about the object of their gaze. This is especially true if accompanied by even a modest amount of detail about the aspects of the object's appearance that generate such a response. The guidance provided by describing the experience of a naive spectator is particularly valuable in the case of novel technological artifacts, which may combine unfamiliarity with complicated operational principles. The spectator in a passage like this is

[136] Quintilian, *Institutiones oratoriae* 8.3.64–5. [137] Scholz 1998, 141.

less familiar with the machine than its designer or builder is, less familiar than an operator, less familiar than anyone except perhaps the reader himself. His viewing experience is therefore the final frontier of the machine's entry into a cultural context; when an author describes the reaction of spectators he is guiding the reader's own reception, telling him quite explicitly how to read the machine culturally. Such an explicit approach is not taken very often, and when it does occur it is a strong indication that the text is meant for an audience that takes primarily a cultural rather than technological interest in a machine. This is the case, for example, with Cicero's use of the *sphaera* in his discussion of how to adapt the Greek intellectual project to Roman needs, and with Plutarch's account of how the power and beauty of siege weaponry glorify a ruler and his state.

A great many of the passages in this section have been from Roman sources. This is not an accident of selection bias; it is a particular characteristic of Roman literary discourse on technology that it focuses not on the mechanical intricacies of the devices themselves, but on the cultural capital they represent. The *sphaera* that so fascinates Cicero's Philus does so not because of its complexity but really in spite of it, in the way it makes the wanderings of the planets comprehensible without requiring specialized astronomical knowledge. The sundials Pliny discusses are, like the *sphaera*, worth commenting on not because of the astronomical knowledge they encode, but because they are the source of stories about Greek invention and Roman acquisition, and about the experiences of the Roman intellectual.

Conclusion

A description of an object in visual terms mobilizes the reader's perspective: a visible object can be encountered from far away or up close, from outside or inside, as a finished product or as it comes together in the process of construction. The reader might be cast in the role of a spectator of a completed device, or as a viewer present at the invention or construction of the device, seeing the parts come together. Incorporating information pertaining to the other senses, particularly the sense of touch, forces the reader to engage imaginatively with the object up close and hands on, as it were. Lists of dimensions or other statistics may lend the text an air of precision, reliability, or administrative control. The confusion and opacity that often result from such aggregations of data can be alleviated by organizational techniques like the walkthrough. Expanding the ekphrasis from a static description to a spatially extended walkthrough or a temporally extended narrative can drastically increase the amount of detail a reader can absorb about the artifact.

The varied layers of knowledge a technological artifact encodes create a wide variety of possible structures for a technical ekphrasis. An author can decide to deploy information about one layer or another, to focus on a machine's theoretical-mechanical principles or the peculiar properties of the materials it is made from, its operational details or its historical background, what it is like to see it or what it is like to build it. The differences between mathematically inclined designer and qualitative spectator, insider and mystified outsider, depend on what information the author chooses to include in a description.

CHAPTER 4

Diagram and Artifact

Introduction

The representational power of verbal descriptions can be supplemented by visual depictions that clarify the description, add detail to it, make it more concrete, or strip it down to its geometrical skeleton. As Athenaeus Mechanicus promises to his addressee Marcellus:[1] "Therefore, if you choose, all the devices will be illustrated (ἐσχηματογραφημένα πάντα ἔσται τὰ μηχανήματα); and what is difficult to explain will be quite clear from them."[2] In late antiquity, the author of the lavishly illustrated *De rebus bellicis* likewise promises his own imperial addressee that

> lest any difficulty arise in creating these types of armaments, I have below appended a color picture of them, not differing at all from the real thing (*imaginem tormentorum nihil a uero distantem coloribus adumbratam orationi subieci*), to simplify the arrangements for imitating them.[3]

He proceeds at once to a kind of ballista, "whose construction, placed before your eyes, an exact picture witnesses" (*cuius fabricam ante oculos positam subtilis pictura testatur*).[4] Adding a visual component to a verbal representation of a mechanical artifact can clarify what the text leaves obscure, and seeing a picture of the device can convince the reader of its feasibility. This is perhaps true especially in the case of the comparatively brief descriptions provided by Athenaeus and the author of the *De rebus bellicis*, but it is hardly limited to them. Illustrations also add at least the impression of easy comprehensibility to technical material: the late ancient text on siege warfare transmitted under the name Heron Byzantius suggests in its preface that the illustrations provided will allow the machines

[1] Most likely M. Claudius Marcellus, nephew of Augustus; on the controversies surrounding Athenaeus' dates, see Athenaeus 2004, 15–25.

[2] Athenaeus Mechanicus, *Peri mēchanēmatōn* 39.

[3] Anonymous, *De rebus bellicis* 6.4.

[4] Anonymous, *De rebus bellicis* 7.1.

described in the text to be grasped easily, in a way that lacks "no didactic or explanatory capacity."[5]

Indeed, easy as it is now to reproduce images accurately, who would dream of representing technological artifacts without at least some recourse to visual depiction? For ancient authors, of course, the challenge was quite different; copying an image is even more difficult to carry out with fidelity than copying text, and provides room for intentional as well as accidental variations. Furthermore, ancient diagrams were without much of the contemporary apparatus of scale markers, multiple perspectives, and exploded diagrams that technical illustration relies on today. Illustrated manuscripts of mechanical artifact texts provoke questions about the interaction of text and diagram, the relationship between the diagram and the object it represents, and the balance of realism and abstraction. These diagrams are sometimes closely bound to the text with letter labels or are themselves described verbally; others are under-determined in the text, which may lead to more variants in their manuscript tradition.

Diagrams of machines and their components occupy a middle ground in the territory between object and text: the diagram more immediately resembles the object it represents, yet remains ontologically distant from it. The diagram's status as an intermediary between a real machine and an abstract verbal representation is wonderfully evoked in a manuscript of a text written much later than those under consideration here, the sixteenth-century *Von Glocken- und Stückgießerei* by the master gunmaker Christoph Seßelschreiber. This work includes an illustration of a hoisting device based on a system of ropes and pulleys, in which the pulleys are drawn on the page, while the ropes are represented by real threads pulled through holes punched in the page.[6]

While the diagrams in the surviving manuscripts of ancient mechanical artifact texts do not connect text and object so directly, they nonetheless employ a variety of mediating strategies. These range from geometrical abstraction to iconic depiction of what a viewer of the real object would see, which may extend to portraying the technology in use. Some of these differences stem from the intended audience of a work: figures in a work intended for an interested layman, particularly for a wealthy patron, may feature rich illustrations of the artifact in use, while a work intended for a technical colleague may adopt clear and simple

[5] Heron Byzantius 1909, 199.4. See also Meißner 1999, 333.
[6] For a discussion of this delightfully inventive manuscript, see Leng 2004.

diagrams. Other differences naturally arise from variations in the type of machine. Ballistic devices are built from many complex components, so that some manuscripts feature diagrams of individual pieces or small subassemblies. Medical equipment like the "Hippocratic bench," on the other hand, is both mechanically simpler than ballistic devices and designed for use on humans, so that the diagrams in Apollonius' commentary to the Hippocratic treatise *On Joints* are shown in the field, being applied to patients.

Diagram and text

Any discussion of ancient mechanical diagrams must begin with a cautionary note on the available evidence. The earliest surviving diagrams are found in Byzantine manuscripts originating centuries after the composition of the texts, and so comparison with their "originals" is simply impossible. The situation is thus slightly different from the related tradition of geometrical diagrams. There, at least a few survive on papyrus; though even these are of course not some notional "original versions," they at least provide a glimpse of an older tradition.[7] Changes in the dominant style of manuscript illustration may trigger the alteration of part or all of a diagram. Sometimes these alterations leave anachronistic clues behind; Weitzmann observes how the garments of figures in astronomical manuscripts change with the prevailing fashion, and the human figures shown in the surviving manuscripts of Apollodorus are outfitted differently in different manuscripts.[8] These anachronistic elements do not necessarily imply that the mechanical artifacts shown in the diagram are suspect, as it is possible that the human figures were added later and the artifacts themselves reproduced more faithfully.

The transmission of diagrams is in many cases somewhat constrained by the text: many texts describe the diagrams or label their features with letters, which provides an in-text mechanism to stabilize the transmission. Saito and Sidoli classify geometrical diagrams as "objects of transmission" within their manuscript traditions.[9] As such, they are subject to variation between manuscripts, requiring careful attention to their individual instantiations but not invalidated as objects of study. Likewise, the images in surviving manuscripts of mechanical artifact texts are part of

[7] On Greek mathematical papyri, a few of which include geometrical diagrams, see Fowler 1995; Fowler 1999, 204–17, 229–78; Jones 2009.

[8] Weitzmann 1947, 157–160.

[9] Sidoli and Saito 2012, 158. For more detailed case studies, see Saito 2009; Saito 2012.

those texts' long and varied lives; we cannot find our way back to the "originals," but we can interrogate the textual references to them, and we can make the varied phenotypes of the surviving copies an object of careful study in its own right.[10]

Diagrams invite further questions about how closely a visual depiction can or should resemble a material artifact. Many of the diagrams in ancient mechanical texts draw upon a tradition of geometrical diagrams in mathematical prose. Netz argues that diagrams in Greek mathematical texts are not meant to stand for an ideal mathematical object outside the text.[11] Instead, the actions carried out according to the text's instructions are, by and large, operations on the diagram. At the same time, the diagram was not seen as "an exact instantiation of the object constructed in the proposition," because even where it is possible to construct the geometrical objects used in the proposition to a very high level of accuracy, some of the intermediary steps required for those constructions are omitted from the proposition.[12] Netz proposes that this quandary may be resolved by regarding the diagram as a "make-believe object," which is similar in form to the referred object, and functionally identical to it, so that it is possible to believe that the circular shape on the page and the perfect circle are identical.

Representing artifacts that (at least potentially) exist materially in the world is a rather different problem from representing mathematical objects, and this shift puts some strain on the established techniques for diagrammatic representation. In a different context, Aristotle grapples explicitly with the imperfect parallel between mathematical-style diagrams and the material objects those diagrams purport to represent. In *On the Motion of Animals*, he argues that if one jointed part of the animal moves, then another must remain at rest. Thus the part containing the joint is "both one and two, both straight and bent, shifting potentially and actually because of the bending."[13] He provides a geometrical diagram

[10] My scope here does not permit detailed analyses of individual mechanical authors' diagrammatic traditions, but the expansion of Saito and Sidoli's approach to geometrical diagrams into mechanical territory is already being paved by work like van Leeuwen's on the Peripatetic *Mechanica* (van Leeuwen 2014).

[11] Netz 1999, 51–6. Netz here mentions a notable exception: instructions to "imagine" objects and actions that cannot be represented diagrammatically.

[12] Netz gives the example of the equilateral triangle in the first two propositions of Euclid's *Elements*. The first calls for the construction of two auxiliary circles, which are then used to construct the triangle. The second proposition, however, elides the step of constructing the circles, calling simply for the construction of the equilateral triangle. The second triangle "acknowledges the shadow of a possible construction without actually performing it" (1999, 54).

[13] Aristotle, *De motu animalium* 698a18–21.

to clarify this argument by means of an analogy to a mathematical construction: "And when [the part of the animal] is bent and moved, one of the points in the joints is moved and one remains in place, just as if Α and Δ of a diameter remained in place, while Β was moved, and the line ΑΓ appeared."[14] However, he is concerned that the analogy is imperfect, because mathematical points are indivisible, whereas the same cannot be said of the joints of an animal:

> But here the center appears in every way undivided (for, so to speak, they fabricate their movements, for none of the mathematical objects moves), whereas in joints the centers exist potentially and actually, sometimes single and sometimes divided.[15]

These misgivings do not stop Aristotle from using diagrams to illustrate several biological phenomena, including the motion of tall men and snakes, the bends in the legs of quadrupeds, the appearance of the male reproductive system, and the features of the baby cuttlefish.[16] These diagrams, so far as the scanty remaining evidence indicates, ranged from geometrical abstractions (some as simple as a straight line with letter labels ΑΒΓΔ) to the finer-featured representation of the cuttlefish.

The surveyor Balbus produced a treatise called *The Description and Explanation of All Figures* (*Expositio et ratio omnium formarum*), most likely in the late first or early second century CE.[17] This text sets forth the various categories of shapes used in the tasks of surveying, using language and structure largely derived from Euclid, along with schematic illustrations. At the same time, language that recalls the *realia* of surveying is still present; for example, he defines a point in Euclidean terms (*signum est cuius pars nulla est*), while simultaneously identifying the point as a tool for the surveyor by saying that the point is what begins and ends each act of measurement.[18] So too he defines the straight line (*rigor*) as follows: "whatever happens in the field because of the surveyor's work to [define] a straight boundary is called a *rigor*; whatever appears on a map

[14] Aristotle, *De motu animalium* 698a21–4. The argument about the center of rotational motion, which he determines to be potentially one but actually more than one, is taken up again at 702b25.

[15] Aristotle, *De motu animalium* 698a24–b1.

[16] Aristotle, *De incessu animalium* 707b17–28, *De incessu animalium* 712a1–13, *Historia animalium* 510a29–35, and *Historia animalium* 550a25ff.

[17] Balbus mentions a Dacian campaign, which Dilke takes to be Domitian's, while Toneatto and Guillaumin believe it to be Trajan's: Dilke 1971, 42; Toneatto 1994, 5; Balbus 1996, 3–5; Chouquer and Favory 2001, 27.

[18] In Lachmann's edition (where this work is entitled *Expositio et ratio omnium mensurarum*) at 97.14–15.

in imitation of these is called a line."[19] That is, the *rigor* comes into being in the act of being measured in the field, and the line in the act of being drawn on the map; so Balbus blurs the boundary between the objects of the surveyor's experience and the mathematical objects for which they stand.

Diagrams of mechanical artifacts exhibit a comparable range of ontological states between mathematical idealization and material reality: some provide a geometrical schema of the artifact, while others show something more like a constructed artifact. The difficulties of representing material objects diagrammatically are as worrying for some mechanical authors as they were to Aristotle, particularly in the Hellenistic period. Later authors, on the other hand, appear to be more comfortable with the ontological gap between material objects and their textual and diagrammatic representations.

Some elements of a "grammar" of technical illustration remain more or less fixed throughout the surviving medieval diagrams of different authors, in different manuscript traditions, from different centuries. These include the representation of ropes and chains by two overlapping sinuous shapes, gears as circles bounded by triangular teeth, and racks as bands attached to triangular teeth. Weights are generally shown as trapezoids. Liquid contents of pneumatic wonders like wine dispensers are frequently drawn as groups of small circles. Color is used in many of the surviving diagrams; many use two or three colors of ink wash or paint over the initial drawing lines. The color sometimes follows the structural lines of the machine; sometimes it fills in features such as a pillar or decorative figure. Alternation of colors may help the reader to distinguish between two or more of a single component type (two sets of pipes in a complex pneumatic device, for example, or two sets of wheels on a siege engine).

Modern techniques for showing what is inside or underneath a solid surface, like the cutaway diagram or the dotted line, are not part of the ancient diagrammatic toolbox. In the absence of such techniques, illustrators of the surviving manuscripts often resort to showing components at odd angles from one another, as though three-dimensional assemblies had been stamped flat, to provide a more or less comprehensible view of their entire depth. For example, a tortoise in Apollodorus' *Poliorcetica* is pictured using this kind of orthographic projection (Fig. 1).[20]

[19] *nam quidquid in agro mensorii operis causa ad finem rectum fuerit rigor appellatur; quidquid ad horum imitationem in forma scribitur linea appellatur.* Balbus 1996, 42–4; 98.12–14 Lachmann. On this passage see also Roby 2014, 47–8.

[20] The diagram that follows is after Paris Gr. 2442, fol. 84v.

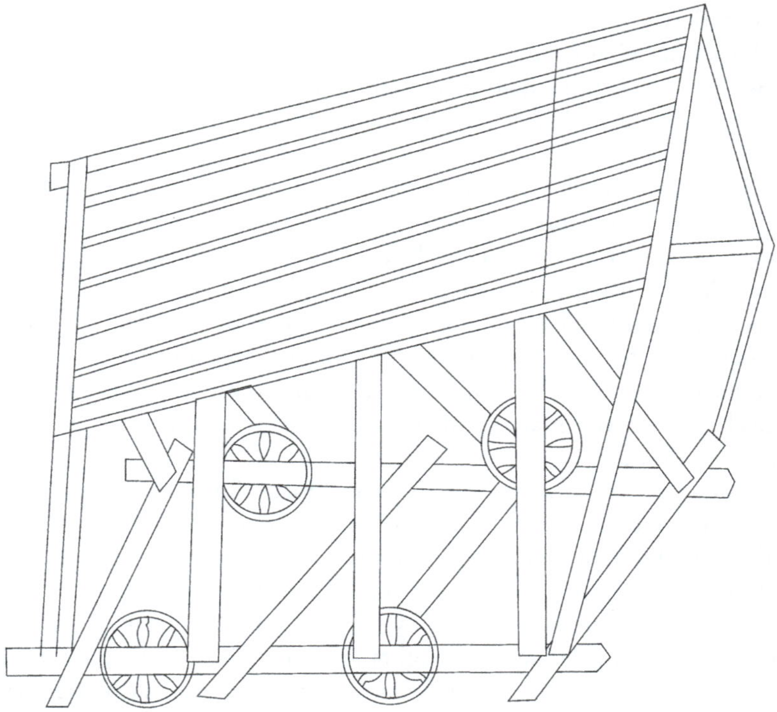

Figure 1. *Tortoise drawn in orthographic projection.*

There are, however, occasional circumstances where a kind of cutaway view is provided. For example, in an illustration of the wall-bursting flamethrower of Apollodorus, the illustrator uses different patterns to show the undamaged exterior of the wall and the view through the sizable hole that the flamethrower has made in it.[21] However, since this illustration shows an actual hole in the wall, rather than the purely notional hole of the modern cutaway view, this is really not an occurrence of the modern technique.

Graphical techniques such as these suggest that while the tradition of mathematical diagrams had a strong influence on many of the surviving mechanical diagrams, some of these texts probably drew on other traditions of illustration as well. The particular challenges of representing a three-dimensional material artifact visually may be better met by techniques like those used to represent architectural constructions, surviving

[21] Paris suppl. gr. 607, 33r.

examples of which date back much further than the mechanical diagrams. The most thorough treatment of these representations is provided by Heisel, whose work covers the surviving architectural plans, engravings, and models from Mesopotamia, Egypt, Greece, and Rome, along with the evidence from literary references to such plans.[22] Like mechanical diagrams, these architectural diagrams sometimes make use of a symbolic "grammar," or what Heisel calls a "code." These codes are culture-specific and chronologically dependent rather than universal and permanent, and must be understood in order for an architectural drawing to be correctly read.

Unlike the mechanical diagrams, the surviving Greek and Roman architectural drawings are not found in textual treatises. The manuscript tradition of Vitruvius, the principal surviving ancient architectural author, is famously devoid of authentic diagrams.[23] At the same time, Gros is surely right to argue that it is extremely unlikely that Vitruvius would have had "l'imprudence ou l'impudence" to offer his imperial addressee a treatise that promised illustrations without delivering them.[24] The surviving architectural drawings are carved on stone or marble, often on the stones of the very buildings they illustrate. The Greek materials are almost all done at full scale, with the exception of the so-called "Pytheos sketch" from the temple of Athena at Priene, which shows a reduced-scale plan of the temple with relatively little detail. Some Greek representations of pillars are compressed in length so as to be easier to represent, but these carvings do not involve true scaling.[25] The same is true for the Corinthian columns depicted on P. Oxy. LXXI 4842, which are drawn with detail and precision at small size, but not truly scaled, as Coulton notes in the appendix to Whitehouse's discussion of the images.[26] The preference for full-scale representations of parts of the building under construction of course considerably limits the objects that can be illustrated, but even these demonstrate a range of symbolic indications of features like the inclination of a roof, the extent of the top edge of an architrave, or the locations of supporting columns.[27]

[22] Heisel 1993. There are several other useful references for this subject, notably Straub 1949; Petronōtīs 1968; Coulton 1977; Université Strasbourg and Centre de recherche sur le Proche-Orient et la Grèce antiques 1985; Gros 1996. However, Heisel's single-minded attention to graphical representations of architectural constructions makes him particularly relevant to this discussion. For an ontological comparison of architectural plans with mathematical diagrams, see Netz 1999, 60–1.

[23] Stückelberger 1994, 96–9; Gros 1996, 23–6; Meißner 1999, 266–8 on the greater emphasis on the graphical in M. Cetius Faventinus.

[24] Gros 1996, 25.

[25] Heisel 1993, 5–6, 154–7, 161–2, 173–4.

[26] Whitehouse 2007, 305.

[27] Heisel 1993, 168–9.

Analogous Roman materials were produced at a variety of scales on a broader range of media, including engravings of city plans (notably the Forma Urbis Romae) and plans of smaller complexes such as aqueducts and graves, as well as a mosaic of the plan for a bath and a marble disc engraved with the lines for the geometrical construction of a volute.[28] With this increased flexibility of media and scale come new possibilities for representation, and the need for a more sophisticated symbology. For example, the engraved plan of a grave complex now held at the National Archaeological Museum of Perugia includes not one but two symbols that appear to indicate stairs: a set of parallel dashes and a rectangle with an inscribed triangle.[29] The former, notes Heisel, was used as far back as Mesopotamian building plans executed in clay, while the latter appears frequently also in the Forma Urbis Romae, where the alignment of the base of the triangles with the entrances to buildings makes quite clear that the triangle points in the direction of ascent or descent. Heisel suggests that the combination of both in a single representation may indicate two different types of stairs.[30] Other symbols in the Roman plans seem to indicate park space, reed beds, trees or columns (these are nearly impossible to differentiate), and aqueducts.[31]

The draftsmen behind the architectural representations seem to have made some use of color as well. In the case of the representations carved on building stones such as those from the renovated temple of Apollo at Didyma, the use of color was limited to applications of red pigment along the main lines of the carving to make it more visible against the background of uncarved stone.[32] However, more extensive and varied coloration seems to have applied to the Roman architectural representations. For example, the bath mosaic of the Via di Porta San Lorenzo matches different colors to different parts of the buildings depicted.[33] Given this evidence, as well as the variety of marking methods used for different surfaces on engravings such as the Forma Urbis Romae and the plan of a grave complex from near the Via Labicana,[34] it appears that color may have been

[28] On the Forma Urbis Romae, see Carettoni 1960; Rodríguez Almeida 1981. The fragments are most easily accessed now through Stanford's Digital Forma Urbis Romae Project; see Koller et al. 2006.

[29] Heisel 1993, 190–1.

[30] Heisel proposes in particular that the parallel lines stand for stone steps (whose weightiness means that their exact extent needs to be specified so they can be properly undergirded with supports), while the rectangle/triangle combination stands for wood steps (which are light enough that they can be freely erected in the space available without this level of planning).

[31] On this symbology see (among many others) Petronōtīs 1968, 29–35; Hesberg 1984; Heisel 1993, 184–207; Stückelberger 1994, 112–20; Nicolet 1991, 63, 71, 99–107; Coarelli 1992; Chouquer and Favory 2001, 45–62, 279–86, 298–303.

[32] Heisel 1993, 75; Haselberger 1999.

[33] Heisel 1993, sec. R4. [34] Heisel 1993, sec. R2.

used widely in these Roman engravings to indicate different surfaces and parts of buildings, though direct evidence for this does not survive.

What is clear is that already by the classical Greek period a certain amount of graphical symbology was already in use in the context of architectural visualizations, and the symbolic "vocabulary" used in this milieu had greatly increased by the Roman imperial period. This symbology represents a mixture of what would in Charles Peirce's terms be called "icons" and "indices": some of the elements of these visualizations (like aqueducts drawn as arches) resemble their referent in form, while others (trees and columns represented by dots) only point to what they stand for.[35] Mechanical diagrams may blend visual elements that conjure up the experience of viewing the object or space in person with those that prompt analysis of more abstract indicators of its components.

Illustrations in texts on surveying are informed by a similar set of challenges: here too, the drawing must often integrate and reconcile material elements such as buildings or landscape features with mathematically idealized elements such as a grid of property line divisions.[36] The illustrations in the surveyors' texts range from the "schematic" (in the sense of resembling a geometrical idealization of the structure) to the "iconic" (in the sense of resembling what the eye sees). On the former end lies, for example, the illustration of a gnomon in Hyginus Gromaticus consisting of a stylized representation of a gnomon and ruler-drawn lines indicating the shadows it casts throughout the day;[37] still more so a geometrized diagram of a range-finding problem in Hero's *Dioptra*.[38] On the latter end is Frontinus' illustration of the technique of cultellation (projecting points on a slope onto a horizontal line) using trees.[39] Many of the illustrations combine elements of both, like Hyginus Gromaticus' definition of maritime and montane *limites*, which shows an "iconic" scene of an ocean and a mountain juxtaposed with the rectilinear grid system on which they impinge.[40] Some of the more "schematic" drawings make use of letter labels that link them to the text like a mathematical diagram. They primarily depict landscape features rather than the mechanical devices used in surveying (though there are a few exceptions[41]).

[35] Peirce 1931, 447.

[36] Dilke 1971, 126–32, reproductions passim; Stückelberger 1994, 112–20; Campbell 2000, xxiii–xxvi, reproductions of illustrations at 277–316; Lachmann 1848, vol. 1; Chouquer and Favory 2001, 49–63, 278–327; list of manuscripts with illustrations at 337–8; Hyginus and Frontinus 2005, 36–40, 237–46.

[37] Cod. Vat. Palatinus Lat. 1564, 94v.

[38] Paris suppl. gr. 607, 66v.

[39] Arcerianus A (Wolfenbüttel, Herzog August Bibliothek, Guelferb. 36.23 A).

[40] Vat. Palatinus Lat. 1564.

[41] For example, Cod. Vat. Palatinus Lat. 1564, 94v (Hyginus Gromaticus, *Constitutio limitum*) depicts a gnomon; Paris suppl. gr. 607, 64r shows Hero's *dioptra* on a column.

Other traditions of manuscript illustration resemble the mechanical diagrams less closely, and may or may not have influenced them. These include the illustrations of plants and animals in Dioscorides, Nicander, and Dionysius of Philadelphia,[42] as well as the depictions of humans and animals in the pseudo-Oppianic *Cynegetica*.[43] Depictions of the varieties of phenomena in the natural world have somewhat different priorities from the mechanical diagrams; they aim in large part to demonstrate the differentiae of phenomena already in existence, as opposed to showing the forms of artifacts to be produced. In this respect certain illustrations in medical texts may be closer to the mechanical tradition; for example, Apollonius of Citium's commentary to the Hippocratic *On Joints* contains several drawings showing the patient and practitioners putting traction devices into action.[44] Here, as in the surveying texts, the mechanical devices are not shown in so much detail as the more prominent elements in the text (in this case, the human body), but they are nevertheless present verbally in the text and may draw on a visual tradition common to mechanical illustrations.

The illustrations found in the surviving manuscripts of Athenaeus and Apollodorus share some significant features with some of the surviving "iconic" pictures in surveying texts: they eschew letter labels in favor of word labels (or no labels), they represent three-dimensional objects using a horizontal orthographic projection, and they depict the exterior appearance of walls or components of the artifact. These visual strategies are strikingly different from the diagrams in texts like Philo's and Hero's works on military engines; these are linked to the text through letter labels and nearly always shown in a top-down view. Other texts exhibit a hybrid approach; for example, the surviving manuscripts of Hero's *Pneumatica* often show both the internal pneumatic workings of these wonders and their carefully crafted outer form. Some of the surviving depictions of alchemical equipment seem to have been strongly influenced by the illustrations of pneumatic devices; in particular, the illustrations in the surviving manuscripts of Zosimus' works on alchemical tools vary strikingly in the degree of precision with which they are executed, but they render vessels such as the *tribikos* quite similarly to those in the pneumatic texts.[45] However, these diagrams use word rather than letter labels, suggesting

[42] Stückelberger 1994, 74–94.
[43] For example, Marcianus gr. 479, whose illustrations are collected and discussed in Spatharakis 2004.
[44] Cod. Laurentianus 74.7.
[45] For an example see Berthelot and Ruelle 1888, 138.

that the texts' verbal style was not influenced by pneumatic works like Hero's and Philo's as strongly as the illustrations.

Since the original illustrated mechanical texts do not survive, the rare textual references to drafting technologies are extremely important. Although different in purpose and form from the mechanical diagrams, architectural drawings provide an ancient textual witness to their theory and practice that may have been valid for the mechanical diagrams as well.[46] Vitruvius, who insists that technical drawing be part of the educational formation of the *architectus*, gives some indication of how the art of draftsmanship was organized. An architect must be able to produce three varieties of visualization of a building (*species dispositionis*, glossed by Vitruvius as equivalent to the Greek *ideae*): *ichnographia*, *orthographia*, and *scaenographia*. Vitruvius defines these as follows:

> *Ichnographia* is a skill that includes moderate use of compass and ruler; from it drawings are taken of the sites of structures on the ground. *Orthographia*, however, is the upright image of its face, colored to some extent according to the plan of the work to be built. Likewise, *scaenographia* is a shaded sketch of the face and receding sides referring all lines to the center of a circle.[47]

Architectural drawings, like the surveyors' illustrations, thus occupy the full spectrum from the "schematic" to the "iconic." *Ichnographia* is on the "schematic" end of this scale, using the tools of geometrical diagrams to produce an image of the building's footprint, which will then be transferred to the site itself. *Orthographia* uses more painterly techniques to indicate what the front of the building will look like. Vitruvius recommends the use of color along the building's structural lines, which is consistent with the approach taken in many of the surviving manuscript illustrations of mechanical artifacts. It appears to be a way of picking out the important structural features of the building rather than a naturalistic portrayal of the finished structure; this projection is still not quite what an actual viewer will see. That role is played by *scaenographia*, which adds in the rest of the building's visible sides, as well as using perspective (the reference to lines converging at the center of a circle seems to refer to the

[46] It should be noted that Vitruvius is not the only literary *evidence* for the use of graphical plans in architecture. A catalogue of passages from authors (including Cicero, Suetonius, Plutarch, and Aulus Gellius, among others) in which graphical architectural plans are portrayed as being deployed for discussion and approval among decision-making groups of various sizes, both on and away from the site of construction, can be found at Heisel 1993, 184–5.

[47] Vitr. *De arch.* 1.2.2.

vanishing point) to project the image in a way that more closely reproduces the experience of viewing the building itself.

Scaenographia in particular brings in additional complications, as accurate drawings of this type require some knowledge of optics (or at least appropriate application of a set of algorithms based on optical considerations). So Vitruvius cautions that column diameters must increase proportionately to the distance between them lest the space seem to "consume and diminish" them. Later he offers a broader discussion of optical illusions, which in the context of building must be remedied by craftiness on the architect's part (*ingeniorum acuminibus*), prefaced by a claim that the same phenomena produce the illusory effect of depth in scene paintings (*in scenis pictis*).[48] In the preface to his seventh book he revives discussion of the techniques used to provide a sense of depth and perspective to theatrical background paintings.[49]

Scaenographia is classified as an optical problem in a passage tentatively attributed to Geminus and included in Heiberg's edition of Hero's *Definitiones*.[50] Though its authorship is difficult to assign, the passage provides some very useful context on *scaenographia*'s disciplinary classification and techniques. *Scaenographia* is first introduced as one of the three components of optics, along with catoptrics and the homonymous subdiscipline of optics.[51] It is later treated in more detail, defined as a task specifically related to architectural drawing, since the architect's goal is to "make the work proportional to the image" (τὸ πρὸς φαντασίαν εὔρυθμον ποιῆσαι τὸ ἔργον) by applying various optical tricks.[52] In particular, *scaenographia* is described as the part of optics that seeks to depict buildings, and in particular to depict them as they appear rather than as they are. The author is quite cognizant of the disconnect between the actual proportions of large objects like buildings and the way those proportions appear to the viewer on the ground, and indeed asserts that the goal of the architect is to use his understanding of optical illusions to make the building appear harmoniously proportioned to the eye (πρὸς ὄψιν) rather than actually (κατ' ἀλήθειαν) being proportional. He goes on to mention

48 Vitr. *De arch.* 3.11.11, 6.2.1–4. On these passages see Büssing 1984.

49 Vitr. *De arch.* 7.pr.11. Tybout compares the principles of perspective outlined in both passages and concludes that Vitruvius is describing two different perspectival systems: the scheme described in Vitruvius' seventh book was developed during the classical period; *scaenographia* proper not until the Hellenistic era (Tybout 1989, 65).

50 On the passage's inclusion in the *Definitiones* text (passage 135) along with some other passages, see Hero 2003, 85.

51 Geminus, *Fragmenta optica*, in Heliodorus, Damianus, and Geminus 1897, 26.1–4.

52 Geminus, *Fragmenta optica* 28.10–30.10.

a few such illusions: a cylindrical column will appear narrower in the middle, while a circle will resemble an ellipse and a square will seem stretched.

Despite the optical challenges it presents, *scaenographia* is valuable both as a representational and as a persuasive tool. Heisel argues that *scaenographia* has a particular cognitive accessibility, suggesting that the *cogitationes et formas*,[53] which Vitruvius says the ambitious architect Deinocrates presented to Alexander, might rather have resembled easily comprehended *scaenographiae* than the sketches and plans used by architectural practitioners.[54] *Scaenographia* has a special persuasive power because it imitates how the object itself would look to a viewer: this striking effect provides even laymen with imaginative access to the building, whereas the other forms may require more technical training and experience for the imagination to reconstruct the building from the drawing.

Vitruvius proceeds to locate the source of all three types of drawing in the intellectual processes of "contemplation" (*cogitatio*) and "invention" (*inventio*).[55] Here Vitruvius might mean that *cogitatio* and *inventio* were required for the initial discovery and development of the drafting techniques he describes, or that they are required for each new application of the techniques to a technical drawing. The latter explanation seems to make more sense in light of the remark's coherence with Vitruvius' other instructions to practitioners of *architectura*, particularly his directive to master both theory and practice, which opens the previous chapter. What is Vitruvius demanding from the *architectus* as illustrator? *Cogitatio* is about technique: draftsmanship executed carefully, with an eye toward the aesthetics of the finished product (as *voluptas* indicates). *Inventio*, on the other hand, concerns the intellectual, analytical flexibility that makes it possible to envision new solutions to complex problems. Technical illustration, from abstract plans to painterly representations, requires essentially the same skill set as Vitruvius prescribes for other technical activities: a combination of analytical problem-solving skills with careful, disciplined craftsmanship.

The same kinds of intellectual and interpretive activity are required for creating technical illustrations, whether they attempt to reproduce what the eye sees in a painterly fashion or represent the form of a structure in an abstract, diagrammatic form. Comparable interpretive work happens on the receiving end of the illustration. Like the purely verbal descriptive

[53] Vitr. *De arch.* 2.pr.2. [54] Heisel 1993, 161.

[55] Vitr. *De arch.* 1.2.2: *hae nascuntur ex cogitatione et inventione. cogitatio est cura studii plena et industriae vigilantiaeque effectus propositi cum voluptate. inventio autem est quaestionum obscurarum explicatio ratioque novae rei vigore mobili reperta.*

techniques used in these texts, the illustrations they contain help the reader to perform the cognitive task of creating a mental representation of the described object in his mind.

The cognitive effect of illustrations' recreating themselves in the mind of the reader is eloquently captured by Peirce's term "icon."[56] He defines the icon most precisely as the image as it exists in consciousness, but the term may also be used for the image in the world that sparks the image in the mind. He includes among icons all kinds of "immediate images," including abstract geometrical diagrams, which he indeed highlights as exemplary icons. Heisel analyzes the process of communication into a source, transmitter, medium, message, and receiver. In the case of communication through diagrams, the diagram is the medium, which channels the concepts (the "source") that exist in the mind of the draftsman (the "transmitter"), translated into graphical form through the drawing process. The observer of the representation (the "receiver") then receives the information it provides (the "message").[57] Cumbersome as this step-by-step explanation may be, it usefully highlights the cognitive processes of translation in the minds of both the draftsman and the reader.

Artifact and diagram

Can looking at diagrams representing an object have the same sort of cognitive immediacy as seeing the object itself? Ptolemy seems to suggest so. In the *Harmonica*, he identifies sight and hearing as the senses responsible for relaying information to the rational part of the soul. These senses acquire this information through verbal sources encountered both visually and aurally. Ptolemy argues that this task is made easier when these sources are augmented by verbal imagery or by diagrams:

> Hearing alone demonstrates things seen by means of explanations; sight alone reports things heard by means of written materials. Both of these together often offer clearer explanations than either on its own, as when knowledge transmitted verbally along with diagrams or figures (μετὰ διαγραμμάτων ἢ χαρακτήρων) is easier for us to learn and remember as we acquire it, and that transmitted by sight appears more faithfully rendered through poetic expression, for example the appearances of waves, topography, battles, and the circumstances of experience, so that our souls are sympathetically affected by the forms of the things described as though we were seeing them.[58]

[56] Peirce 1931, 447. [57] Heisel 1993, 4–5.
[58] Ptolemy, *Harmonica* III.3.64–71.

Ptolemy here emphasizes the duality of sight and hearing, and so he links diagrams (given the phrasing he chooses here, he is probably thinking of lettered diagrams in particular) explicitly with oral explanations. His own extensive use of lettered diagrams in his written works suggests he thought that they are at least as effective as explanatory tools when they form part of a text designed to be read rather than listened to.[59] These labels, a defining attribute of mathematical diagrams, serve as the anchor points between the verbal and visual components, allowing them to work in harmony. While most mechanical authors who refer to diagrams use letters to label some elements of a machine, Ptolemy is unusual in rigorously labeling each point-like component, such as corners and the attachment points of strings.

Ptolemy's comparison between the effects of poetic imagery and diagrams on the rational soul suggests he perceives some common ground between their cognitive effects. Poetic imagery brings the soul into a condition sympathetic with the form of the object being described, the same condition it would be in if the eye actually encountered that object. Diagrams, meanwhile, make it easier for the rational part of the soul to internalize verbal information: visual diagrams and verbal description are partners in transmitting knowledge from the world to the understanding.

Diagrams mediate between text and concrete object in a variety of ways, and the way the diagram is referred to in the text can indicate some fine but important distinctions in the ontological relationship between the object and its diagram. Hero, for example, alludes to his extensively letter-labeled diagrams in such a way that the elements of the diagrams are assimilated completely to the components of the machine they represent, and Ptolemy takes a similar approach. Earlier authors, by contrast, struggle more visibly in making an equivalency between mechanical artifacts and their diagrammatic representations. It appears that the descriptive strategy of adapting mathematical conventions like the lettered diagram to mechanical artifacts was a textual technology that developed only gradually.

Philo refers to diagrams much less frequently than Hero or Ptolemy, and when he does discuss them, he does not equate the diagram to what it represents. Instead, he asserts that the diagram "looks like" (ὁρᾶν)[60] the

[59] Netz hypothesizes the development of mathematical propositions accompanied by diagrams as a comparable kind of interaction between the verbal and the visual. This development begins with the drawing and letter-labeling of a diagram, a process "accompanied by an oral dress rehearsal – an internal monologue, perhaps" (Netz 1999, 86). This oral component provides the outline of the argument, which is then codified in written text. On the interaction between the visual and verbal in mathematical texts, see also Netz 1999, 26–9, 86–8.

[60] Philo, *Belopoeica* 64.16.

machine, or something along these lines. Due to an unhappy accident of the manuscript tradition, Philo's surviving texts are devoid of diagrams; only empty spaces remain in a pair of manuscripts,[61] and otherwise they have vanished completely. Judging by these spaces and by the scattered references in the text, Philo used diagrams to enhance his descriptions much less often than Hero. He does appeal to the power of visual reasoning, as when he claims regarding the problems he has just outlined with current ballistic technology that "one could best display these to view (ἐπὶ τῆς ὄψεως) one piece at a time."[62] Yet he does not apparently seek here to attempt such a visual demonstration through diagrams; perhaps Philo recommends visual access to the machine itself as a first resort.

On the few occasions where he does refer to diagrams, his remarks are nearly always limited to saying that a diagram follows, which will clarify the preceding verbal description. The only part of this work in which he does refer in any detail to a lettered diagram is during his discussion of the improvements he is proposing to the frame of the catapult. Here he uses letter labels to denote components A through M, which are sufficient for him to portray the whole frame and shooting mechanism of the catapult. His references to the lettered points of the diagram, at their most detailed, are as follows:

> Let there be the visible part of the side support, conforming in shape to the drawing below, where [the point] A is (ἔστω δὲ ὑπὸ τὴν ὄψιν κείμενος ὁ παραστάτης τῷ σχήματι ὑπάρχων ὁ ὑπογεγραμμένος, ἐφ' οὗ τὸ A). Imagine its stated length is that from one line all the way to the other, where [the points] A and B are … we said the parts extending above the line, where [the points] Γ are, were like a double tenon.[63]

Another passage using letter labels follows, and then Philo largely dispenses with them. Why does he choose to use them here? Here he is describing his own design, which the reader is encouraged to reproduce, rather than the older designs of Ctesibius and his predecessors. So, while Philo devotes extensive verbal description to the legacy designs, his own frame design perhaps warrants enhancing its verbal description with detailed diagrams, not only because of its novelty, but because this is the design the text actually suggests the reader build.

Philo does not describe the contents of the other diagrams to which he alludes in such detail. However, he does discuss how these diagrams

[61] These are Paris gr. 2442 (P), and Vat. gr. 1164 (V), both dated to the eleventh century.
[62] Philo, *Belopoeica* 58.35. [63] Philo, *Belopoeica* 63.29–40.

can assist the reader with the acts of imagination needed to visualize the assembled device:

> Imagine the spaces in between the side supports and center supports being closed off, so that the spring is hidden and cannot be harmed from any direction it faces, while the frame, having neither washers nor under-washers nor pawls lying on top of it [to be damaged], nor any other roughness, appears elegantly simple and presents an attractive sight. We will make its appearance clear to you at the end by drawing a diagram.[64]

The diagram is evidently intended to show the covered exterior rather than the interior components, and is perhaps given as a guide to achieving the right aesthetic effect, a recurring concern of Philo's. The reader's imagination functions here to add the cover to a pre-existing visualization of the interior of the catapult; Philo reminds the reader of the elements, like washers, that would be seen without it, and then hides them from view. Philo often invokes the imagination in this way, to add a new element to a visualized subassembly, or to slightly change an element that has already been described. A few of these instances will suffice to give an idea of how he puts the imagination to work:

> Imagine that you are looking at the frame of one of the old models, and that the side supports and center supports are similar to those in ours, but in place of the hole carrier something like an architrave rests on them.[65]
>
> Imagine that [the tenons on the connecting-rods] are polished and rounded underneath so as not to hurt the springs.[66]
>
> Imagine that there are also some posts worked from oak, having the same width and length as the connecting rods, but a width of 1½ dactyls.[67]

In all these cases, Philo asks the reader to imagine editing some element of a mental image he has already constructed. As the latter examples show, he does not usually provide a diagram to assist with the operation; instead, he uses verbal description of how components must fit and move together to define their forms. The outer housing is a special case; because it is more aesthetic than functional, its form is not sufficiently determined by the verbal description, and perhaps this is why Philo chooses to illustrate it visually. Netz describes a comparable phenomenon in mathematical texts, where forms of νοεῖν are used when objects are not included in a diagram for some reason, or where they are perceived as being inadequately indicated in the diagram.[68] This occurs particularly in the case of

[64] Philo, *Belopoeica* 62.15–25. [65] Philo, *Belopoeica* 62.2–6. [66] Philo, *Belopoeica* 65.14–16.
[67] Philo, *Belopoeica* 65.20–2. [68] Netz 1999, 52–3.

three-dimensional objects and in "applied" mathematics, where an element of a geometrical diagram is supposed to stand in for a material object like a balance. It is not surprising, then, to find that Philo adopts this same lexicon for his own descriptions of material objects.

Another type of interaction between an imagined visualization and a diagram accompanies Philo's more detailed description of the lettered diagram that originally illustrated his frame design:

> νόει δὲ τὸ σχῆμα συντεθέντων αὐτῶν ὁρᾶν τὸ ἐφ' οὗ τὰ Α·ἔστω δὲ ἡ δίοπτρα μὲν ἐφ' ἧς τὸ Γ, ὁ δὲ ἁρμὸς ὁ τῆς συμβολῆς ὁ Δ ...
>
> νόει δὲ ταύτας εἶναι κατὰ τὰς γραμμὰς ἐφ' ὧν Ε. ὑπόλαβε δὲ καὶ τὸ προειρημένον σχῆμα κατὰ κρόταφον θεωρεῖν· καὶ ἔστω τὸ ἐφ' οὗ τὸ Ζ. ἐχέτω οὖν ἐκτομὰς πεποιημένας ἐξ ἑκατέρου μέρους τὰς ἐφ' ὧν Η, ἕως κατὰ τὴν ἐπιτομὴν πιπτούσας ἀκριβῶς, ἐχούσας δὲ πλάτος δακτύλων β· ἀπέχεις οὖν καὶ τὴν τῶν μεσοστατῶν κατασκευήν.

> Imagine the form of these once they have been put together looks like that where [the points] A are. Let the sight be at Γ, and its joint of connection Δ ...
>
> Imagine [the excisions in the center support] are along the lines where E is. Now suppose you are seeing the form previously described on its side, and let it be where Z is. Then let it have excisions made in each part, which are where the letters H are, up to the point that they fall precisely on the other incision, having a width of 2 dactyls; thus, you have the construction of the center supports.[69]

Here the imagination is explicitly invoked to bridge the gap between the constructed machine and its representation in a diagram, but this bridge remains incomplete in Philo. The machine "looks like that where [the points] A are," but it is not to be identified with what is shown there. The reader imagines that the rabbets that hold the center support in the frame are "along the lines where E is," but they are not taken to be identical with those lines. Even where the connector Δ seems to be identified with its diagrammatic counterpart, this cannot be the case since Δ is masculine (i.e. it refers neither to a point or a line, but most likely simply recalls the *harmos*). The sight (*dioptra*) probably makes the same kind of reference back to the material component, though in this case it could grammatically also refer to a line, which would make spatial sense here. The rather awkward grammatical constructions to which Philo resorts in this passage consistently call attention to the effort required to impose the imagined machine on the visible diagram. Philo's approach is quite unlike that of Hero and Ptolemy, who draw

[69] Philo, *Belopoeica* 64.16–30.

an equivalency between diagram and machine without requiring this kind of linguistic and conceptual effort.

Philo was not the inventor of this kind of circumlocution. The language he uses to establish a non-equivalent relationship between the points of a diagram and components of a machine had been used already by Aristotle, in the biological texts referred to earlier. He narrates the diagram illustrating the concavities in the gait of a tall man or a snake by saying, "Let the point at the right front side be where A is, and that on the left where B is, and the point at the left rear be where Γ is, and that on the right where Δ is" (σημεῖον δεξιὸν πρόσθιον ἐφ' οὗ Α, ἀριστερὸν ἐφ' οὗ Β, ὀπίσθιον ἀριστερὸν ἐφ' οὗ Γ, δεξιὸν ἐφ' οὗ Δ).[70] By contrast, the diagram he uses at the beginning of *De motu animalium* to describe the function of the joints is presented as a hypothetical mathematical abstraction, containing purely geometrical objects: "just as if A and Δ of a diameter remained in place, while B was moved, and AΓ appeared" (ἂν εἰ τῆς διαμέτρου ἡ μὲν Α καὶ ἡ Δ μένοι, ἡ δὲ Β κινοῖτο, καὶ γίνοιτο ἡ ΑΓ).[71] Here he refers to the points simply as ἡ A, for example; their labels agree in gender with καμπή, although they have already been conceptually separated from the joints by the use of the geometrical hypothesis that they are parts of a circle.

The pseudo-Aristotelian *Mechanica* provides another point of comparison. The author of this text uses both approaches, combining direct identification of diagrammatic and material elements with the periphrasis using *epi*.[72] For example, a set of dentist's pliers is described using letter labels, some of which are identified directly as parts of the pliers, and some of which are identified periphrastically as in Philo's text:

> ἔστω γὰρ τῆς ὀδοντάγρας τὸ μὲν ἕτερον ἄκρον ἐφ' ᾧ τὸ Α, τὸ δὲ ἕτερον, τὸ Β, ὃ ἐξαιρεῖ· ὁ δὲ μοχλὸς ἐφ' ᾧ ΑΔΖ, ὁ δὲ ἄλλος μοχλὸς ἐφ' ᾧ ΒΓΕ, ὑπομόχλιον δὲ τὸ ΓΘΔ· ὁ δὲ ὀδοὺς ἐφ' οὗ Ι σύναψις· ὁ δὲ τὸ βάρος.
>
> Let one end of the pliers be where [the point] A is, the other B (this is the one that extracts). One lever is at AΔZ, the other at BΓE, and the fulcrum is ΓΘΔ. The tooth is at I, the juncture; this is the weight.[73]

70 *De incessu animalium* 707b26–8.

71 *De motu animalium* 698a23–5.

72 The *Mechanica* is no longer usually attributed to Aristotle, but is widely agreed to be the work of some member of the Peripatetic school. A concise recent overview of claims about the text's authorship is provided at Coxhead 2012, 302. Bottecchia Dehò provides much more extensive discussion of its authorship, concluding that the text is not likely to postdate Aristotle by long and might well be genuine: Aristotle 2000, 27–51. Berryman argues that Strato of Lampsacus (who succeeded Theophrastus as the head of the Peripatetic school) is the most likely author (Berryman 2009, 106–8). Drachmann notes that in any case, a pre-Archimedean date is almost assured by absence of references to the screw or to centers of gravity (Drachmann 1963, 10).

73 *Mechanica* 854a25–30.

In the very next problem, a nutcracker is analyzed using just one point on a diagram (the fulcrum), and this point is indicated using *epi*: "the device comprises two levers that have the same fulcrum, the juncture where [the point] A is" (τὴν συναφὴν ἐφ' ἧς τὸ Α).[74] These two problems, which are mechanically very similar and yet manage to use different strategies to establish a relationship between a device and its diagram, show a linguistic flexibility that is not found in later texts seeking to establish a relationship between an object and a letter-labeled diagram.

There is, of course, a conceptual difference between the mechanical elements referred to in the Peripatetic *Mechanica* and those in Philo's work. The former are notional beams, oars, and rudders, used to illustrate mechanical principles in a stripped-down, idealized form. Like the frictionless plane of today's physics texts, these objects exist somewhere in between purely mathematical objects and the imperfect and variable material objects that Philo grapples with. Philo's reluctance to identify components of the diagram directly with components of the machine perhaps owes something to this distinction. Despite these differences, however, Aristotle, the author of the *Mechanica*, and Philo are all engaged in the same experimental process of adapting the techniques of mathematical prose (themselves still fluid) to a new class of material objects. These objects may be closely assimilated to their textual and diagrammatic representations, or may instead be kept at a certain conceptual distance. In the texts from the early stages of this interdisciplinary borrowing of mathematical techniques for mechanical applications, use of the periphrasis with *epi* prevails.

The use of *epi* in textual references to lettered diagrams of material objects prompts comparison with its use in the mathematical texts to which the genre of mechanical prose owes so much. Mathematical authors, from the early Hippocratic fragment to later authors, frequently use *epi* to denote points of diagrams. Archimedes and Apollonius, for example, use *epi* with the dative or genitive to perform the task of placing points on a pre-established geometrical object like a section of a cone.[75] *Epi*, along with other prepositions, is used by these authors to "describe various *spatial* relationships between the letters in the diagram and the objects referred to."[76]

[74] *Mechanica* 854a38–9.

[75] See, for example, Archimedes, *De conoidibus et sphaeroidibus* 1.162.6–16; Apollonius, *Conica* 2.44.1–4; Apollonius, *Conica* 3.13.6–7, etc.

[76] Netz 1999, 47.

Netz argues that the letters of mathematical diagrams should generally be interpreted, in the terms of Peirce, as indices rather than icons: they signify by means of a deictic relationship with their object, by standing next to it, rather than by similarity to it. The consequence of this is that the text is very strongly focused on the diagram as a spatial object, "as spatial as the world of plates and cupboards."[77] If the letters function as indices, then representations using the letters must necessarily refer to the diagram.[78] Federspiel notes in a detailed analysis of the language of Euclid 1.4 that the formulation τοῖς ἐφ' ἑαυτῆς σημείοις for "the points of a line" is quite unmarked in mathematical prose.[79] Indeed, he proposes that the usage is something like the Latin *pugna ad Cannas* meaning "the battle of Cannae." Yet there is a complication in this example; *epi* used in this way may be an unmarked genitive in *mathematical* prose, where the spatial context is pre-established, but this is not true in ordinary Greek usage.

As we have already seen, Aristotle is another relatively early source for the use of letters as references to points of diagrams. As Netz points out, however, he uses letters as well to denote relationships between conceptual objects that are not associated with diagrams, and where the relationship cannot be taken to be spatial. Netz cites a passage from the *Prior Analytics* as an example, arguing that Aristotle is making use of "a bold *metaphor*, departing from the spatial mathematical usage."[80] Aristotle is seen here in an experimental mode, expanding textual techniques that lend themselves well to representing a certain class of objects (geometrical forms staked out by lettered points) to a new class of objects that share some, but not all, of the features of the original class. Seen in this light, Aristotle's challenge in the *Prior Analytics* has much in common with his struggle to represent the components of an animal using a toolbox developed for the components of a mathematical construction, and with Philo's struggle to apply this same toolbox to mechanical problems.

The mathematical texts Netz analyzes provide a verbal narrative of mathematical operations that are taken as being performed on the visible diagram. By contrast, Philo and Aristotle (when Aristotle is talking about concrete objects like limbs) are trying to perform a slightly different task, establishing a kind of metaphorical mapping between locations on a diagram and components of a material body (a catapult or a person), which really are part of Netz's "world of plates and cupboards." For these material objects, two kinds of mapping must be taken into account: points

[77] Netz 1999, 47. [78] Netz 1999, 51. [79] Federspiel 1991.
[80] Aristotle, *An. Pr.* 30a28–30; Netz 1999, 48.

A and B have a spatial distance from each other within the diagram, while the diagram itself is a kind of projection into the text of the material objects labeled as A and B.

This second kind of mapping highlights the ontological distance between the object and the diagram that represents it. This distance is necessary in the case of the mechanical diagram (its two-dimensionality alone separates it from the object), but not in the case of mathematical texts. It is theoretically possible to make an accurate diagrammatic construction of certain geometrical forms, such as the equilateral triangle (constructed using compass-generated circles) mentioned earlier. Some mathematical diagrams do not allow for perfect representation of geometrical forms, and in some other cases the steps required to achieve such precision in the diagram are elided. Even in these situations, however, the operations the text instructs the reader to perform are carried out, as Netz argues, *on the diagram*. In the mechanical texts, the instructions are in most cases meant to be carried out not on the diagram but on the machine itself. Even when the diagram is used as a template, the operations are carried out on the diagram only in order that they may be transferred to the machine itself. The operations are no longer operations with lines and points, but with beams and pegs.

This calls for a change in the rules of the game of "make-believe" Netz associates with Greek mathematical diagrams.[81] Such "make-believe" activity succeeds for mathematical diagrams because the diagram is both similar in form and functionally identical to the object it points to (in the sense that the operations described in the proposition can be carried out on the diagram). In the case of mechanical diagrams, this is no longer true, and it is not possible to reconcile the object and diagram so neatly. This gap prompts Philo to hedge his references to the diagram with the use of *epi*, with claims that the object "looks like" the diagram, and with instructions for the reader to imagine that elements of the machine stand in the place of A or B.

As a textual tradition of mechanical prose based in part on the conventions of geometrical prose developed, the challenge of representing material objects using the language of mathematical diagrams seems to have become less problematic. After Philo, the next extant link in the chain is probably Biton's *Kataskeuai*. Biton does not use any kind of periphrasis while introducing the mechanical elements that appear in lettered diagrams following each description, or say that the machine "resembles" the

[81] Netz 1999, 51–6.

figure. Instead, he directly identifies the components of the machine with the lettered components of the diagram. For example, he introduces the beams of a stone-throwing catapult simply by saying that "it has straight beams AB" (ἔστι γὰρ ἔχον ὀρθοὺς κανόνας τοὺς AB).[82] He uses this approach throughout his text, wherever he labels mechanical components with letters.

This direct approach is also, of course, used in mathematical texts. Netz raises an important cautionary point about constructions such as Euclid's ἔστω εὐθεῖα ἡ AB.[83] This phrase is conventionally translated as "let AB be a straight line," which Netz argues suggests a *correlation* between some AB and a straight line. The problem with this translation is the Greek article; the use of the feminine gender means that AB must already conceptually *be* a straight line, so that the conventional translation of ἔστω leads to the reading "let the straight line AB be a straight line." Netz recommends the alternative translation "Let there be a straight line, [viz.] AB"; this phrasing, rather than claiming some relationship between AB and an object (a straight line), takes the quite different action of assuming a certain straight line, labeling it AB, hence locating it in the diagram.

Biton, like the other mechanical authors who identify mechanical objects with their diagrams, uses ἔστω in this same way, to establish an object rather than to ascribe attributes such as straightness to it. Biton does not make a scrupulous effort to label each component mentioned in the text. He normally labels the larger components such as the main beams, and sometimes other principal components such as the bow, but leaves many components unlabeled, even if they do appear in the diagram. He uses letters to label many different kinds of components (for example, the sling of a stone-throwing catapult), not just those that correspond more or less to individual points (for example, the ends of a beam). In just one passage of his *Kataskeuai* Biton varies his conventions for labeling beams several times: the catapult he is describing has a pair of beams referred to collectively as AB (in the diagram A corresponds to one end of each beam, and B to the other), two more beams referred to as ΓΓ, and yet another pair of beams which are first referred to as ΔE, and then changed to Δ when Biton focuses closely on the upper side. Biton's labels are thus quite fluid, capable of shifting from labeling point-like features of the catapult like the end of a beam, to line-like components like the beam itself, and even components that do not closely resemble any basic geometrical object, like the sling.

[82] Biton, *Kataskeuai* 2.4. [83] Netz 1999, 243.

Biton's approach directly identifies material components with their counterparts in a diagrammatic representation, without any indication of the conceptual struggles of identification characteristic of Philo and the *Mechanica*. However, this direct approach probably does not represent a fully thought out adaptation of mathematical prose to the mechanical arena so much as it reflects Biton's typically casual descriptive approach. (Consider, for example, his concluding advice: "Do not be confused because we have used set measurements; you will not need to use the same measurements. If you want to build larger ones, do it, and likewise smaller ones; just try to maintain the proportions."[84]) Biton's fluid letter labeling relies on the fact that he is not performing any rigorous geometrical operations on the labeled components. The unlikelihood of Biton's having conscientiously shaped his language around a newly developed theory of the ontological status of the objects he describes, however, suggests that by the time of his writing, the same direct approach to labeling diagrammed components might already have become a common feature of mechanical texts.

Hero adapts the techniques of geometrical texts to mechanical topics in a more complex way than Biton. Where Philo is sparing in his allocation of letter labels to components of the catapult frame, Hero labels nearly every feature of the machine in his *Belopoeica*. Here, for example, he describes the components of the *gastraphetēs*:

> Let there be the bow mentioned ΑΒΓΔ, having curved ends ΑΒ and ΓΔ too firm to be wound by the hand of man (Ἔστω γὰρ τὸ εἰρημένον τόξον τὸ ΑΒΓΔ, ἔχον τὰς ἐπικαμπτομένας ἄκρας τὰς ΑΒ, ΓΔ βιαιοτέρας τῆς διὰ τῆς χειρὸς τοῦ ἀνθρώπου γιγνομένης καταγωγῆς), and the bowstring ΑΔ. Connected to the bow, below the middle of the concave edge, let there be a beam ΕΖΗΘ, having on its upper surface a dovetail-shaped channel ΚΛ.[85]

The labels used in this passage correspond to points on a diagram, and the frequent appearance of letter labels in the verbal instructions strongly suggests that the diagram is intended to be used together with them to help the reader visualize the bow.[86] Like Biton, Hero identifies the mechanical components with their representations in the diagram, rather than resorting to the use of *epi* or ὁρᾶν as Philo does. With the phrase "Let there be the bow mentioned ΑΒΓΔ," for example, he introduces points ΑΒΓΔ as demarcating the form of the bow, not the form of a diagram resembling

[84] Biton, *Kataskeuai* 8.3–5. [85] Hero, *Belopoeica* 5.1–10.

[86] In Paris suppl. gr. 607, this diagram is found at fol. 47v.

the bow.[87] The letter labels denote objects that, far from being geometrical abstractions, have tangible physical qualities: ΑΒ and ΓΔ are "too firm to be wound by a man's hand," while ΑΔ is a bowstring, not a line connecting two points that represents a bowstring. Tybjerg observes of other works by Hero that "Hero's diagrams simultaneously act as a geometrical diagram where the diagram *is* the object manipulated and a technical illustration that *represents* the object."[88] In contrast to Philo's approach, the letters in the text now mark out material objects with material properties like elasticity, no longer just diagrammatic elements that "look like" those objects.

Hero's direct approach to verbalizing the content of diagrams is not confined to the *Belopoeica.* He uses the same easy equivalency when using lettered diagrams to describe other types of mechanical artifacts. Consider, for example, a device in the *Pneumatica* that uses bubbling water to simulate the songs of mechanical birds, prompting an owl to turn and look at the birds, which then fall silent.[89] The components of this device, from the outer vessel to the concealed siphons and weighted chain, are all identified with their equivalents in the diagram using letter labels. Likewise, in what remains of the *Mechanica*, mechanical components of all kinds are identified by letter labels, as in this explanation of a hauling machine with a given mechanical advantage:

> τῷ δὲ αὐτῷ ἄξονι τῷ ΚΛ συμφυὲς <ἔστω> τύμπανον τὸ ΞΟ ἔχον ὁμοίως τὴν διάμετρον πενταπλασίονα τῆς τοῦ ΜΝ τυμπάνου διαμέτρου.
>
> On the same axle ΚΛ, let a drum ΞΟ be attached, likewise having a diameter five times the diameter of the drum ΜΝ.[90]

Hero draws the same equivalence between machine and diagram whenever he refers to diagrams of material objects in the *Cheiroballistra*, the *Automata*, and the portions of the *Dioptra* in which he discusses artifacts such as the odometer and the *dioptra* itself. In other texts such as the *Metrica* or *Catoptrica*, which are concerned with more purely geometrical problems, he uses the same approach. By Hero's time, then, it seems that material, technological artifacts had become quite fully assimilated into the tradition of diagrammatic representation that began with mathematical authors describing strictly geometrical objects.

Ptolemy, like Hero, identifies components of an artifact with their counterparts in a diagram. At the same time, however, there is nevertheless

[87] This latter approach is often used by Ptolemy, as when he introduces an instrument represented in a diagram as a "rectangle."

[88] Tybjerg 2004, 50. [89] Hero, *Pneumatica* 1.16. [90] Hero, *Mechanica* fr. 1.1.32–5.

a subtle but important difference in the ways these two authors carry out the identification. In Ptolemy's approach, as in Hero's, elements of the artifact are mapped onto elements of the diagram without the kind of struggle we witnessed in the Hellenistic authors. However, Ptolemy normally assigns these components a somewhat idealized geometrical identity, rather than foregrounding material attributes like toughness as Hero does. Consider, for example, Ptolemy's description of the harmonic *kanōn*:

> Let a *kanōn* be imagined on the straight line ΑΒΓΔ, and bridges near its ends, everywhere equal and alike, forming under the strings surfaces nearly spherical (σφαιρικάς, ὡς ἔνι μάλιστα, ποιοῦσαι τὰς ὑπὸ τὰς χορδὰς ἐπιφανείας). [Let one of these, ΒΕ, be constructed] around the center Ζ of the surface mentioned, and ΓΗ similarly around the center Θ, with Ε and Η the points obtained by bisection of the convex surfaces. Let the bridges have such a position that the lines extended through the bisection points Ε and Η and the centers Ζ and Θ, that is ΕΖΒ and ΗΘΓ, be perpendicular to ΑΒΓΔ. If now we stretch a string of similar length from the points Α and Δ, as ΑΕΗΔ, it will also be parallel to ΑΒΓΔ, because the bridges have equal height. And let the points at Ε and Η be taken as the beginnings of the plucked part of the string; for on these it will make its connections to the convex surfaces, because ΕΖΒ and ΗΘΓ are also perpendicular to it.[91]

Ptolemy's language here is highly mathematical. Although he is discussing a material object, the *kanōn*, he describes it in idealized geometrical terms: bisected straight lines, parallel and perpendicular pairs of lines, the centers of spheres, strings that contact their bridges and supports at single points Κ and Λ, respectively designated as the bisection of the string as a whole and the bisection of the half. The illustration of the *kanōn* is drawn in the surviving manuscripts with ruler and compass, and the points referred to in the text are labeled (Fig. 2).[92] It is much closer to a geometrical diagram of a mathematical object than a representation of a material object as it would appear to the eye.

The components of the *kanōn* are laid out, as it were, on a normalizing geometrical diagram, a process highlighted by the initial command to imagine the *kanōn* as "on (κατὰ) the straight line ΑΒΓΔ."

Ptolemy, in this sense, takes the identification between object and representation even further than Hero. Whereas Hero is more comfortable than Philo in establishing a conceptual identity between a component of an artifact and its counterpart in a diagram of the artifact, Ptolemy assigns the geometrical attributes understood from the diagram (perpendicularity,

[91] Ptolemy, *Harmonica* 1.8.31–43.

[92] The diagram is after Paris suppl. gr. 59, fol. 169v.

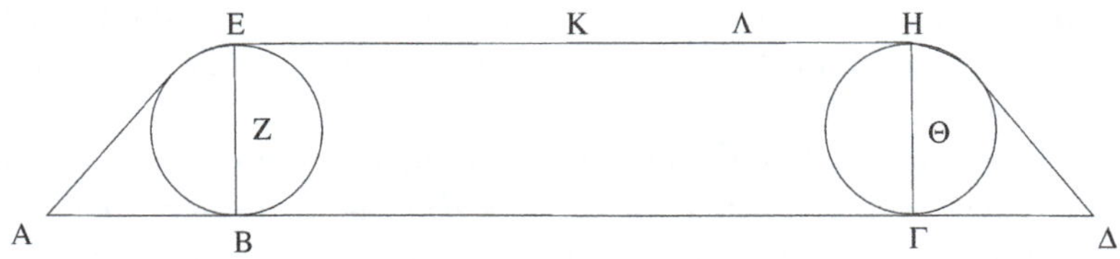

Figure 2. *Diagram of the* kanōn.

say, or perfect bisection) to the artifact itself, locating the artifact in a mathematical space. In Creese's insightful formulation, "the *kanōn* is itself a diagram of the arithmetical relationships which underpin the phenomena. The diagram of *Harmonica* 1.8 is the diagram of an instrument of a diagram."[93] This is possible because Ptolemy introduces the *kanōn* diagrammatically (as opposed to the way he describes the astronomical instruments in the *Almagest*). Creese argues that Ptolemy's use of highly mathematicized language and diagrams "suggest[s] the *ekthesis* of a mathematical proof," but that this suggestion is at the same time undermined by the reminder of material imperfections – the surfaces of the bridges are only "nearly" spherical. This creates the tension or "collision" Creese observes between the mathematical and the material.

The only exception to the assimilation between the *kanōn* and its diagram is the action of stretching the string between A and Δ. Rather than defining the string simply as AEHΔ, Ptolemy provides an instruction to stretch "a string of similar length from the points A and Δ, as AEHΔ" (χορδὴν σύμμετρον, ὡς τὴν AEHΔ). This phrasing signals the assignment of the particular attribute of length measurement to the string, where AEHΔ serves as an example of a string with this attribute. A comparable turn of phrase occurs in Hero's *Belopoeica*. In his discussion of the early history of torsion catapults and their frames, Hero says "they constructed a frame from four strong beams, like ABΓΔ" (πλινθίον ἔπηξαν ἐκ κανόνων δ ἰσχυρῶν, οἷον τὸ ABΓΔ).[94] Here, as in the case of Ptolemy's *kanōn*, the turn of phrase seems to combine with the introduction of material components in a particular way, identifying a part of the artifact as having a particular material, structural attribute (such as being of a specific length, or being constructed from four strong beams), as well as

[93] Creese 2010, 70–1. This view is echoed in Creese's approach to the *Sectio canonis*: "harmonic diagrams are only hummable once one has the monochord in mind. The reader who follows the construction of propositions 19–20 on a monochord redraws their diagrams acoustically. The monochord is thus an indispensable diagram-reading instrument in the *Sectio*" (2010, 175).

[94] Hero, *Belopoeica* 8.16–17.

a punctuated moment of creation ("they constructed" the frame, "let us stretch" the string). Both approaches serve to reify the artifact as a material construction.

In the second book of the *Harmonica*, Ptolemy proposes an instrument that could serve as an alternative to the *helikōn*, an instrument that he says was promoted by theorists to replace the eight-stringed *kanōn* but difficult to use in practice. He never gives this alternative instrument a name, but introduces it immediately as a geometrical construct (a diagram of a rectangle), into which material elements are gradually introduced:

> Next to this instrument, if we simply lay out (ἐκθώμεθα) a rectangle, as ΑΒΓΔ, and we imagine the lines ΑΒ and ΓΔ at the plucked part of the string (νοήσωμεν τὰς μὲν ΑΒ καὶ ΓΔ κατὰ τὰ ἀποψάλματα τῶν χορδῶν), and ΑΓ and ΒΔ on the boundary tones of the octave. Then if we extend [a line] ΔΕ equal to ΓΔ and cut the side ΓΔ, [measuring against] rulers, at the proper ratios of the genera, placing the limit of high pitch as Ε, and through the cuts that emerge on it, we extend strings parallel to ΑΓ and equal-toned to each other, and when this has been done we place what will become the common bridge of the strings underneath them, at the position linking the points Α and Ε underneath, as ΑΖΕ.[95]

The instrument as a whole is strongly subordinated to the mathematical diagram by its initial identification as a rectangle, which is to be placed next to the previous instrument; this form is even "laid out" using a verb that recalls the *ekthesis* of a mathematical proof. Identifying the instrument with its geometrical representation retrospectively recalls the *helikōn*, next to which it is suggested the rectangle be placed.[96] As the passage continues, the relationship between the diagram and some material components of the instrument (the strings) is again given a distanced structure by the use of *kata*, as in the description of the *kanōn*. The brief material reality of the strings quickly evaporates as their terminal points are reidentified as the bounding notes of the octave – here the arithmetical abstraction of harmonic ratios compounds the geometrical abstractions already performed.

Finally, the common bridge is placed under the strings; this is a physical action on a material instrument, which nevertheless calls attention to its geometrical "shadow" here through the explicit identification of Α and Ε as "points" (σημεῖα). The description thus weaves nimbly between the

[95] Ptolemy, *Harmonica* II.2.35–43.

[96] As Solomon notes, the παρὰ which opens up this description could also mean "in addition to" (Ptolemy 2000, 67 n. 35). Even so, the identification of the new instrument with the geometrical construction remains.

material and the geometrical, but the emphasis remains consistently upon the latter. Ptolemy's liberal use of geometrical terminology (this passage alone includes rectangles, parallels, and points, as well as a host of letter labels), and the ease with which he identifies components of the instrument with their abstract mathematical equivalents, demonstrate his preference for analyzing these devices in distinctly geometrical terms. Likewise, as Creese argues, the simpler string diagrams in the *Harmonica* are used as tools of proof alongside the appropriate mathematical arguments.[97]

For both Hero and Ptolemy, then, the material artifact and abstract diagram are relatively easily reconciled. Ptolemy's descriptions of artifacts do occasionally separate the two using *kata*, but for the most part the artifact and diagram are identified. Hero does not use any such language of separation. This represents a significant change from the textual references to diagrams of material objects in Aristotle's biological texts, the pseudo-Aristotelian *Mechanica*, and Philo's *Belopoeica*. These earlier authors struggle (explicitly in Aristotle's case) with the problems of identifying elements of technological artifacts with their representations in the form of lettered diagrams.

This shift appears to mark a textual-historical process in which the geometrical diagram and the language associated with it were gradually appropriated into texts on other technical subjects. The objects that the conventions of the geometrical-style diagram were now called upon to illustrate shared some characteristics with their mathematical relations, and could be introduced to the reader similarly piece by piece through a process of "construction" (*kataskeuē*). On the other hand, they are concrete objects, possessing material as well as abstractly spatial attributes. They are subject to irregularities of components and construction: a bowstring is not a line, and one bowstring is not interchangeable with every other. These differences problematized the wholesale adaptation of conventions that had been developed for mathematical objects to descriptions of artifacts. The circumlocutions of Aristotle and the later Hellenistic authors can thus be seen as a kind of experimentation, a tentative process of appropriating elements of mathematical prose to serve the task of describing material objects.

Diagrams can also intervene more directly between text and object. Hero, in particular, sometimes recommends the use of diagrams as templates:

[97] Creese 2010, 71.

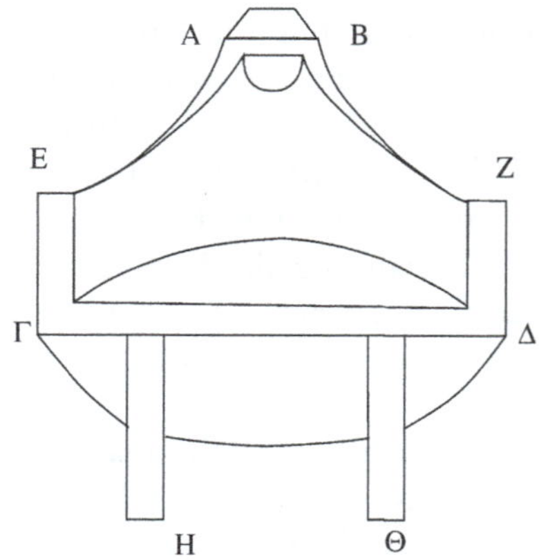

Figure 3. *Washer template.*

> The washer is made in this way: it is necessary to construct a template similar to ΑΒΓΔΕΖ depicted below, which has rounded edges ΑΕ and ΒΖ, and straight ones ΕΓ and ΖΔ, and ΑΒ equal to the diameter of the hole; and then for the washer to be carved out upon this. If it is going to consist of bronze, it is necessary to make it by molding a circular cast from beaten bronze, having thickness sufficient for the force of the engine.[98]

The user is meant to reproduce the diagram, scaled appropriately, upon some separate material suitable for sizing the hole. The washer could either be rectangular and made of wood plated with metal (to avoid friction wear), or round and made of cast bronze (Fig. 3). Where a wooden washer was used, the "template" for the washer hole was inserted into the hole during the chiseling process to check its diameter.[99]

A different use of the diagram as template occurs in Hero's discussion of the peculiarities of the straight-spring (*euthytone*) engine as compared to the V-spring (*palintone*) engine.[100] In this kind of catapult, the two springs are shorter and mounted on a single frame, separated by the breadth of the slider, a design which is simpler than the V-spring but requires accurate

[98] Hero, *Belopoeica* 20.1–13. The word ἐμβολεύς translated here as "template" has a primary meaning of "piston" in mechanical contexts (Hero uses it with this meaning extensively in the *Pneumatica*), but it is also used, as here, to denote templates of various kinds. Philo uses it for a shaping block used to impart curvature to a slab of bronze, and Anthemius (*Fragmentum mathematicum Bobiense* 56.27) appears to use it in the context of constructing one mirror so as to be identical to another.

[99] Marsden 1971, 53 n. 29. The diagram is after Paris gr. 2442, fol. 76r.

[100] For technical and historical discussion of the differences between these types of catapult, see Marsden 1969, 16–33. Marsden's work is updated with further archaeological evidence in Baatz 1994; Rihll 2007; Campbell 2011.

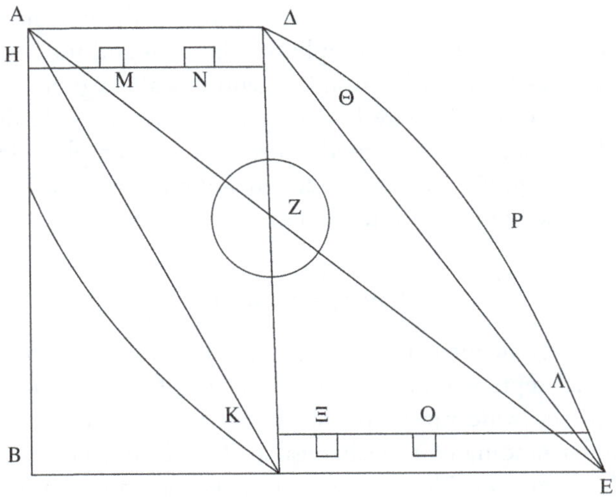

Figure 4. *Hole carrier template.*

placement of the hole carriers. In order to achieve this, Hero recommends transferring the design directly from a diagram in the text:

> Add together the widths of the side supports and center supports, which we called "counter-supports" in the palintones, and also the diameters of the holes, and the width of the barrel (that is between the center supports), and construct on a board a line this long. Draw [lines] ΑΓ and ΒΔ perpendicular to it, each of which is equal to the diameter of the hole, or a little larger equal to the width of the side support, and then draw around the circumference ΓΕΔ of the circle through Γ and Δ. And the figure ΑΒΓΔΕ will be your underside of the hole carrier in the euthytone that consists of two hole carriers (καὶ ἔσται σοι τὸ ΑΒΓΔΕ σχῆμα τὸ κάτω μέρος τοῦ περιτρήτου ἐν τῷ εὐθυτόνῳ τῷ συγκειμένῳ ἐκ τῶν δύο περιτρήτων).[101]

The construction described in this passage is then matched with another one that can be used to construct the top piece once the side supports and center supports are installed (Fig. 4).[102] The transition from drawing to machine is seamless; Hero says that the figure (σχῆμα) *will be* the underside. Unlike the passages above where the identification of diagrammatic element is a convention Hero uses, here the two are really the same object.

These two templates allow the designer to construct certain elements of the catapult directly from the text, in two different ways. In the first case,

[101] Hero, *Belopoeica* 26.16–28. Wescher's edition of the text replaces ΑΓ and ΒΔ with ΑΔ and ΒΓ.

[102] The diagram is after Paris suppl. gr. 607, fol. 50v.

the diagram of the washer assembly is used to size the hole in the washer. The second template is more complicated in form and use, requiring the designer to transfer a fairly complex geometrical diagram onto boards of wood. Both cases illustrate how diagrams can help to bridge the gap between the artifact and its textual representation. Templates leverage the text's capacity to guide the work of design and building, involving the diagram itself directly in the production of new artifacts.

Layered diagrams

The way a diagram mediates between text and object depends on what kind of visual representation of the object it offers. The previous chapter outlined the wide variety of descriptive strategies used to shape the reader's understanding of a mechanical artifact. It may be described from an "insider's" or "outsider's" perspective; its mechanical features or its aesthetic effect on the viewer may be emphasized, and so on. Diagrams of these artifacts require the author and illustrator to make similar decisions: a diagram can show the object as a whole or in part, as a painterly representation or a geometrical abstraction, from the long view of a layman or in meticulous mechanical detail.

The surviving mechanical diagrams are generally "iconic" in Peirce's sense, visually resembling at least some components of the object's physical form. However, depending on which components are illustrated, the diagram may or may not visually resemble what a viewer of the completed object would see. The visible exterior surface of an artifact, after all, is only one piece of it, and the most interesting parts may in fact be inside. Since ancient mechanical diagrams lack the common modern drafting technologies of exploded diagrams or multiple perspectives, the depiction of these artifacts frequently requires a choice to be made between showing the interior and exterior.

The diagram might show the artifact as a constructed whole, in parts, or both. Naturally, this has something to do with the kind of verbal description of the machine that the diagram accompanies. Texts that present the machine in the process of construction are, not surprisingly, more likely to describe individual subassemblies of the device one by one. In the surviving manuscripts, this approach is correlated with the depiction of subassemblies in separate diagrams, which may or may not be shown alongside a diagram of the complete machine.

Hero's *Cheiroballistra* exemplifies the latter approach. This work is a truncated account of an artillery weapon; it has no preface, and a few components are missing.[103] However, the text is still complete enough to provide a detailed, component-by-component account of the device. Unlike his *Belopoeica*, where he describes devices that were already well known in Philo's time, this text details a device likely to have been contemporary with Hero himself. Marsden points out several distinctive features that the device described here shares with those portrayed on Trajan's column, including a wide spacing between spring frames (which are separated by the curved strut called the *kamarion*, and which have a distinctive circular shape), and an added lower strut called the *klimakion*.[104] If Hero's floruit was indeed in the middle of the first century CE, the Dacian wars are not far in the future.[105] For all its incompleteness, the text of the *Cheiroballistra* provides a highly detailed description of the components of a device that perhaps held considerable interest as a technological novelty.

The surviving text begins *in medias res*, with the stock. Here, and throughout the text, Hero identifies components individually using letter labels. He does not consistently provide a letter label for each vertex of the component, but he provides enough to identify the parts of each component unambiguously. For example, the stock includes two dovetailed boards, which are at first identified with two letters corresponding to diametrically opposed corners (ΑΒ and ΓΔ), and then later acquire an additional label on the side with the channel or pin (they become ΑΖΒ and ΓΕΔ).

He very frequently refers to accompanying diagrams. Immediately following his description of the stock, he identifies each of the several pieces of the trigger assembly with groups of letter labels: a handle ΑΒΓΔ, a tenon ΕΖΗΘ, a claw ΚΛΜ, and so on. A diagram (which he calls both *schēma* and *theōrēma*) follows. As he refers to this illustration several times in this passage as the "first diagram," this reference must be understood to include the diagram of the stock assembly, because he also refers to a diagram in that passage. In the surviving manuscripts, this diagram (which could alternatively be considered a set of diagrams) includes drawings of all the component pieces of the stock and trigger assemblies, as well as both completed subsystems (Figs. 5 and 6).[106]

[103] Marsden identifies the base, universal joint, ratchet gears and their corresponding pawls, and the winch as the components that are certainly missing from the part of the text that survives (Marsden 1971, 210).

[104] Marsden 1971, 208.

[105] On the debate over Hero's date, see Hero 2003, 8–30.

[106] The diagrams are after Paris suppl. gr. 607 (M), 56r–56v; Paris gr. 2442 (P), 68v–70r. The diagrams of the stock assembly are similar in content, but there are some differences: P is drawn in a blocky

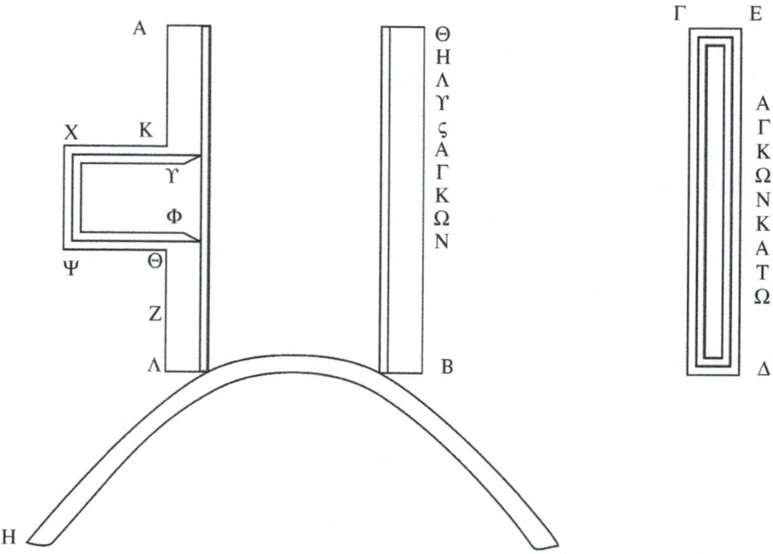

Figure 5. *Layout of the* cheiroballistra's *stock assembly.*

This technique, which clarifies both the appearance of individual components and how they are to be combined into larger systems, is still used today in mechanical diagrams. It emphasizes the construction process, forcing the reader to think not only about the appearance of the completed machine, but about its composition from interworking subassemblies. Diagrams organized in this way are relatively easy to comprehend, for several reasons. First, they allow the illustrator to represent components at different scales, so that a complex subassembly that is small compared to the size of the whole device can be given enough visual room to draw all its components in sufficient detail. Second, they tend to follow the text more closely. In many detailed descriptions of complex machinery, authors choose to walk through the machine descriptively, building it up piece by piece. This style of diagram provides a graphical analogue to the textual walkthrough; this makes it easy to proceed through the two

style suggesting three-dimensionality while M is an orthographic projection; M includes more letter labels, as well as word labels, which P does not have. The differences between the diagrams of the trigger assembly in the two manuscripts are minimal: the diagram in P includes word labels as well as the same letter labels as that in M, the form of the claw KTM is slightly different, and the trigger HΞ is more robust.

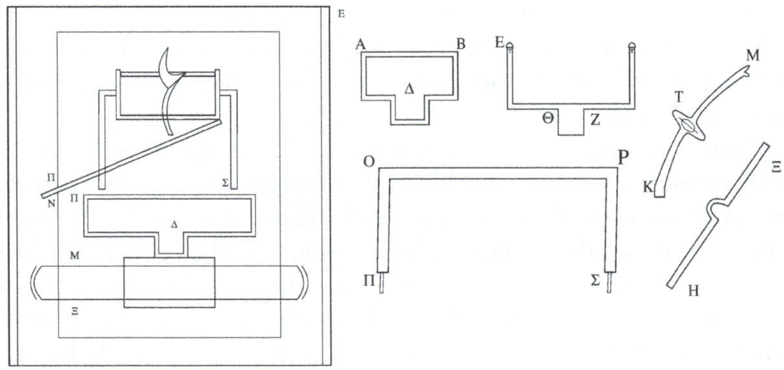

Figure 6. *Layout of the* cheiroballistra's *trigger assembly.*

representations together, gradually composing a complete mental picture of the machine.

The alternative approach is to provide a single diagram depicting the completed device. This may be achieved by showing the machine from a single perspective and allowing features on its far side to be obscured by elements closer to the viewer. Apollodorus strongly favors this approach; he is also the author who most often describes artifacts in the field, deployed against enemy towers, which are also sometimes featured in his illustrations.[107] A more comprehensive view can be provided in a single diagram by abandoning the notion of a fixed point of view and showing components from all planes of the machine as though projected onto a single plane.

This approach has the advantage of showing each component, but the viewer must learn to "read" these diagrams, to reconstruct imaginatively a three-dimensional form from the oddly angled components shown on the page. Biton's *Kataskeuai* and Hero's *Belopoeica* generally feature these complex single diagrams, together with letter labels and some word labels (ΔΙΩΣΤΡΑ, for example, or ΤΟΝΟΣ). The diagrams in the manuscripts of Athenaeus often also have word labels, but no letter labels are found either in the diagrams or in the text itself. The presence of word labels in these illustrations might be contrasted with the name labels given to animals in the late ancient mosaics analyzed by Leader-Newby. Rather than

[107] See for example Apollodorus, *Poliorcetica* 172.16, where Apollodorus explicitly calls for the diagram to show the ram "together with the upper floor of the tower described previously" (προγεγραμμένου). This tower was both verbally described and pictured earlier in the text, so Apollodorus might be referring to either one with the adjective προγεγραμμένου.

"[turning] the general into the particular" as Leader-Newby argues is part of the function of these labels, they might instead be seen to turn a particular representation of a mechanical component into a more general case of a *diōstra*, for example.[108]

A further feature that may be noticed in these diagrams is the presence of ornamental elements, such as decorative capitals on pillars and patterned surfaces. Sometimes, these patterns clearly represent structural elements so common that they do not need to be mentioned in the text, like the bricks or stones in a wall under siege. Sometimes there is a less clear-cut visual allusion to structural features of a machine, such as the patterns of orderly dotted circles that decorate the surfaces of several devices, including rams and siege towers, in a manuscript of Apollodorus.[109] From the text it seems that these represent pegs or nails, such as those attaching the four-dactyl wide boards he specifies as a covering for the exterior of a tortoise. A more elaborate version adorns an illustration of an offensive tower incorporating a ram, a ladder, and a great many protective coverings.[110] Apollodorus recommends that the tower be covered with hides to make it fireproof, and that these are to be attached with flat-headed nails. These nails are shown in the diagram as a pattern of diamonds running along the forward struts of the tower, in contrast with the more common pattern of circles, also present in the diagram, which appears to show a different set of fasteners.

In other cases, decorative elements in diagrams correspond to components of the device that are explicitly mentioned in the text as important to the appearance of the artifact. For example, Hero specifies that the stand of the *dioptra* is to be finished like a Doric column, "for the sake of good looks" (εὐπρεπείας ἕνεκα).[111] This decorative element is quite faithfully reproduced in the existing manuscripts, though the style of the column may vary.[112] Similarly, Philo specifies that the frame of the catapult is to have an exterior piece designed in a certain way, with a wave-shaped molding along it, for aesthetic reasons.[113] He specifically notes that he will provide a diagram of its appearance later on, but this has been lost. Ancient mechanical diagrams thus seem to have provided a visual transmission not only of the mechanical features of these artifacts, but also, on many occasions, of their decorative and aesthetic elements as well.

This is particularly true of the pneumatic wonders, many of which are crafted to look like figures such as birds or people, while concealing

[108] Leader-Newby 2007, 198. [109] Paris suppl. gr. 607, 34r.
[110] Apollodorus, *Poliorcetica* 173.1–174.7. [111] Hero, *Dioptra* 3.9–11.
[112] For example, Paris suppl. gr. 607, 64r. [113] Philo, *Belopoeica* 62.6–15.

their mechanical workings within. The interior and exterior are both potentially interesting, both are usually described in the text, and so in some ways it is up to the illustrator to determine which should be foregrounded. A self-trimming lamp described in Hero's *Pneumatica* provides a striking example of how a visual representation can differ between manuscripts. In one version of the diagram, the lamp is drawn in the shape of an elaborately decorated bird, upon which the mechanical components of the self-trimming mechanism have been superimposed, along with three letter labels (Β, Γ, Ζ).[114] By contrast, another manuscript shows only the mechanical components, along with some additional letter labels (Α, Β, Γ, Δ, Ε, Ζ, Θ).[115] The former version, in other words, foregrounds the decorative exterior of the lamp, while the latter focuses on the functional elements on the interior.

The text associated with this diagram reads as follows:

> Let the lamp be ΑΒΓ. Through its mouth (διὰ δὴ τοῦ στόματος) let an iron bar ΔΕ be pushed, moving freely around the point Ε. Let the wick be wrapped loosely around the bar. Let a toothed gear Ζ be placed next to it, moving freely around an axle whose teeth touch the bar, so that when it turns the wick is pushed through the teeth. Let the lamp have a mostly open valve [for oil] (ἐχέτω δὲ ὁ λύχνος ἀνεῳγότα τὸν ὀμφαλὸν ἐπὶ πλέον).[116]

At first glance, the animal figure of the lamp might seem to be referenced in Hero's text, which uses terms like *stoma* and *omphalos* to identify components of the device. However, these are in fact typical terms for parts of a lamp, and do not indicate that the artifact in question has an animal shape. There are, of course, many devices in the *Pneumatica* that Hero does identify in the text as incorporating animal or human figures. He describes devices that include figures of people playing instruments or pouring libations on an altar, as well as singing birds, hissing snakes, and even an (unspecified) animal that drinks after having its neck cut by a knife-wielding herdsman. The illustration in which the self-trimming lamp is given the shape of a bird therefore harmonizes with the overall subject matter of the *Pneumatica*, even though Hero does not specify an animal shape for this particular device in the text. Still, Hero's text does not prompt the reader to imagine the lamp as shaped like a bird, so the diagram from the Marcianus manuscript (and the later manuscripts that follow the same tradition) shows the autonomy of the illustrator,

[114] This version appears in Marcianus gr. 516, 181v.
[115] Paris gr. 2515, 15v. [116] Hero, *Pneumatica* 1.34.

destabilizing these diagrams as evidence of "what Hero would have done" as an illustrator.

At the same time, the illustrations' flexibility (as inconvenient as it is to the project of understanding precisely what they might have looked like in Hero's own time) is an essential part of the living history of these texts, of the way they actually propagated. It is likewise an important reminder that the relationship of text and diagram is not entirely harmonious. There can be a kind of struggle between them to dominate the understanding of the artifact that the reader walks away with, in this case depending on whether he privileges the mechanical information provided in the text, or the whimsical design suggested by the illustration. The way this struggle plays out in illustrated technical ekphraseis serves as a reminder that this is just one part of a broader set of problems having to do with the relationship of ekphrasis and illustration. Squire asserts that ekphrasis and image each influenced the other's reception; as an ekphrasis can be read in different ways, "responses to it were also mediated by images themselves; by the same logic, ecphrastic descriptions might actively alter the way in which viewers responded to images."[117] Squire draws on studies of medieval manuscript illustrations and the few illustrated papyri surviving from antiquity to show how ancient illustrations can "engage with a text by *not* corresponding with its literal details."[118] Though it comes from a very different context from the one Squire is concerned with, the autonomy and contradiction between these illustrations of the lamp raises the same point: the variation tells us more, not less, about how the ekphrastic text works.

Conclusion

Scarry contrasts visual to verbal representations by classifying the former as "immediate sensory content," confronting the reader with a textual element that resembles the object itself, rather than the latent resemblance encoded in text that requires the reader to construct his own picture imaginatively.[119] In fact, diagrams need not resemble what someone viewing the completed artifact would see; they can show the interior workings of a device in addition to or instead of the outer surface, or show components separately. They may instead show something more closely resembling what the artifact would look like in the world, by portraying it complete: perhaps in action, perhaps alongside its human users. Diagrams can emphasize the mechanical interworking of the various components

[117] Squire 2009, 145–6. [118] Squire 2009, 133–9. [119] Scarry 1999, 5.

of the device, the geometrical symmetries in its design, or the aesthetic appeal of its decorative elements.

The considerable room for variation in the diagrammatic representations of mechanical artifacts is a reminder of the multilayered complexity of the artifacts themselves. These objects are, at one and the same time, both crafted artifacts and embodiments of mechanical principles. For some types of artifact, the aesthetic component is of primary importance, resulting in a diagram that foregrounds the object's elaborately worked exterior rather than its mechanical elements (though these may still be included in the diagram). Other authors, like Apollodorus and Apollonius of Citium, highlight the use of artifacts more than their principles of design and construction, and the diagrams in the text may follow suit by showing the artifact in a context of use. The illustrations in Apollonius' text are particularly remarkable for their inclusion of human figures. Such diagrams, particularly when humans are depicted, are a reminder of the social context in which mechanical technology is inevitably embedded. Such illustrations stand in stark contrast to the geometrical style of Ptolemy's diagrams, which cohere with his usual textual assimilation of the artifact to a set of mathematical objects.[120]

Choices about how to represent an artifact visually are made by the author (and are often reflected in the verbal description), and then mediated by generations of illustrators. This aspect of the text reproduction process interferes with the goal of grasping exactly what an author had in mind to show visually, but it is nevertheless an important part of the life and use of these texts. At the same time, the verbal component of the text serves to rein in the whimsy of the illustrator, often guiding the production (and subsequent reading) of the diagram quite closely. The result is a complex cooperation between the verbal and visual. On the one hand, the relationship between the two is sometimes overtly tense, as when the Hellenistic authors hesitate to identify the diagram and the object it represents. On the other, these diagrams give authors an opportunity to provide the reader with an experience of greater conceptual closeness to the object described in the text than can be provided by verbal description alone.

[120] This variation in representational style may be usefully compared to the sculptural formations of walls on a few important buildings included on the Forma Urbis Romae. Although the reduced scale of the city plan of course makes it impossible to include all the features that an observer of the buildings would see, in a few cases the designer of the plan has prioritized the representation of sculptural elements over the more abstract representational style more typical of the plan. See Heisel 1993, 193.

CHAPTER 5

Description and Instruction

Introduction

Machines are material constructions, but they are also embodiments of knowledge – mechanical, physical, mathematical, material, cultural, historical, etc. – and ekphrastic texts selectively explicate the knowledge encoded in the machine. We might think of the encoding of knowledge by the machine as analogous to the semiotic aspects of an artwork. Ekphraseis on such objects may well describe the physical attributes of the object itself, but will often also elucidate the meaning and cultural significance of, for example, the mythological background of the scene it portrays.

The artificial restriction of ekphrasis to works of art, so common among modern critics, owes much to the semiotic richness of those objects. Heffernan's dictum that ekphrasis is a "verbal representation of visual representation"[1] was a particularly unrealistic limitation, which he later adjusted to the more useful distinction that ekphraseis on art objects are marked by the interpretability of the artistic object, and are hence akin to art criticism.[2] The criterion of interpretability highlights the effort of encoding human knowledge in the artifact and decoding it through the text, rather than the arbitrary criterion of whether a decodable artifact is made in a form that imitates something else.

[1] Heffernan 1993, 3.

[2] Heffernan 1999. In emphasizing the common ground between ekphrasis and art criticism, Heffernan is preceded by others, notably the late nineteenth-century work of Bertrand and Bougot, described at Webb 1999, 15–16. The interpretation of works of art, representative or otherwise, is of course fraught with its own enormous set of complexities and problems, which I do not mean to minimize but will not touch on here. Starting points particularly relevant to the questions here can be found at Mitchell 1994, 28–34, 88–107; Tanner 2005 is the most thorough guide to these problems in the context of ancient art, particularly 1–21, 45–55, 70–91, 158–70, 220–34; Squire 2009, 77–89, 120–46, 241–62, 322–33, 340–56, provides a much broader view, though still grounded in the ancient context.

Textual representations can draw on and reinforce the reader's understanding of the aesthetic, physical, literary, and historical principles that underlie the physical form of any kind of man-made object, whether that object is worth the making principally because of its aesthetic or its technological value. Jaeger observes, for example, the "discourses" associated with the two astronomical *sphaerae* that Cicero describes in the *De re publica*. The discourse associated with the solid *sphaera* is connected to a lineage of astronomical knowledge involving Thales, Eudoxus, and Aratus, and to this particular artifact's history as loot from Syracuse. In its identity as a beautiful, portable object whose transfer of ownership plays an important role in its story, Jaeger compares it to Homeric mixing bowls and the poetic discourse in which they participate. The mechanical *sphaera*, on the other hand, features in a story that begins with Archimedes, and which in Jaeger's view is not completed until Philus recounts Gallus' explanation, which provides it with "layers of significant associations."[3] The verbal disclosure of structural, functional, and semiotic information takes on a special immediacy in texts that set out explicitly to instruct the reader about the principles underlying the construction, use, or history of an object. The instructional mode is characterized by the creation of a particular kind of cooperative relationship between author and reader: instructions directly invoke a reader's capacity to imagine the results of following the instructions, or even to actually follow them.

The connection between descriptions and instructions is particularly clear when the instructions are explicit and extended, as they are in construction manuals (which I here term *kataskeuai*) framed as instructions to the reader to build an artifact for himself. Instructions are expressed in a wide range of voices: imperatives address the reader directly, creating the impression of a collaborative workspace, whereas impersonal forms of instruction present the process of construction as disengaged from human hands.[4] Even in a single text an author can traverse the entire range from pure description to explicit instruction. Philo's *Belopoeica*, for example, uses a verbal range from declarative (ἐστι, 63.11), to third-person imperative (ἔστω, 52.7 and νοείσθω, 63.32), to second-person imperative (νόει, 62.15 and λάβε, 57.24). The middle ground here, between declaring that something *is* and commanding the reader to take hold of it physically, is in the territory of imagination, where the reader is asked to create a

[3] Jaeger 2008, 56.

[4] These questions have of course been explored for other types of ancient texts both "technical" and otherwise; particularly useful studies include Risselada 1993; von Staden 1994; Gibson 1998; Hine 2009; Nutton 2009.

mental image of some part of a machine. An author often strengthens this impression of instruction with explicit commands to the reader to "picture" this or "imagine" that.

Instructions for seeing

Epigrams are a natural place to begin looking at the relationships between texts and objects, authors and readers, since they are focused on an object (real or imagined) and often incorporate an address to the viewer or reader, sometimes even featuring imperative instructions. Tueller analyzes the speakers and addressees of a diverse catalogue of epigrams dated from the seventh to the fourth centuries BCE.[5] He finds that during the Hellenistic period, authors of epigrams began to experiment playfully with the identity of the speaker and the addressee of the poem. The speaker of a sepulchral epigram, for example, had already changed in the classical period from defaulting to the object to defaulting to the deceased.[6] Hellenistic authors like Callimachus experimented with new possibilities for the speaker's identity, such as a three-way conversation between a passerby, the deceased, and the grave-marker.[7] This kind of play between object, addressee, and speaker directly challenges the reader to contemplate subtle variations in the relationships between them. The reader is consequently implicated in creating a shared imaginative space via what Goldhill calls "directive reading," a phenomenon he associates particularly with Hellenistic literature.[8]

Texts that guide the reader to visualize the construction of mechanical artifacts create the same sense of shared space: they share with epigram the central goal of vividly representing an object, and concomitantly the different ways in which the reader can be guided to "see" it correctly. Such "virtually inhabited" spaces are a place for the reader to engage actively with questions presented by the textual representations of real and imagined objects: what exactly am I picturing? What is my relationship to the object, and to its creator? What is the voice of the author prompting me to imagine or to do?

Texts that create a space in which the reader is guided in an imaginative encounter with an object are, of course, not unique to Hellenistic Greece. The relationship between description and instruction is a central concern,

[5] Tueller 2008, 16–35. [6] Tueller 2008, 20.
[7] Callimachus, *Epigrammata* 13 Pfeiffer.
[8] Goldhill 1994, 212–14.

for example, in the *Imagines* of Philostratus. The *Imagines* obviously does not speak for ekphrasis in general, particularly as it owes so much to its own cultural environment.[9] However, this text blends theoretical explication and descriptive practice so as to illuminate with particular clarity the intersection between description and instruction.

The *Imagines* comprises a series of ekphraseis on paintings depicting mythological and literary subjects, framed by Philostratus' story about a visit to Naples (a city singled out for its abundance of Greek learning), where he happened upon a portico decorated with a tasteful variety of paintings. While Philostratus contemplates giving a lecture praising the paintings, he is approached by the son of his host, who asks him to explain all the paintings. Philostratus of course obliges, first waiting for a crowd of listeners to assemble, and the resulting ekphrastic explanations furnish the bulk of the text. The boy, who is given the task of active questioning as well as passive absorption of information, is the focal figure for these explanations: "I said, 'Since the boy has come, let him be placed in front, and the effort of my speech be directed to him, but all of you follow along, not just agreeing but asking questions in case I say something unclear.' "[10] The proposed interactive lecture recalls the play of relationships between author, fictional spectator, and reader characteristic of Hellenistic epigram. The goal of the discourse is, similarly, not merely to describe the paintings (which, of course, are visible to the assembled listeners) but to teach the listeners how to be good spectators, how to see and interpret the paintings.[11]

Philostratus first addresses his young protégé before the painting of the Scamander:

> Boy, do you know that this comes from Homer, or have you not realized this at all because you are contemplating the marvel of how fire could live on the water? Let us figure out what it means: look away to see what it is based on (συμβάλωμεν οὖν ὅ τι νοεῖ, σὺ δὲ ἀπόβλεψον αὐτῶν, ὅσον ἐκεῖνα ἰδεῖν, ἀφ' ὧν ἡ γραφή).[12]

In order to really *see* an information-bearing artifact like a painting, the viewer must look past its physical features to its underlying principles. The viewer must decode the artifact, and a good ekphrasis will guide him

[9] On the *Imagines* and the "Second Sophistic," see for example Goldhill 2001, 157–9; Webb 2006, 116–17; Elsner 2009, 10–11; for a broader contextualization see Bowersock 1989. On *enargeia* in the *Imagines*, see Squire 2009, 243–4, 419–27; Baumann 2011, 18–23, 168–71.

[10] Philostratus, *Imagines* 1.pr.3. On the dramatic shift of perspective that occurs when the address to the boy begins, see Webb 2006, 118–19.

[11] Philostratus, *Imagines* 1.pr.5.

[12] Philostratus, *Imagines* 1.1.

through this process. There are a variety of ways to do this: a painting may, for example, be explicated as a reflection of a literary text (as with Philostratus' "Scamander"), or a moral or philosophical lesson (as in his "Komos"), or through the principles of the painting's spatial arrangement (for example, the deceptive effect of perspective that the artist achieves in his "Menoeceus"), or specialized knowledge about the subject matter (such as the miraculous sounds made by the statue of Memnon in Ethiopia, or the types of soil preferred by pine, fir, and cypress trees in the "Marsh"), or the circumstances of its creation by an artist. Philostratus' ἐκεῖνα ... ἀφ' ὧν ἡ γραφή is potentially a very broad designation.

Throughout the text, Philostratus uses imperatives to frame these descriptions as instructions for seeing (e.g. ὅρα, 1.4, σκόπει, 1.7) or imagining its other sensory effects (e.g. ἄκουε, 1.6).[13] Philostratus blurs the sensory encounter between viewer and painting into an encounter between the viewer and the objects portrayed in the painting. Often the reader is encouraged to imagine perceiving these objects in ways that would not literally be possible for a viewer, as when he is told to hear the castanets of the "Komos," or to fully experience the roses crowning Komos, which Philostratus praises as being "painted along with their scent."[14] This educational tour is not aimed directly at the reader by a disembodied instructor; Philostratus himself, already identified in the proem as the leader of the dialogue, injects himself into the task of interpretation (e.g. οἶμαι, 1.2, ἐμοὶ δοκεῖν, 1.7) and joins his protégé in the imaginative effort (e.g. δεξόμεθα, 1.4, ἴδωμεν, 1.6).

Meanwhile, the enthusiastic young boy and the crowd of other listeners stand in for the reader as target of those instructions.[15] Webb observes of this double-layered perspective, "as we receive information about the larger pictorial content of the painting, we also simultaneously share in the new clarity of the Boy's perception."[16] This approach is hardly limited to Philostratus; authors in a wide range of genres, particularly those whose purpose is somehow didactic, use such intratextual surrogates for author

[13] On *paideia* and the *Imagines*, see Elsner 1995, 28–37; Newby 2009; Squire 2009, 243–4; Baumann 2011, 23–30, 150–5.

[14] ἐπαινῶ καὶ τὸ ἔνδροσον τῶν ῥόδων καὶ φημὶ γεγράφθαι αὐτὰ μετὰ τῆς ὀσμῆς, 1.2. On this kind of imaginative supplementation as a feature of ancient ekphrasis, see Webb 2009, 21. On synaesthesia in Philostratus, see Manieri 1999; Webb 2006, 120–1; Platt 2011, 3. Squire focuses on descriptive clarity as a way of compensating for the fact that "the actual physical image of what was described is unavoidably written out of the analysis" (Squire 2009, 244); Webb makes of this a positive feature, the convincing fictionality of the work that stems from the power of ekphrasis (Webb 2006, 119).

[15] On the didactic value of this embodiment of teacher and student in the text, see Webb 2006, 123–5.

[16] Webb 2006, 125.

and audience to lead the reader to think about the subject matter in a certain way.[17]

It should, finally, be emphasized that the goal of using ekphrasis to decode an object down to its principles is explored by many authors besides Philostratus, in many cultures besides that of Second Sophistic Greece. For example, Faber singles out Hellenistic ekphraseis as challenging the reader's understanding both of technical details like the materials an object is made from and their cultural associations (such as the differing status of wood and metal vessels), and the various literary predecessors their description is meant to summon up.[18] Männlein-Robert likewise argues that Hellenistic ekphrastic epigrams are oriented more toward spurring interpretation than providing exhaustive description. She draws on Goldhill's characterization of Hellenistic culture as a "culture of viewing"[19] to argue that the primary work done by these poems is not the imitation of an artwork but "the poetic *mise en scène* of an important interpretative pronouncement on it."[20] Indeed, the connection between representation and the dissemination of knowledge arguably goes back at least as far as Aristotle's comparison of *mimēsis* and *mathēsis*.[21]

Descriptive text thus often serves a didactic function; the reader learns to see the object through its underlying principles, and so to understand it more deeply. Graf argues that works of art, though not specifically identified in the rhetorical handbooks as a category of subjects for ekphrasis, must nevertheless have been fair game for ekphrastic description because they can be "read" and interpreted by a viewer who has learned how to see them correctly. Graf bases his analysis in part on the late ancient rhetorician Nicolaus of Myra, whose method of viewing advocates beginning at the beginning and proceeding methodically from one part to another. This advice may seem trivial, but in fact it is vital to the central didactic

[17] On this see, for example, Gale 2004. [18] Faber 1995.

[19] Goldhill 1994. On the juxtaposition of the "connoisseurial eye" and "imaginative eye" see Platt 2011, 174. Gow offers a more negative reception of the intellectual challenge these epigrams present, describing them as "a field for the exercise of the intellect rather than display of the emotions" (Gow and Page 1965, 592). Tanner, discussing this reaction, characterizes the epigram as "a very suitable vehicle for aesthetic response within a rationalist culture of viewing," which may nonetheless strike the modern reader as an "inadequate [response] to art" (Tanner 2005, 262).

[20] Männlein-Robert 2007, 252.

[21] Aristotle, *Poetics* 1448b. For discussion of how this connection, and Aristotle's view of the "cognitive pleasures of viewing," should affect our understanding of *enargeia*, see Walker 1993, 365 n. 24. Tanner, by contrast, focuses on the problems presented by the passage in the *Nicomachean Ethics* (1117b28–1118a25), arguing that the pleasures of vision are intermediate between purely sensuous bodily pleasures and purely rational intellectual pleasures in which the body plays no part, so that the pleasure of mimesis is at best rather problematically linked to intellectual activity (Tanner 2005, 257 n. 192).

task: rather than gravitating immediately to whatever appears strange, the student is supposed to learn to see systematically.[22] "Systematic" seeing is, for Nicolaus, a matter of seeing the described object piece by piece, and so coming to understand it as a greater whole; he cites the example of a bronze statue of a man, which is described limb by limb until the object "obtains life from all sides."[23] Properly disciplined description provides the reader access not only to the visible qualities of a statue, but (by approaching them in the correct order) to the sensible arrangement that makes the object a statue of a living man.

Scarry argues more radically that the relationship between description and instruction in fact has a practically unlimited breadth, that a written description of anything is really a kind of instruction set for recreating that object in the reader's imagination. She distinguishes three ways in which an object can communicate to an audience: the first of these she calls "immediate sensory content," such as we obtain from a painting; the second is "delayed sensory content," a set of instructions (perhaps encoded in a nonverbal form, like a musical score) which lead to the production of sensible content in a fully determined way. The third is "mimetic content," which is never directly available to the senses, and comes instead from underdetermined stimuli such as the words in a verbal description.

In order to find his way through the mimetic content to a mental image of the object being described, the reader must interpret the components of the description as instructions for seeing the object.[24] Because verbal descriptions usually carry only "mimetic content," cognitive work is required for the reader to get from there to the mental image that "immediate sensory content" provides directly. Scarry cites as an example a passage from *Tess of the d'Urbervilles* that guides the reader to think in a particular way about a man walking, but of course every verbal image requires its own kind of work.[25] Each description guides the reader to assemble imaginary components into a working complex of form and principle.

[22] Graf 1995, 148.

[23] Nicolaus, *Progymnasmata* 69.16: οὕτω γὰρ πανταχόθεν ἔμψυχος ὁ λόγος γίνεται.

[24] Scarry 1999, 5–6.

[25] "[*Look closely at the walker's legs.*] The pair of legs [*now picture their work of weight bearing*] that carried him [*assess how well they hold that weight*] were rickety, [*and how that affects his motion*] and there was a bias in his gait [*watch which way the load leans*] which inclined him [*superimpose a geometric figure into the midst of this representational picture*] somewhat to the left of a straight line" (Scarry 1999, 37).

Instructions for building

Texts that include instructions for constructing an artifact (*kataskeuai*) make the connection between description and instruction much more explicit even than texts that give instructions for seeing. *Kataskeuai* illuminate the reciprocity of description and instruction: passages of text describing artifacts can help explain their structural and functional principles to the reader, while passages of explanatory text can teach the reader how to see the artifact in a disciplined, well-ordered way. These texts often strongly emphasize knowledge embodied by the artifact above and beyond its structural details, sometimes explicitly providing the required background knowledge to the reader before proceeding to the description of the artifact itself.

When a text supplies instructions for building an artifact, the relationship between text and object undergoes a radical change from the description of a specific machine already embodied in a particular historical context. There, the reader may venture into imagined contact with the machine, its structure, and activities, through the author's detailed account; he may picture it quite fully in his mind, but once that part of the text is done, so is the machine. Instructional texts, on the other hand, can provide the means for the machine to come to life not just in the reader's mind, but in the world. This kind of text may demand different kinds of knowledge from the reader; beyond describing the outward appearance of a machine, instructional texts draw upon and add to a complex body of mechanical knowledge and craft skills.

These differences drive a variable quality of descriptive texts that might be labeled "generativity." Highly generative texts come in different forms, such as a narrator's personal account of building a device, an impersonal passive description of an artifact's construction, or building instructions addressed to the reader. They are distinguished from other texts neither by reliance on any one particular form of authorial voice (though a particular set of voices are most often used in these texts), nor by the assumption that the reader is necessarily intended actually to go forth and follow the instructions. The crucial difference is rather that the artifact is described through instructions that show it coming together before the reader's eyes. Such texts enhance the *enargeia* of the descriptions they offer by encouraging the reader to imagine performing the manual processes of manipulating and assembling components of the device, not just looking at it. Verbal descriptions of this kind of manipulative activity have an extraordinary cognitive effect on their audience, appealing to the same areas of the brain that are engaged during actual manual activities.[26]

Imagined manual engagement with an artifact in the making recalls what Kittay calls a "parcours," a trip made by the eyes over a visible surface (or its analogue in writing).[27] The integration of the reader's point of view into the "parcours" itself reflects the shift from description to narrative, "an interplay, represented or understood, between the limits of perception and the limits of body movement."[28] The continuity between highly "generative" texts and ekphrastic texts that encourage the reader to imagine looking, but not touching, is illuminated by the "ekphrasis of the *tropos*"[29] – that is, ekphraseis on actions, including construction. Aelius Theon defines the term by example:

> There are also ekphraseis of the *tropos*, such as those of equipment, arms, and *mēchanēmata*: in what way they were prepared, as the armor-making in Homer, the circumvallation of the Plataeans in Thucydides, and the preparation of the *mēchanēma*, the great beam which they halved and hollowed out.[30]

There has been some debate over which passage of Thucydides is referred to here,[31] but Webb argues that Theon's use of the term *periteichismos* suggests construction in progress more than a completed artifact, and so 2.75–8 fits the bill better.[32] Theon is not the only author to describe this phenomenon; Doxapatres' commentary on Aphthonius draws a distinction between "ekphraseis of the finished appearance" (*schēma*) and ekphrasis of the *tropos*.[33] The distinction between the artifact the reader encounters in finished form and the artifact he encounters through a narrative of its creation is precisely the quality of "generativity."

The forms of address and other verb forms used in a *kataskeuē* play a crucial role in determining whether the artifacts are presented as the immediate results of actions on the reader's part or as objects detached from human interference, as future possibilities or past achievements. Authorial voice is of course an important consideration for other types of didactic and technical text as well. The use of the first person has

[26] Martin and Wiggs 1996. This experiment is referenced briefly at Scarry 1999, 146.

[27] Kittay 1981, 228. [28] Kittay 1981, 235. [29] Webb 2009, 69–70.

[30] Aelius Theon, *Progymnasmata* 118.21–5; see in general 68.6–10, 118.21–32.

[31] The passages in question are 2.75–8 and 3.21; respectively a longer account of the process of constructing the fortifications at Plataea and a brief account of the finished fortifications. As Webb notes, the translations of Hermogenes by Patillon and Kennedy indicate that they believe the latter passage is referred to, but Webb argues otherwise (Hermogenes 1997; Hermogenes 2005; Webb 2009, 69).

[32] Webb 2009, 69. [33] Webb 2009, 69–70.

attracted particular attention. Lloyd discusses authorial "egotism" and self-presentation in Greek literature as early as Hesiod; these continue in the lyric poets and the natural philosophers, and achieve a spectacular degree of intensity and variation in the texts of the Hippocratic corpus.[34] Von Staden describes the "scientific self" that Celsus projects in his medical work, differentiating several types of *ego* that appear over the course of an impressive 240 occurrences of the first-person singular.[35] These include the "nomenclative" *ego*, used where Celsus wants to emphasize the precision of his quantitative terminology; the "*ego* of *dispositio*," which adumbrates the text's structure and weaves together cross-references; the "autoptic-empirical" *ego*, which makes claims to personally witnessing clinical practices; the "author as reader" who engages with the writings of his predecessors; and the "*ego* of scientific independence," which settles conflicts between scientific views. To this varied list Hine adds a variation of the "autoptic-empirical" *ego*, which he calls the "autobiographical" *ego*, in which authors (he cites Seneca and Pliny) refer to personal experiences that do not fall into the purview of the work at hand, as well as the "*ego* of the self-conscious stylist," as when Seneca draws attention to his use of a legal term.[36]

The multiplicity of types of *ego* bespeaks a complicated set of authorial priorities, which, as von Staden notes, do not always correspond to "the typical explanatory mechanisms of claiming originality or engaging in polemics."[37] In fact, von Staden observes that Celsus often uses impersonal constructions to explain his own theories and clinical experiences. Instead, Celsus' "I" often indicates an author taking charge of rhetorical rather than medical concerns, as he "performs moderate modulations" of prior rhetorical formulations, reworking them for his own authorial ends. Von Staden is careful to warn against oversimplifying Celsus' "divergent authorial intrusions."[38] Celsus, an author writing in a discipline we would call "scientific," thus projects himself as engaging personally with rhetorical concerns in a sophisticated and multifaceted way.

Galen is positively notorious for the intensity of the first-person presence in his work. This presence manifests both as explicit claims to personal expertise and as personal appearance of the author in his text through first-person verbs and personal pronouns. Nutton summarizes the persuasive intent of these appearances by saying that "what Galen says must be right, in part because he is Galen, the 'I.'"[39] Such a

[34] Lloyd 1987, 56–70. [35] von Staden 1994, 106. [36] Hine 2009, 18.
[37] von Staden 1994, 106. [38] von Staden 1994, 117. [39] Nutton 2009, 61.

characterization will not surprise the reader who has taken even the most casual of glances at Galen's prose. However, to make a reliable test of the comparison against Celsus, avoiding the possible problems of changing languages, Nutton uses Niccolò da Reggio's fourteenth-century Latin translation of Galen's *De motibus dubiis* to test his assertions about Galen's egocentricity. He confirms his assertions about Galen's unusual reliance on first-person forms by comparing the density of first-person pronouns and verbs to Celsus and to an anonymous pharmacological text written in Greek by a Roman contemporary of Galen's, also translated by Niccolò da Reggio.[40] Galen also uses first-person forms to define a term, to testify to something he himself has witnessed (Asper labels this the "Ich empirischer Beglaubigung"), or to engage in polemical debate.[41] Asper distinguishes as well a "disponierendes Ich" and an "Ich des Wissenschaftlers," which respectively handle the tasks of discussing the arrangement of the work itself (similar to von Staden's "*ego* of *dispositio*") and building the author's high scientific profile.

The first-person plural has its own set of uses, though it is often interspersed with uses of the first-person singular. Asper identifies the "integrative we" as having a particular power to elide the distance between author and reader and create a sense of collaboration.[42] In referring to it as a "Konsensappell,"[43] Asper draws on Usener, who describes the distinction between first-person singular and first-person plural in the context of Athenian oratory.[44] The choice of authorial voice may thus have as much to do with the circumstances surrounding the text – in this case, principally Galen's work to establish himself against the background of his textual and practical competitors – as it does with the task of finding a voice appropriate to the level of the material in the text.

Of course, for most ancient texts it is not possible to say much for certain about those circumstances, and one runs the risk of engaging in circular argumentation or unjustified "biographical" readings of the text. We know a great deal more about Galen's life, about the context of competition and knowledge transmission in which he carried out his medical and textual work, than we do about any mechanical author.[45] Where it appears possible to draw connections between a particular instructive voice and its historical context, I shall try to do so. For the most part, however, the

40 Nutton 2009, 58–9. 41 Asper 2007, 333. 42 Asper 2007, 333.

43 Asper 2007, 128. 44 Usener 1994, 32, 135–6.

45 For a richly detailed reconstruction of the details of Galen's life and working environment, see Boudon-Millot 2012; for a lively and dramatic portrayal of his professional life at Rome, see Mattern 2013.

focus here will mostly remain on the intratextual environment of instructive address.

This environment is populated by something like what Volk calls, in the context of didactic poetry, the "constellation" of teacher and student.[46] The reader can be invoked in the text as a listener, a viewer, or an active corporeal entity. He may or may not be asked directly to perform visual, imaginative, or manual tasks (whether a particular reader actually carries those tasks out is immaterial to how the text is presented). The many forms this instruction can take produce a diverse set of "constellations," each inducing the ekphrasis to operate on the reader's imagination in its own way, each with its own way of positioning the artifact vis-à-vis the reader.

The *addressee* of the text, named or not, is often unknown, and may not feature in the text beyond the preface. Asper, for example, suggests that Teuthras, the addressee who appears at the very beginning of *De pulsibus libellus ad tirones*, was probably not known to the actual reader.[47] Meißner argues that Philo of Byzantium and Aeneas Tacticus both address their instructions on siege warfare to any potential besieger or defender, whereas Onasander seems to assume a reader with some technical knowledge and sufficient political power to take on the tasks of military leadership, even though his preface appeals to a less expert readership.[48] That is to say, the level of political power and technical expertise possessed by the nominal addressee of a text need not reflect the qualities the text actually requires of its reader.

Von Staden's sobering reminder about the dangers of "reductive generalizations"[49] in assessing authorial voice applies just as well to mechanical artifact texts. Even a single author may deploy the "authorial self" in many different ways, and the possible variations are hopelessly multiplied for different authors working on different types of project at different times. Some types of authorial voice which have previously been identified in other technical texts are also present in texts on mechanical artifacts. Others change profoundly in a context where material objects are brought into being (in imagination or reality) through the actions indicated in the text – as it were, before the reader's eyes.

The verb forms chosen by the author of a *kataskeuē* shape the encounter between author, reader, and object. One approach is of course to address the reader directly, framing the instructions as second-person imperatives

[46] Volk 2002, 37–9. [47] Asper 2007, 334–5. [48] Meißner 1999, 164, 189.
[49] von Staden 1994, 105.

and perhaps also using personal pronouns to situate the figure of the reader in the action of building. This approach encourages the reader to engage imaginatively with the manual tasks of building, to picture himself as involved with that work. Nutton observes that Galen most often addresses the reader directly using the second person when he describes the anatomical experiments of the *De anatomicis administrationibus*, or for similar passages in other works like the *De motibus dubiis*.[50]

First-person plural verbs may suggest collaboration between author and reader, an impression Lehoux describes as the author's putting "his arm around the reader's shoulder."[51] The sense of active engagement of author and reader may be strongest in the case of the first-person plural subjunctive, which Risselada argues "is only used when the speaker somehow joins in the action involved."[52] However, a first-person declarative verb can also function implicitly as a directive form, suggesting that the reader should do as "I" do.[53]

Alternatively, an author may include statements in the first-person singular to highlight his particular achievements, innovations, or past constructions. These first-person singular formulations do not so strongly induce the reader to picture himself involved in these activities as the second-person or first-person plural directives that appeal to him directly, but they do create a "lived-in" space of construction, where the artifact's construction is depicted as the result of a particular person's building activity. I take the concept of a "lived-in space" of technical discourse from Clarke, who develops it in the context of geographical periplus texts.[54] This lived-in space is not only place (that is, not only the stops on the periplus that are described and linked by narrated travel), but also inhabited narrative: "the 'lived-in' nature of both space and time *is* brought out by the use of the first person to refer to those participating in the voyage."[55] Hine makes a similar observation of Mela's geography: "Formally these are indefinite second persons, but the first, *credas*, the main verb, directly engages the reader, and the second and third vividly treat the reader as a traveler who is on the spot, and is invited to imagine sailing along the coast of Spain, or looking at the river Araxes in Armenia."[56] First- and second-person grammatical constructions in texts that give the

[50] Nutton 2009, 58. [51] Lehoux 2013, 137. [52] Risselada 1993, 163.

[53] Gibson notes this approach in the case of Latin didactic literature, describing the effect as "I do this (and you can/should too)" (Gibson 1998, 71).

[54] Clarke 1999, 37–42, 201–7. Compare Platt on Pausanias' *Periēgēsis* as a text "requiring readers to conduct their own imaginative journey," constructed in a way that encourages "a 'phantastic' mode of reading" (Platt 2011, 251).

[55] Clarke 1999, 201. [56] Hine 2009, 27–8.

reader instructions for building an artifact create a similar kind of inhabited workspace.

The alternative to this "lived-in" space is a depopulated or impersonal construction space, in which the artifact seems to come together without human intervention. Passive verbs and *dei* + infinitive constructions are commonly used Greek verb forms that do not require the reader to imagine himself (or anyone else) participating in the work of construction. Each of these forms allows the reader to picture the artifact coming together, or the way it ought to come together, without imagining any personnel directly involved with that process.

Another such form is the third-person imperative, conventionally translated in English as "let there be ..." and so forth. This form is especially common in mathematical prose, where it seems to separate the objects upon which they operate (mathematical objects in this case) from author and reader alike; they are pre-existing, independent of the act of reading. Fowler and Taisbak describe the unseen agent of these activities as the "Helping Hand"; as they put it, "Greek geometry is not about what we can do, but about what can be done."[57] Creese observes further that "while there is indeed a Helping Hand in Greek mathematics, there seems not to be a Helping Imagination."[58] This latter distinction is to say that present passive imperative forms of "draw" (γράφειν) are rare, while the opposite is true for "imagine" (νοεῖν). Creese takes this to indicate that two-dimensional objects can be "drawn for us," while three-dimensional objects for the most part require some purely imaginative activity.

Asper suggests that the detached "Helping Hand" might serve to augment the author's perceived objectivity and hence his persuasiveness, as well as to reject the influence of individual perceptual experience in mathematical claims.[59] However, this form of directive is also commonly used, sometimes coupled with other markers of "mathematical" prose, in instructions for building mechanical artifacts, for whose separate pre-existence it is obviously not possible to make such a case. Asper contrasts the use of these "impersonal" imperatives (most notably in Euclid) with the "personal" imperatives of the "Rezeptstil," a set of textual markers he associates more closely with instruction in practical mathematics than geometry.[60]

57 Fowler and Taisbak 1999, 362. 58 Creese 2010, 68–9.

59 Asper 2007, 126–8. For a cautionary note on the tendency to homogenize mathematical prose see Netz 2009, 102–3.

60 Asper 2007, 101–2, 125, 197–202.

Authors of directive texts in Latin of course have access to a different range of possible structures.[61] Risselada's study of directive expressions in Latin describes a wide variety of grammatical structures that communicate instructions to the reader, including genuine imperatives, subjunctives, and various periphrastic expressions.[62] The hortatory first-person plural subjunctive is quite commonly used in these texts, and functions like the "integrative we" seen earlier. Risselada argues that this form has "commissive" aspects (meaning that the speaker's actions are involved) as well as directive, so that the first-person plural subjunctive implicates both author and reader in the activity described.[63]

Besides the familiar simple imperative, Latin also has the imperative in *-to*, which is sometimes referred to as the "future imperative," though Risselada prefers the term "imperative II." The "imperative II" differs notably from the simple imperative in that the realization of the action specified in the content of an "imperative II" directive is envisaged as non-immediate, either in a temporal or in a conditional sense.[64] Cato makes unusually liberal use of this form, for example, in giving instructions for building the components of a pressing room:

> Have the wheel made (*facito*) with Punic-style joints, 4 feet across and 6 fingers thick, and oak dovetails inserted (*subscudes iligneas adindito*). When you have fastened these together, have them closed with cornel wood pins (*eas ubi confixeris, clavis corneis occludito*).[65]

This pattern gives Cato's work the "non-immediate" effect that Risselada describes.[66] Hine observes as well Cato's unusually liberal use of third-person jussives; these might similarly suggest an impersonal directive mode.[67] This is not to say that the instructions presented in the book were not intended to be carried out by anyone, but the expectation might have been something along the lines of the model suggested by Reay, in which slaves are regarded as "prosthetic tools."[68] However, we do not have to know how much time Cato might himself have spent engaging in these activities to appreciate the distance built into his instructions.

Varro's *De re rustica* is, of course, a very different sort of text from Cato's; it is elaborately framed as a dialogue on farming, perhaps even written (as Kronenberg suggests) with satirical intent.[69] Whatever the author's intentions, the text is certainly structured very differently from Cato's relentless

[61] Nutton notes some of the challenges and pitfalls associated with this shift at Nutton 2009, 55–6.
[62] Risselada 1993. [63] Risselada 1993, 158. [64] Risselada 1993, 122–9.
[65] Cato, *De agricultura* 18.9.1–4. [66] Risselada 1993, 128. [67] Hine 2011, 629.
[68] Reay 2005, 333. [69] Kronenberg 2009, 7–20, 73–129.

instruction. However, certain passages do include recommendations for how certain structures are to be put together, often providing guidance using impersonal directive expressions like *oportet*. A good place to see these structures in action is the description of the typical farm's aviary. These practical structures, built *fructus causa*, are sharply distinguished from the peculiar aviary built *animi causa*: that one is described as a finished artifact, so we do not see it come together.[70] The first step in making the more utilitarian aviary is the construction of a building (described as a *testudo* or *peristylum*) capable of holding several thousand birds; this construction is treated in the text by a simple declarative *fit*. Then follows a series of directives using *oportet*, each instance of which may govern several different infinitives:

> It should have (*habere oportet*) just enough space for light so that the birds can see where to settle, where their food is, and where their water is. It should be bordered (*tacta esse*) with light stucco around the doors and windows, so that no mouse or other creature can enter. There should be (*esse*) many poles around the inside of the structure's walls where the birds can settle, as well as inclined rods going from the ground to the wall, with transverse perches fixed on them at measured intervals, looking like the enclosures at the theater.[71]

Varro also makes use of the occasional second-person subjunctive, as when he warns that you should take care (*caveas*, III.V.4–5) to provide the aviary with recessed shelves where birds that die in the aviary can be stored for later accounting. These constructions are quite rare, however; it is the impersonal *oportet* that is most common in these scenes of construction. Gibson observes that *oportet*, like the gerundive, appears to be used in a more "binding" sense than "oblique" forms of directive like the third-person indicative (which can be construed as a directive in the comparatively weak sense that the reader ought to consider imitating what these other people are doing).[72] However, Gibson notes that at the same time the lack of a pronoun object for this impersonal form gives the reader license to choose whether to identify himself with the addressee of the directive form. Hine adds that in the context of Varro's lively dialogue, "the generalizing nature of the second-person singulars is clear."[73] As I read Varro, I acquire the impression that *someone* ought to build an aviary like this, but I am not compelled to imagine myself doing it.

[70] Varro, *Rust.* III.V.1–5. [71] Varro, *Rust.* III.V.3–4. [72] Gibson 1998, 93–6.
[73] Hine 2011, 634.

Vitruvius imposes a similar grammatical disconnect between the reader and the mechanical artifacts he describes in the *De architectura*. He does sometimes allude to the workmen who operate a machine, or describe how buildings both public and private serve the needs of visitors and inhabitants; accounting for these needs is part of the knowledge the architect must master. However, his apparent concern for the needs of users contrasts with an almost complete absence of the figure of the user or builder when he describes machines in action or in the process of construction. In these passages, Vitruvius tends to elide the human figures that are so important elsewhere in the work. This helps to simplify the description, but perhaps more importantly establishes a separation of the reader from the notional figure of the artifact's builder or user.

For example, the user is absent from Vitruvius' detailed description of the water organ. Instead, the external applied forces are described using passive verbs, while the consequent internal activities of the machine are active:

> So, when the levers are pushed up, piston rods draw down the bases of the cylinders to the bottom; the "dolphins," which are mounted on hinges, driving the disk valves into them, fill the space of the cylinders with air, and the piston rods, pushing up the bases in the cylinders with a forceful frequency of pulses and closing off the upper openings with valves, force the air enclosed there into pipes, squeezed by the pressure.[74]

The water organ is a very complex device consisting of several sections. A description of a person playing it would require the reader to visualize the player and the air-compression mechanism as well as the sounding part of the organ described here. Vitruvius all but removes the figure of the user from the description, reducing him to his contact with the machine (*tactus*) and replacing his actions with their passive effects on the stops of the organ.[75] He thus confines the reader's view to the interior of this section of the organ.

Vitruvius normally provides guidance about building artifacts using passive verbs, jussive subjunctives, and gerundives, never referring to the figure of the builder or identifying the reader as a builder.[76] This choice may reflect the distance between his readers and the craftsmen who might actually build these artifacts, the "deferential didacticism" that Whitehead

[74] Vitr. *De arch.* 10.8.5.

[75] *quarum pinnarum tactus motiones efficit regularum continenter*; Vitr. *De arch.* 10.8.4.

[76] He does use the first-person singular frequently, and the first-person plural more rarely, as tracked in Hine 2009, 24. However, these refer to his authorial work, not to accounts of building, except for the case of the basilica he designed at Fano (5.1.6–10), which he prefaces with *collocavi curavique* before returning to the usual passive and impersonal forms.

attributes to Apollodorus.[77] This explanation, though often applied to figures like Cato and Vitruvius, is certainly too reductive, however, and does not explain the fluctuations over time that Hine observes in the use of the second person in Latin technical texts even within the first century CE.[78] We can only extract from the surviving evidence the field of possible forms of address, and position authors within that field of possibilities. Vitruvius, though a particular (imperial) addressee for the work has already been established, nevertheless communicates information without addressing the reader personally. It is therefore possible to speak, as Meißner does, of humanity as a collective, generic *indirect* addressee.[79] Addressing a work to a generalized, indirect audience results not in the elimination of technical detail from the work, but in the imposition of a certain conceptual distance between the reader and the construction activities described.

It is rarely, if ever, possible to say anything certain about the historical circumstances that might have led the author of any particular text to use a certain instructional mode, and it is not my intention here to attempt any kind of totalizing claims about what instructional texts were like in the Hellenistic era and so forth. The chronological ordering of the following case study does, however, make for a convenient way of grouping together texts that might have been written under similar cultural circumstances, and which might have thus faced some of the same pressures on their form from the prevailing literary climate, both in "technical" texts and otherwise. It is in this spirit as well that I mention some elements of those literary environments, such as the "directive reading" and fascination with epigrammatic encounters between reader and object that were notable features of Hellenistic literature, or the emphasis in the early Empire on creating texts that could attract and hold the interest of a broad audience, and so forth. It is scarcely inconceivable that the authors of mechanical artifact texts might have consciously availed themselves of descriptive techniques used in broader contemporary literature.

Case study: military machines

The important collection of texts on military machines by Philo, Biton, and Hero provides a convenient opportunity to compare these various

[77] For discussion of how this picture changed throughout the imperial period and from one discipline to another, see Horster 2003, 176–83, 191–2. On "deferential didacticism," see Apollodorus 2010, 33.

[78] Hine 2009; expanded in Hine 2011.

[79] Meißner 1999, 178. Meißner is here referring to Vitruvius' well-known commendation of his work to *omnibus gentibus* (Vitr. *De arch.* 9.pr.2).

forms of directive. This collection has special value because of its cohesiveness. Although the texts themselves were authored in different periods, they have largely traveled together as a "collected works" since the eleventh century, often copied and bound alongside related authors like Athenaeus Mechanicus and Apollodorus.[80] Analyzing this group of texts, which have quite similar subject matter and so much of their textual tradition in common, allows us to explore some chronological and cultural variations in the way these authors treat their subject, while keeping other variables as stable as possible.

In Philo's *Belopoeica*, the question of the intended audience for his instructions is immediately raised by the reference in the prologue to an addressee, a certain Ariston (who is otherwise unknown).[81] Philo claims here to have addressed other works in his *Syntaxis* to Ariston as well (here he specifically identifies his treatise on harbor building), so we must not assume that Philo wrote for a reader planning to specialize in the design of military engines. The other branches of practical mechanical knowledge that seem to have featured in the *Syntaxis* do not figure in the *Belopoeica*, which keeps a tight focus on the problems and skills associated with the design and construction of catapults. Philo does cite results from the treatise on levers that was apparently part of his *Syntaxis*; however, these results are theoretical generalizations rather than concrete problems from another domain of engineering.

Philo approaches the task of describing the engines to the reader, and teaching him to think about and even build them for himself, by connecting geometrical abstractions and mechanical theorizing with concrete problems of craftsmanship. He has a particular interest in novel invention and refinements of design and craft, making him attentive to even minute differences between engine designs. Passages of directive text addressed to a single reader (nominally Ariston, though he is explicitly named only at the beginning of the work) alternate with passages describing Philo's own prior activities and didactic intentions. A few other passages use impersonal constructions with *dei*, or passive constructions where the machine itself or its components serve as subject. The overwhelming majority of the instructions in the text, however, are delivered through personal directive forms that create the sense of a participatory "lived-in space" of construction.

[80] Wescher 1867, xxxvii–xliv; Marsden 1971, 13. Philo's text is missing from the collection in one branch of this manuscript tradition, but included in others.

[81] Marsden 1971, 156 n. 1.

We can identify a few consistent figures in this "lived-in space." Philo uses the first-person singular to describe his designs for siege engines, particularly when he emphasizes his own innovations. The first-person plural encompasses both an editorial–authorial Philo who describes his decisions in shaping the text, and a construction mentor Philo who intervenes to guide the reader through acts of building. The second person introduces the student–addressee figure to whom that guidance is directed. Some didactic maneuvers require Philo to put together compound structures involving different persons and tenses. For example, "I attempted to remove [the hole carriers] entirely from the system, so that we could insert connecting rods underneath of any magnitude we liked in both depth and breadth ... we will stretch out the springs ..." combines a designer in the first-person singular with a construction collaborator in the first-person plural.[82] Philo's precise modulations of authorial voice demonstrate his awareness of the need to bring together individual innovations and their propagation to students, authorial work and craftwork, manipulating objects and theorizing the principles behind them.

Philo uses the first-person singular when he describes his own process of learning about and improving upon the designs of Ctesibius. For example, he describes how he eliminated a flaw in the original design by making some radical structural changes:

> I saw that in pre-existing machines the springs were set over against one another, and that most of the engineers considered that this was harmful to the firing, but were unable to change its position because this was the natural result of the arrangement, and it was not possible to change course. Because of this and the rest of the inconvenient attributes that presented themselves in the arrangement, I tried to change the design and the whole structure, so that I could re-imagine whatever I wanted in every respect, nothing getting in my way.[83]

Philo then goes on to list the many improvements he has made to the traditional design, beginning with the installation of springs that run parallel to one another rather than coming into contact, thus radically improving the machine's range. The forms Philo chooses here are reminiscent of the "autoptic-empirical" *ego* that von Staden observes in Celsus; there too an author emphasizes that he personally witnessed technical practitioners engaged thoughtfully with their work. However, the aggressive intervention Philo portrays himself as making with the design is quite different from Celsus' kind of *ego*, perhaps closer to the emphatically subjective

[82] Philo, *Belopoeica* 60.29–35. [83] Philo, *Belopoeica* 59.34–45.

"Ich des Wissenschaftlers" that Asper identifies in Galen's polemics about the improvements he makes to his discipline.[84]

Combining personal pronouns with first-person singular verbal forms intensifies the effect of Philo's personally stepping in to refine the design of the machine. He uses this approach when he describes an improvement he has made in the method for stringing the catapult:

> I assert that, in the time it takes for one of the other machines to be readied for shooting by means of the stretcher, I will prepare mine using the same assistants, but without this device (ἐγὼ τοῖς ἴσοις ὑπουργοῖς <μὴ> χρησάμενος ἐντονίῳ), wrapping the spring around harmlessly by hand, applying neither packing bolts nor needles, nor clamps, nor anything of the kind, by which the spring is usually harmed.[85]

Third-party personnel appear here, in the form of the assistants required to perform the mechanical adjustment, along with the spring-damaging tools they typically employ. Philo presents his design as a performance of efficiency, himself at the helm: once he comes on the scene, he will complete the task with no damage to the spring. He proceeds in this same vein in the remainder of this discussion, detailing his own construction activities as improvements to legacy designs, emphasizing his contributions with first-person verbs and participles. He is assertive about the advantages of his innovations; the current approach is not just inefficient but actually harmful to the structure of the spring. In these and other cases where he presents himself as an innovator, *ego* and first-person verbs are predominant; his language suggests his presence in workspaces of design and construction.

When Philo speaks about his authorial work, about his intentions for the text he is composing and his hopes to clarify the material for the reader, he uses a kind of editorial "we." For example, he says at the beginning of his book, "If it had happened that everyone previously involved with this part [of mechanics] had used the same approach, then perhaps we would need nothing except to explain the arrangements of their machines, since they would be the same."[86] However, this did not happen, and instead "we see" (ὁρῶμεν) a patchwork of different designs, from which Philo promises to select the latest and greatest to describe for his reader. Though Philo uses a plural first-person form for this rather than the singular, this kind of authorial voice resembles the "*ego* of dispositio" and the "*ego* of the self-conscious stylist" identified by von Staden and Hine.

[84] Asper 2007, 333–7. [85] Philo, *Belopoeica* 61.10–16. [86] Philo, *Belopoeica* 49.6–17.

Later in the treatise, when Philo is guiding the reader through the tricky steps of constructing a device, a new kind of "we" makes its appearance: for example, "we are not putting washers on" (63.7), "we add sufficient thickness to the center stanchions" (64.8). Here Philo's "we" is not serving in the editorial capacity, telling the reader what the text will explain, but rather that of a guide who accompanies the reader through the work of construction, Lehoux's "arm around the shoulder." Hence this text creates two types of "lived-in space": a space for teaching and a space for building.

Both these "collaborative" uses of the first-person plural recall the "integrative we" that Asper connects to Galen's work. In keeping with this collaborative spirit, the editorial "we" is often found next to references to the reader as *se* or *soi*. This is especially common when Philo is discussing the text itself as a learning tool:

> We will record for you (ἱστορήσομεν οὖν σοι) what we ourselves were taught at Alexandria, spending a great deal of time around the craftsmen who were involved with this sort of thing, and at Rhodes, having become familiar with many master builders and learning from them that the most highly esteemed machines fell into a category quite similar to the method that is going to be explained here. (51.15–19)
>
> Since we have disclosed to you the methods of the craft … (ἐπεὶ οὖν δεδηλώκαμέν σοι τάς τε μεθόδους τῆς τέχνης) … (56.13–15)

The emphatic repetition of personal verb forms and pronouns gradually solidifies the impression the text gives of a "lived-in space." This space contains not only the authorial "I," but also a "you," embodied in direct commands to see and even take hold of parts of the machine:

> Hold a hole carrier before your eyes, detached from the frame and not yet nailed together or prepared. What kind of sight will it present to you, hollowed out and see-through from all directions, thickly studded with the holes that border the circles? Contemplate and calculate how much force it must withstand, and thus its design will seem contemptible to you (θεώρει δὲ προσεπιλελογισμένως, ἡλίκην αὐτὸ δεῖ βίαν ὑπομένειν· οὕτως γὰρ εὐκαταφρόνητον φανεῖταί σοι τὸ σχῆμα). (57.24–30)

The last of these passages indicates an approach, common in Philo, of developing the reader's ability to see the device as well as the core principles of its design. Philo's instructions are devoted to teaching the reader to see the machine in his mind's eye and to think about its mechanics (*theōrein, logizein,* etc.) all the while he is learning to build it. The hole carrier is described as a material, crafted object to the reader, who is instructed to handle it and hold it up to his own eyes. If the reader ends up following Philo's construction directions, the result will be a combination of literal and imagined visibility.

Philo makes his reader see the arithmetical, geometrical, physical, and material relationships in his mind's eye, even if he never builds the machine that would embody them in the world. The imaginary vision Philo seeks to create is underscored by the repeated use of *emphainein*, very often paired with *soi*:[87]

> We shall explain to you the result discussed above [i.e. that Philo's novel machine will shoot further than an old-fashioned machine of the same size], as we say, framing the discussion through mechanical demonstrations and physical theories; you will do well to give heed and apply yourself to these. (59.12–16)

Philo's text deploys the language of visualization and persuasion, along with frequent use of personal pronouns, to create a "lived-in space," an imagined space shared by author and reader where the artifact comes together.

The juxtaposition of "we" and "you" suggests a direct and personal transmission of information, mirroring in text Philo's own acquisition of information from the protégés of the famous Ctesibius, and from craftsmen at Alexandria and Rhodes. He twice (67.44, 72.36) mentions those who transmitted Ctesibius' ideas to him, specifying that those who had watched Ctesibius "quite inquisitively" (περιεργότερον) "gave him reports" (ἀνήγγειλαν) on Ctesibius' designs. This adds not just to Philo's cachet as a technological practitioner, but to the status of his textual work as well.[88] Such self-positioning is, of course, also a persuasive technique, as Philo invokes the power of in-person knowledge transfer.

It would be too much to read the hands-on approach to technological patronage that Diodorus Siculus claims for Dionysius of Syracuse as a direct parallel to the process Philo describes, but the emphasis on collaborative mechanical creation within a populated space is nevertheless distinctively Hellenistic.[89] Asper refers to a "simulated closeness" that creates the same sense of security and authenticity for the knowledge being transmitted as a process of personal transmission would

[87] This form of φαίνω is used principally of indicators of information, such as medical symptoms (see especially LSJ, defs. 3 and 4). See also *Belopoeica* 72.6: "We will explain about their utility" (προσεμφανιοῦμεν περὶ τῆς εὐχρηστίας αὐτῶν); 73.36: "We will write about [a repeating catapult] for you, explaining its parts one by one with all due accuracy" (περὶ ἧς σοι γράψομεν ἐμφανίζοντες τὰ κατὰ μέρος μετὰ τῆς ἐνδεχομένης ἀκριβείας).

[88] Meißner proposes that these are mutually beneficial (Meißner 1999, 165).

[89] Diodorus Siculus, *Bibliotheca historica* 14.42.1–2.

provide.[90] Although he did not have the chance to witness Ctesibius directly, Philo emphasizes the direct contact he makes with those who did. Mediators like these are the next best source of the tacit, practical knowledge that is so difficult to convey precisely in text.

The forms of address Philo chooses for his own text suggest an attempt to collapse the distance from the workspace that text tends to impose, to recreate an environment of personal instruction in textual form. In addition to including the reader (personalized as *se*) in scenes of calculation and construction, Philo also invokes him with second-person singular imperatives. Sometimes the reader surrogate appears embodied, instructed to handle pieces of the machine.[91] Sometimes he is instructed just to think about it, and to make calculations.[92] In all of these activities, the addressee is at the center of the text, constantly tended to as he learns the principles of catapult design.

Philo marshals this busy multiplicity of voices carefully, in order to bring together the disparate types of knowledge that go into designing and crafting a catapult. The reader is constantly urged to take an active role in engaging with the mathematical, mechanical, and material layers of knowledge about the machine. On the builder's part, this kind of active engagement demands a constant commitment to tinkering with the parts, while the reader's perspective is built upon language of physical involvement. Eyes and hands are prominent players in Philo's text, examining and handling pieces of machines, pushing a ruler around to make a calculation on the fly, fine-tuning the aesthetics of the machine's appearance.

Active attention to both the physical form and the cultural significance of an artifact is a central feature of Goldhill's model of the Hellenistic "culture of viewing" as focused on fostering and describing interactions between objects and viewers.[93] The mathematical methods presented at the beginning of the text are likewise freighted with the expectation that the reader will engage with physical as well as geometrical knowledge. Philo's frequent inducements to the reader to see, handle, imagine, and reckon with the components of the catapult are a way of strengthening the engagement between reader, text, and object, and hence between author and reader. Every link in this chain is fortified by the methods Philo chooses: he forges a strong connection between himself and the object by

[90] Asper 2007, 43.

[91] E.g. λάβε γὰρ περίτρητον πρὸ ὀφθαλμῶν, (57.24), ἔλαυνε τοὺς σφῆνας· τοῦτο δὲ ποίει, ἕως ἄν σε τὸ μῆκος τῆς τοξείας ἐκτίνῃ (66.21).

[92] E.g. θεώρει δὲ προσεπιλελογισμένως (57.29).

[93] On this aspect of Hellenistic visual culture see also Zanker 2007; Goldhill 1994.

emphasizing the work he himself has done, illustrates the object's structure with highly specific geometrical terms fleshed out by intensive descriptions of the real materials from which it will be built, and connects the object to the reader by direct address and commands to imagine and handle it.

Biton's *Kataskeuai* takes a slightly different approach. Like Philo, Biton has a single addressee, but in this case the addressee is the king Attalus, who does not fit so neatly into the student role as Philo's Ariston apparently did. Biton's text does maintain a vestigial addressee throughout, as σοι appears formulaically at the beginning of each description in Biton's survey, and there are as well a few second-person admonitions to think ahead (4.6) or not to forget something (4.39). Biton himself is visible at times throughout the text, presenting himself as a kind of intermediary between the technology and the king, whose primary responsibility is to choose a canon of weapons that will be maximally useful. He occasionally uses the first person to emphasize his own broad technical expertise, as when he claims that "there is also a theoretical approach to this, which I have discussed in my *Optics*, for matters of dioptrics lie within my grasp."[94] Yet Biton does not advertise himself as an expert innovator when it comes to the mechanical subject matter of the *Kataskeuai*, and Rihll suggests that Biton often describes technology that had probably been in place for at least a century.[95] Indeed, he introduces each engine with a famous historical exemplar of the genre, identifying the designer and location of each one.

Biton's building instructions are far less detailed than Philo's, although he indicates at the beginning that this work was accompanied by a list of dimensions (*logothesia*; the word is also used for the reckoning of an accountant), which could have compensated somewhat for his relatively low level of descriptive detail. Even if this were the case, however, the way in which the reader encounters that information would have been quite different: a list of dimensions to be referenced after reading the text is a far cry from Philo's long-form, detailed description of a smaller group of machine types. In the text itself, Biton's instructions use both absolute dimensions and letter labels; the labels correspond to diagrams following each passage of text.[96]

94 Biton, *Kataskeuai* 4.10.

95 Rihll 2007, 164–5. Claims like this should be treated carefully because of the controversy over Biton's dates; however, the inclusion of descriptions of equipment from a certain era in a text is not a reliable reason to date a text to that era (this is true for Hero as well). Certainly Biton appeals to more securely dated historical exemplars from quite a wide chronological range.

96 The dimensions recorded were connected to the historical exemplars rather than being Biton's own proposals; as Gille observes, "Biton donne bien les dimensions de ses machines, mais elles lui viennent d'ingénieurs qui ont réussi" (Gille 1980, 147).

Biton's main text is divided into passages on six engines: two stone-throwing catapults, a *helepolis*, a *sambuca*, and two types of "belly-bow" (*gastraphetēs*). Each description follows a fairly formulaic pattern: Biton begins by naming a famous exemplar of the machine type, its inventor, and place of deployment; he then describes its structure, and finally closes by alluding to a following diagram. In each case, the description of the historical object is in the aorist; when it comes time to dispense instructions for building a new machine of the same type, Biton gradually moves out of the aorist and into a combination of (first-person plural) futures and (third-person) imperatives. In this way the reader receives an impression of a machine that is firmly rooted in its successful, famous past, which he can make real for himself by following Biton's directions. By combining these aorists with imperative and future verb forms, Biton shifts away from the distancing effect the aorist has in less "generative" texts like histories. Here, the aorists instead bring the past to bear as a supporting argument for choosing a particular design for a machine yet to come.

Although Philo's text and Biton's share this quality of "generativity" to some extent, the two authors create very different environments for information transfer in their texts. Philo praises his predecessor Ctesibius as both knowledgeable and ingenious, but invariably hastens to add that his own designs represent a step forward. In doing so he himself embraces the role of a singular inventor communicating his ideas to a student colleague: as he learned from the colleagues of Ctesibius, so shall the reader learn from him. Though the learning process will necessarily be different, mediated as it is through text rather than hands-on transmission of practical knowledge, Philo does not highlight the difference but instead contrives novel ways of eliding the textual distance between himself and his reader. Claims to technological innovation are intertwined with a program of textual innovation: Philo's innovations are reified through their integration into texts, which themselves become part of the mechanical tradition. Philo's achievement, as practitioner and as author, is peculiarly his; it is his individual contributions to the field that count most of all.

Biton, by contrast, downplays his direct involvement with the task of building. He makes the occasional first-person appearance in the text, and is not shy about asserting his expertise in other fields. However, he does not present the building process as an intensive, sustained interaction between author and reader, but rather the predictable result of following a set of easily transmitted instructions. Meißner proposes that the structure of Biton's text indicates that the engineer was meant to provide the ruler or patron with sufficient preliminary data (e.g. notes, drawings, and building plans) to be able to hand off the project to craftsmen, rather than participating personally in the work of construction.[97] The presence of the list of dimensions alongside the text would serve this effort, while Biton's breezy assurances that the machines described could be resized to suit with relative ease would surely please a patron more than the unfortunate craftsman tasked with carrying out the rescaling.

Biton's text, unlike Philo's, resembles in many respects the approach Asper refers to as the "Rezeptstil." This form of information transmission is defined in part by its use of instructions delivered through imperatives, reference to accompanying diagrams, and discrete sections of text with well-marked transitions.[98] Texts belonging to this category thus collapse the distance between author and reader, creating a kind of "lived-in space" somewhat like Philo does. However, Philo's text does not share most other characteristics of the "Rezeptstil" works. Those texts are "discrete," rather like a catalogue of problems, while Philo's is a continuous, ever-deepening examination of the principles according to which a catapult functions. A "discrete" text structure encourages formulaic repetition like Biton's: each of the problems in the texts Asper mentions (including Babylonian and Egyptian mathematical texts, as well as Greek texts in stereometry) is structured the same way as its text-mates. The overall effect is of a standard procedure reliably providing linear, step-by-step instructions for a technical task. The authors of these texts often claim to enable the reader to grasp the method of solution of a particular type of problem, generalize that solution, and move on to the next problem model; Biton does the same.

Biton's text meets the "Rezeptstil" criteria of including references to accompanying diagrams and imperative guides to action, though his imperatives are third-person rather than the direct second-person commands that seem to be characteristic of the texts Asper describes. Despite this difference, Biton's text comes closer than any other surviving mechanical artifact text to meeting Asper's criterion of linguistic standardization.[99] Each machine entry in Biton's text is structured in the same way. The passage begins with a programmatic statement in the first person

[97] Meißner 1999, 161. [98] Asper 2007, 198.

[99] Hero's *Mechanica*, which survives primarily in Arabic, does appear from what survives to display a relatively high degree of linguistic standardization. It is not, however, a text on mechanical *artifacts*; while he does discuss the simple machines in this work, they are not concrete artifacts but representations of mechanical principles. This of course is not to say that the knowledge of theoretical mechanics this text explicates was not itself informed by the work of mechanical practitioners; on this see Schiefsky 2008.

(usually plural), which makes a future-tense promise of the description to come (ὑπογράψομεν[100]), a present-tense characterization of the act of description (ὑποτάσσομεν[101]), or an aorist account of the act of writing (προκεχείρισμαι ἀναγράψαι[102]). This remark is followed by a short account of the machine's historical exemplar. The description follows, with some variation in length and content. Finally, every description is closed with a formulaic reference to a diagram: "its appearance is like that found below" (τὸ δὲ σχῆμα οἷον τυγχάνει ὑπόκειται).

Like the problems in the "Rezeptstil" texts, Biton's descriptions provide absolute dimensions, but unlike the values in the texts Asper describes, these are not confined to integer values. Far from being a minor qualification, this points to a more profound difference between Biton's text and the mathematical texts that incorporate the "Rezeptstil." Those texts are identified by Asper as problem texts for teaching future practitioners; hence they use repetition with slight variations to inculcate into the reader a skill that he can then apply to any appropriately structured future problem. Biton's text does not seem to be intended this way. For one thing, it is nominally addressed to a king who is probably not intended to develop any skills at all from the text. More importantly, however, the "standard procedure" or "general case" to which Asper refers has no real analogue in Biton's text.

This important difference has to do with the degree of flexibility in the relationship between the problems or designs described in the text and those the reader will face in the future. The texts Asper catalogues under the "Rezeptstil" heading provide "model" problems with sample quantities. The reader who learns to solve these problems can go on to solve other problems that have the same structure as the model, involving a potentially unlimited range of quantities. In Biton's text, each machine type is given a "model" embodiment, a particular historical exemplar, which every machine to be built on that model must follow more or less precisely. The idea that following the text's directions is a process of duplication rather than redesign for a potentially unlimited range of ballistic requirements would have been reinforced by the attached list of dimensions. Referring to a historical exemplar ties the artifacts described quite tightly to the "original version," unlike the practical-mathematical problems of the "Rezeptstil" texts, which are not tied to any artifact at all.

It might be suggested that the unlimited flexibility of application that the problem texts implicitly offer the reader–practitioner is echoed in

[100] Biton, *Kataskeuai* 3.1. [101] Biton, *Kataskeuai* 4.1. [102] Biton, *Kataskeuai* 6.2.

Biton's explicit assurance that the engines he describes could be scaled up or down to meet any possible needs. Even though the actual requirements for such a process of resizing to be successful would have been much more complex than Biton suggests, the promise of the text closely resembles that of the mathematical problem texts: the text contains model instructions that can be returned to and followed with new numbers for new situations.[103] As Meißner notes, the devices described are in general expected to be similar rather than identical to those whose future construction the text encourages.[104]

However, if one intends to produce a predictably functioning ballistic device out of conventional materials, the flexibility and scalability of its dimensions is in fact quite limited compared to the theoretical flexibility of the problems in the mathematical texts Asper cites. Contrast this approach to solving new ballistic problems with Philo's: rather than a model to be emulated, Philo spends most of his time giving the reader the mathematical-physical principles that underlie the design of the engines he describes. The resizing of new engines is not so much a turn-the-crank repetition of steps learned by rote, but a more complex system of calculations starting with the basic principle of the relationship between the spring hole and the missile size.

The surviving technical prose from the Hellenistic period is, unhappily, quite limited.[105] However, Philo's and Biton's texts on artillery suggest a few likely features of the strategies used by authors of *kataskeuai* in this period. Both works mention a sort of mechanical canon of catapult types, which Philo is intent on improving with his own innovations, and which Biton uses as models for the instructions he addresses to his royal addressee. This authorial stance might correspond to a broader tendency in Hellenistic technical literature: the works of previous authors are treated first and foremost as material to be improved upon, broadened and enriched with one's own contributions.[106]

Both texts are addressed to specific readers, though the instructional relationship between author and addressee takes two very different paths.

[103] The challenges of this particular task are demonstrated, for example, by Vitruvius' sobering story of Callias' scale-model siege engine, which proved a failure at full scale (Vitr. *De arch.* 10.16.3–5). The moral of the story is that material constructions are subject to material constraints, and do not allow for the predictable linear scaling that might naively be expected.

[104] Meißner 1999, 162. [105] Meißner 1999, 169.

[106] Naturally, not all suggested innovations turn out to be successful; while Philo argues for the advantages of the wedge-stretched catapult over those made with bushings, Meißner points out that most of the archaeological finds of catapults from Hellenistic and Roman times are of the latter type (Meißner 1999, 166).

Framing the text's work as a direct communication between author and reader recalls the structure of Hellenistic literary epigrams that address the reader directly with instructions for seeing an object.[107] It is also connected to an epistolary culture in which technical knowledge is propagated through missives addressed to particular readers, such as Archimedes' letter to Eratosthenes and Eratosthenes' letter to the king Ptolemy.[108] All these texts serve in their various ways – from Philo's frequent references to the reader's hands and eyes, to the "simulated closeness" of Biton's "Rezeptstil"-like work, to the artificial intimacy of the epistolary form or the "directive reading" of the epigram – to build a connection between author and reader, reader and artifact.

The differences between the two texts are equally important. Philo likely wrote his treatise during a time very rich in technological developments; he was quite possibly a near contemporary of Archimedes, as well as of numerous fellow designers of siege engines like Charon of Magnesia and Zopyrus of Tarentum.[109] He lived close enough to Ctesibius' time to observe those who had worked with him in person. His text participates actively in a lively, multivocal world of technological development by making constant claims to improve on the existing body of knowledge, improvements rooted in the mechanical and mathematical principles that Philo insists his reader should learn to grapple with as well.

These figures are all history to Biton, who writes about their achievements as something past, and who does not propose mechanical improvements to their designs. Biton is tied to his discipline's past by the ubiquitous textual pointers to his royal addressee, to whom he presents machines with a proven track record. His text is much less detailed, more like a field guide to the various kinds of military machinery than a detailed education in the minute details of their actual construction, and he often uses hands-off grammatical forms that separate the reader from the activity of construction. Instead of presenting himself as a technological innovator, Biton casts himself in the role of technologically savvy anthologist, collecting and reshaping technologies devised by others and presenting them in a text that increases their accessibility and utility for his own audience.

107 On Hellenistic innovations in this respect, see Tueller 2008, ch. 3.

108 On this subject see Archimedes and Eutocius 2004, 12–15; Asper 2007, 75; Taub 2008.

109 For a recent set of arguments on the dates of these figures, see Keyser and Irby-Massie 2008.

Hero's *Belopoeica*, probably written in the first century CE, differs strikingly from Philo's work.[110] Where Philo is constantly concerned with issues of invention and innovation, representing his own designs as cutting-edge technology, Hero restricts his discussion to older mechanisms (perhaps predating him by as much as three centuries), acknowledging right off the bat that he will describe machinery that is not current.[111] Hero's text is framed not as a technical manual addressed to a particular person or even a particular class of practitioner, but as a general introduction to a subject of importance for all.[112] He announces his intention to fix a particular problem – and it is a textual problem rather than a technical one: this material, important as it is, has not previously been explicated in a way that makes it accessible to everyone.

Hero begins by attempting to claim some philosophical ground in the name of mechanics: he explains that technological excellence, particularly when it is directed toward the creation and development of military technology, is the surest route to *ataraxia*.[113] The text as a whole is framed as an evolutionary history of the development of various kinds of catapults, a contrast to Philo's layers of mathematical, mechanical, and craftsman's knowledge and to Biton's "discrete" catalogue of ballistic successes. Hero's philosophical prologue, statement of intent to write a treatise suitable for laymen, and historical organizational scheme are all part of an approach that is somewhat disengaged from the devices themselves. However, this text is larded throughout with vignettes of the construction process that open a window onto the practical problems of building a catapult. Hero's text is a fascinating hybrid of subject matter both general and specific, sketchily abstract and grittily hands-on.

In order to make his subject matter accessible, he promises to speak about complete engines as well as their individual parts, to discuss technical vocabulary and practical details like fitting the spring, and to discuss how the engines are used. All this information will, moreover, be encapsulated in remarks about the taxonomy and development of every

110 Here, for expediency's sake, I do not engage with the question of Hero's dating. Giardina provides an excellent guide to the controversy at Hero 2003, 5–30. For additional analysis pointing toward a first-century CE date, see Keyser 1988; Asper 2001; for a more cautious perspective, see Sidoli 2011.

111 Hero, *Belopoeica* 2.12. On Hero's acknowledged intention in this text to present older technology in a new way rather than innovating, see also Marsden 1969, 44 n. 2; Tybjerg 2000, sec. 2.1; Cuomo 2002, 167–70. For an extensively documented argument that Hero draws from many sources representing a broad chronological span, see Schellenberg 2008.

112 The embodied σε addressed by Philo and Biton rarely appears in Hero's works. A specific addressee (Dionysius) is mentioned in the proem of *Definitiones*, but does not appear in the rest of the text.

113 Hero, *Belopoeica* 1.1–15.

type of engine.[114] Hero does not correlate each device with a named inventor or patron, as Biton did. Instead, he proceeds through an evolutionary history without inventors or other particular personnel, except for the vague mention of "they" who demanded this evolution. The *gastraphetēs* was devised because "they" needed more range than handbows could provide; torsion engines were developed, again, because of "their" desire to increase the weapon's range and missile size, and so forth.[115] This generic, impersonal account of technological evolution is very different from Philo's claims of having participated in a collaborative process of learning and development alongside Ctesibius' protégés, and even from Biton's quick-fire references to the inventors associated with every engine.

The same impersonal approach pervades the instructions for building the engines. Philo's and Biton's treatises featured a varied cast of characters: builder, patron, mentor, student, and so forth. By contrast, Hero's discourse on the design and layout of the engines he describes is provided mostly through constructions that involve no particular person, though the anonymous "they" are occasionally identified as responsible for naming components. Hero usually uses δεῖ + infinitive constructions and third-person imperatives; a typical passage looks like this:

> Imagine (νοείσθω) the edge of the barrel ΕΖΗΘ along the length ΩF; riveted to this a toothed rod ϙϡ; to the barrel along ͵ΑΒ let a hook ͵Γ͵Δ be attached, moving around a rod. The barrel having been drawn down, it would happen that the hook ͵Γ͵Δ (which they call the pawl) was carried obliquely along the teeth.[116]

Hero's grammatical constructions, as well as elements of his vocabulary such as νοείσθω and προσκείσθω, are characteristic of mathematical prose.[117] His careful mapping of points on the machine to points on a diagram also suggests an assimilation of mechanical design to geometry; sometimes he indeed walks the reader through the geometrical

[114] Hero, *Belopoeica* 2.15–24.

[115] On Hero's explanation of surveying as likewise arising in response to changing human needs, see Tybjerg 2004, 31–3.

[116] Hero, *Belopoeica* 7.7–19.

[117] A few words about what it means to be "characteristic" are undoubtedly called for here. This assertion is based on the "full corpus counts/author" information available from the TLG for the verbs of these forms that Hero uses (for example, γεγενήσθω). Overwhelmingly, these verbs are most commonly used by mathematical authors; they are also, with varying frequency, used by philosophical, mechanical, and medical authors, and in later eras by theological authors. The earliest uses are almost exclusively in Hellenistic mathematical authors, though Polybius (alone among historians) shares a fondness for third-person imperatives as well.

construction.[118] This approach is very different from Philo's, which exploited geometrical methods and a loosely defined set of mechanical laws where possible, but acknowledged the frequent need for a trial-and-error, craft-based approach to design. The differences between these two authors also highlight different ways a mechanical text can be "mathematical." Hero's descriptions of machines regularly assimilate elements of the machine to points on a diagram or in a geometrical construction, and use grammatical constructions common to geometrical texts. However, he postpones instructions for calculating component sizes to an appendix at the end, rather than emphasizing in the main text the need to calculate and recalculate to be assured of an accurate result, as Philo does.

Hero is not carrying out proofs, but laying out a field of design types, so he is not engaged in quite the same activity as a mathematical author. He is, however, disconnected from the creation of his mechanical objects in the same way Asper observes of geometers: Hero makes no claims to be either their designer or builder, and the reader is not, in these passages, directly enjoined to construct them either. Of course these disconnections exist for different reasons: the objects mathematical authors study might be viewed as eternally pre-existent, whereas the mechanical objects Hero discusses were man-made (albeit many decades before him). Still, the similarity is remarkable for the frequent instantiations of directive structures that Asper argues are so rarely encountered outside mathematical texts that they mark a language specialized to the discipline.[119] Hero is not really doing mathematics (at least outside the appendix), but his choice of these imperative forms seems to signal that he is staking a claim on that territory for mechanics.

What happens when Hero moves from describing mental constructions of mathematicized artifacts to describing manipulations of objects freighted with material complications? In passages concerned with the hands-on process of construction, Hero typically abandons the impersonal constructions he uses when discussing the design and layout of the machines, and shifts to a descriptive mode populated by a collaborative "we" and a second-person addressee of imperatives and directive future indicatives. For example, Hero uses this combination of forms to describe how the spring (for the *palintone* catapult, or the frame for the *euthytone* model) is tightened:

> Then whenever we want (βουλώμεθα) to stretch either the half-springs of the *palintone* or the frame of the *euthytone* … we will attach (ἐξάψομεν) one

[118] Hero, *Belopoeica* 26.23–26.39. Tybjerg addresses a similar approach in Hero's *Catoptrics*, where "he simultaneously constructs a diagram that is the site of the geometrical proof and a working device that can produce certain effects" (Tybjerg 2004, 49).

[119] Asper 2007, 125–7.

> end of the spring to one pin ... When the holes seem to you to have been filled up (ὅταν δὲ καλῶς δόξῃ σοι πεπληρῶσθαι τὰ τρήματα), then as for the remainder of the spring, if it is considerable, you will cut it off short (ἀποκόψεις); but if it is only a little, you will let it go, and wrap it around half of the spring (ἐάσεις καὶ ἐπειλήσεις). Then push the arm through and carry out the next step (τὰ ἑξῆς πρᾶττε), as was described earlier. If because of the frequent removals the spring takes up some slack, you will give the washers an extra turn (ἐπιστρέψεις), as was mentioned before, using the iron bar holding the ring as a lever.[120]

Throughout the *Belopoeica* Hero distinguishes between generally impersonal passages describing the aspects of the device that can be reduced to geometry, and passages describing hands-on craft activities, where he uses directives in the first and second persons. Nor is this pattern peculiar to the *Belopoeica*; other texts by Hero such as the *Cheiroballistra* and the section of the *Dioptra* devoted to the odometer exhibit the same distinctions.

The *Dioptra* mostly uses impersonal forms to describe the odometer's construction. Personal verbs are largely reserved for ruminating on his authorial choices and for describing manual engagement with the artifact. He introduces the machine as follows:

> Since some of those who came before us laid out some methods by which this is done [i.e. measuring distances on land using some type of cart-mounted odometer], it will be possible to judge both the machine written by us and those written by my predecessors (ἐξέσται δὲ κρίνειν τό τε ὑπὸ ἡμῶν γραφόμενον ὄργανον καὶ τὰ ὑπὸ τῶν προτέρων). First, let there be a frame (γεγονέτω οὖν πῆγμα) like a small box, in which will be all the equipment that is going to be described.[121]

When the design of the machine is under discussion, Hero relies on third-person imperatives and passive constructions. However, when the text turns to a tricky piece of craftsmanship requiring hands-on testing, Hero switches to personal verbs:

> ἀλλ' ἐπεὶ ἡ ἑκάστου κοχλίου στροφὴ οὐκ ἀκριβῶς οὐδὲ μεμετρημένως τοὺς παρακειμένους ὀδόντας στρέφει, ἡμεῖς τῇ πείρᾳ ἐπιστρέφομεν τὸν πρῶτον κοχλίαν, ἕως οὗ τὸ παρακείμενον αὐτῷ ὀδοντωτὸν τύμπανον μίαν ἀποκατάστασιν λάβῃ, μετροῦντες ὁσάκις αὐτὸς ἐπιστρέφεται.

[120] Hero, *Belopoeica* 28.21–29.5.

[121] Hero, *Dioptra* 34.7–11. The phrase *graphomenon organon*, the eponymous "written machine," is especially interesting; it appears to suggest something more comprehensive than mere description, indeed a real textual encoding of the machine itself by instructions on how to build and operate it. The cover of this book shows the marriage of word and image in the version of the accompanying diagram from Codex Parisinus Suppl. gr. 607.

> But since each turn of the screw does not turn the connecting teeth precisely or according to a stated measure, we turn the first screw as a test, until the toothed drum connected to it makes one revolution, measuring how many times it is turned.[122]

Despite Hero's apparent preference for impersonal grammatical constructions when discussing his catapults' theoretical, structural, and functional properties, he regularly populates his descriptions of craft activities with participants.

Unlike geometrical constructions, manual craft activities involve a significant component of tacit knowledge that makes them difficult to represent completely in text. Personal verbal constructions may bridge the gap between the text and the physical object by imaginatively integrating the reader into the populated, hands-on environment of construction. Hero's *Belopoeica* is, like his other mechanical works, written largely in a "geometrical" style: he makes extensive use of lettered diagrams, rarely mentions absolute measurements, and uses grammatical forms and vocabulary reminiscent of mathematical texts.[123] The "geometrical" features of Hero's texts are reminiscent of the texts Asper calls "Elementaschriften," highlighting their Euclid-like use of impersonal directive forms.[124] The author of these texts does not appear personally, except for the occasional formulaic "I say" (λέγω ὅτι), nor is the recipient addressed directly. The effect of impersonality is fortified by reliance on the third-person imperatives typical of mathematical prose.[125] The form thus presents a strong contrast to the "Rezeptstil," in which Asper notes the persistent use of "personal" imperative forms.

However, Hero is not providing a rigorously mathematical approach to mechanical engineering: he only occasionally refers to actual geometrical constructions in his mechanical texts. Instead, he is carrying out a rhetorical maneuver to assimilate his mechanical work to mathematical prose, perhaps to enhance its cultural cachet.[126] Cuomo argues that Hero aims to present his reader with a story about the evolution of new and improved technological solutions to practical problems, "and of how those solutions themselves have been retrospectively established on solid geometrical foundations."[127] Hero claims value for this kind of knowledge based

122 Hero, *Dioptra* 34.68–73.

123 On these elements, see especially Netz 1999, ch. 4; Tybjerg 2004, 35–9.

124 Asper 2007, 125.

125 Asper's examples include ἔστω, εἰλήφθω, and ἐπεζεύχθω.

126 Tybjerg 2004, 51–4. On the rhetorical power assumed for "mathematical" argumentation particularly by Hero and Ptolemy, see Feke 2014.

127 Cuomo 2002, 170.

not on his advanced, specialized technical know-how, but on a more general understanding of the contribution ballistic technology makes to civic *ataraxia*.[128]

Compared to Philo's emphasis on innovation and his creation of a sense of collaboration between mentor and student even in text, and to Biton's businesslike instructions for recreating proven mechanical success stories, Hero's *Belopoeica* uses hands-off discourse more suited to an amateur audience of armchair mechanics. He takes his reader on a high-level tour of the activities of the mechanical engineer, sampling everything from geometrical design to the material challenges of craftsmanship. In doing so, he recasts pre-existing mechanical devices in quasi-geometrical terms, but at the same time omits or dilutes much of the mathematical calculation that was central to Philo's treatise. He does maintain a link to the mathematical theory underlying catapult design, but this material is removed from the main action of the text to an appendix on finding the size of the spring hole. Hero's simplified and decidedly non-cutting-edge introduction to ballistics combines with his extensive use of impersonal verb forms to create a sense of disengagement between the reader and the objects he is nominally being instructed to build.[129]

Vitruvius and Athenaeus Mechanicus, who were probably close predecessors of Hero, as well as the somewhat later Apollodorus, use the same types of grammatical construction as Hero does. They also situate their subject matter in a similarly broad context, less concerned with arguing for their own innovations than with providing a survey of technological exemplars suitable for non-experts. The mathematical background, downplayed in Hero's text compared to Philo's, is more or less completely absent from these authors. Vitruvius, for example, does not address the important task of finding the spring hole size suitable for a particular missile weight by providing a variety of calculation techniques like Philo, or even a geometrical appendix like Hero. Instead he provides an abbreviated list of corresponding weights and diameters (which Philo also does, but accompanied by mathematical techniques allowing the calculation for any weight). Vitruvius rationalizes this decision by saying that he does not want his less mathematically inclined readers to be "detained by deliberation during wartime danger."[130] This is perhaps a bit disingenuous; there

[128] On likely candidates for the historical circumstances that might have prompted this stance see Cuomo 2002, 170–3.

[129] Asper describes this type of effect as "simulierte Distanz" (Asper 2007, 43).

[130] Vitr. *De arch.* 10.11.2.

are no catapult designers in foxholes. In fact, Vitruvius provides instructions suitable for an audience less interested in a full mathematical and mechanical description of the catapult's operation, more eager to move ahead quickly to a survey of the design of the machine itself.

Vitruvius depersonalizes the construction process by delivering his instructions in passive verbs, mostly subjunctives; for example, "let the cross-piece [of a catapult] be described (*describatur scutula*)."[131] Even when he briefly discusses the hands-on process of stringing and tuning the catapult, he does not introduce personal verbs:

> Then the ends of the ropes are inserted through the holes of the cross-beam; they are drawn across to the other side, and then they are cast and wrapped around a capstan, so that when the ropes are stretched on them by levers, when touched with the hand, each makes an equal correspondence of sound … And so, blocking them in with wedges, catapults are tuned to their sound by musical observations (*ita cuneorum conclusionibus ad sonitum musicis auditionibus catapultae temperantur*).[132]

No personnel are present in this scene of construction, not even in the tricky manual task of adjusting the tension in the ropes so that it sounds right: the adjustment is accomplished by "musical observations" (*musicis auditionibus*), not by a person listening, and not by a set of hands. This passage is quite typical of Vitruvius' instructional approach, and makes a remarkable contrast with Philo and even with Hero.

Vitruvius' artifacts seem to come together autonomously, without human intervention; the reader is not encouraged to imagine himself participating in the craft process, as he was throughout Philo's work and sometimes in Hero's. Vitruvius issues nearly all his directives without using second-person forms, and the exceptions are confined to his prefaces.[133] In Meißner's view, Vitruvius' preference for impersonal forms coheres with his claim to make architectural knowledge accessible to everyone (*omnibus gentibus*), to systematize an entire discipline for future generations of practitioners and laypeople.[134] Asper suggests that such impersonality may indicate a choice to pursue consensus over competition.[135] Vitruvius' choice to avoid second-person references is echoed by Celsus (who likewise completely avoided them) and Frontinus (who used them extremely sparingly). As Hine observes, later Latin authors of instructional texts (like Columella, Vegetius, and the late ancient medical writers) revive the use

[131] Vitr. *De arch.* 10.11.4. [132] Vitr. *De arch.* 10.12.1–2. [133] Hine 2009, 28.
[134] Meißner 1999, 178. [135] Asper 2007, 44.

of second-person forms; hence, Vitruvius and Celsus made a "conscious choice" to avoid them.[136]

Athenaeus Mechanicus was likely a close contemporary of Vitruvius, and the two have much in common despite working in different languages. They mention many of the same figures from the history of the field, such as Geras of Carthage and Pyrrhus of Epeirus, and draw heavily for their material on Agesistratus (mentioned by name only once by Vitruvius, but much more often cited by Athenaeus).[137] Their organizational and descriptive approaches are likewise similar: they both proceed encyclopedically through a variety of devices used in siege warfare, rather than focusing intensively on a few devices. Athenaeus claims that his concision is meant to improve the book's comprehensibility for a wide audience. For example, he provides an introductory parable relating a speech of the Indian gymnosophist to members of Alexander's retinue:

> Hence Calanus of India appears not to have gone wrong in saying to them that "we are not like the philosophers of the Greeks, among whom many words are expended on small matters. We," he said, "are used to say the least about the greatest things, so that it will be easy for everyone to remember."[138]

Calanus' appearance here is perhaps intended to remind the reader of the Hellenistic legacy of military machinery, while his emphasis on brevity sets the stage for Athenaeus to claim higher value for mechanical work than for lengthy philosophical discourse. Indeed, Athenaeus specifically rejects the strategy of incorporating Greek philosophy into his discussion of mechanics, saying that authors from Archytas to Aristotle have wasted their readers' precious time with theoretical digressions.[139] It might be argued that his rambling introduction merely wastes the reader's time in a different way, but the part of the text in which machines are actually described is nothing if not concise. The uncomplicated brevity of Athenaeus' mechanical descriptions seems intrinsic to his attempt to secure high status for mechanics as a practical and efficient discipline.[140]

Athenaeus almost always describes machines using passive verbs taking the machine itself as their subject, though like Vitruvius he also uses active third-person verbs for historical vignettes involving the machines'

[136] Hine 2009, 29. [137] Athenaeus 2004, 25ff.

[138] Athenaeus Mechanicus, *Peri mēchanēmatōn* 5.7–11.

[139] The prologue of the Latin version of Philo's *Pneumatica* contains a similar promise to avoid time-wasting philosophical digressions.

[140] Meißner highlights the parts of Athenaeus' text that draw most heavily from Agesistratus (Meißner 1999, 192).

construction, use, or destruction. Some of his descriptions are billed as direct quotations (normally signaled by φησί) from technical treatises, largely from the Hellenistic era. Within these quoted passages, constructions like ἐχέτω sometimes occur, but Athenaeus does not use the third-person imperative outside these quotations, and he does not use second-person imperatives at all. He does sometimes insert editorial comments in the first person; most of these are about the arrangement of the work, while a few are critical judgments favoring one type of design or another, though he does not make technical mechanical arguments for his preference.[141]

Athenaeus is, in other words, highly visible as an editor engaged with the material he is presenting to the reader, but largely disengaged from the technology he describes. Even on the one occasion where he presents himself as personally involved with an invention (a device to keep siege engines steady on the deck of a moving ship), he devotes nearly all of the passage to a vivid description of a ship at sea, and explains his innovation only as attaching "the so-called little ape" (τὸ λεγόμενον πιθήκιον) to the deck.[142] He does not claim to have invented the device itself, and he describes it in so little detail (here and on a few other occasions when he mentions it) that its evocative name is the only indication of its form.[143]

Despite his own desultory descriptions of mechanical technology, Athenaeus makes an explicit claim that the discipline of mechanics should provide status opportunities not just for those who design machines well, but also for those who explain them well:

> For when a work is well conceived, the good reputation of its architect is secured; when it is set forth in words, the arrangement will have the greatest fame in the commentaries (τοῦ γὰρ ἔργου καλῶς διατυπουμένου τούτου, τοῦ ἀρχιτέκτονος εὑρίσκεται εὐδοξία· κατὰ δὲ λόγον ἐκτιθεμένου, τὰ συντάγματα μέγιστον ἕξει κλέος ἐν τοῖς ὑπομνήμασιν).[144]

Just what kind of activity confers this fame? The verb διατυπόω refers to giving form to an object by engraving it, or to a mental image by conceiving of or representing it graphically, or to an abstract system by arranging it. The closely related noun διατύπωσις is used in a text long attributed to Hero to refer to a system of devices used in measuring

141 E.g., respectively, κατετάξαμεν, 31.5; ἀρέσκει δέ μοι καὶ τὸ καρχήσιον, 35.5.

142 Athenaeus Mechanicus, *Peri mēchanēmatōn* 32.11.

143 Athenaeus is not unusual among his contemporaries in his lack of emphasis on personal innovation; in any event, as Meißner observes, Athenaeus made more of an effort than Onasander to include devices identified as new (if not of his own design) in the text: the above-mentioned "little ape," small modular siege engines that could be assembled at the destination, a tortoise with a steerable wheel, and a wall-mounted covering for a ram (Meißner 1999, 193).

144 Athenaeus Mechanicus, *Peri mēchanēmatōn* 15.2–4.

water;[145] however, in rhetorical terms, this term means something quite like ekphrasis,[146] another kind of "leading before the eyes."[147] The ἔργον καλῶς διατυπουμένον referred to here might therefore be a well-designed machine, a well-depicted machine, or a well-described machine. As Platt observes, "the language of *tupōsis* ('impression') implies a chain of images emanating from a prototype, a direct link between an object and its representation."[148] Whitehead and Blyth render συντάγματα as "instructions."[149] Certainly this idea might be incorporated here, but more inclusively we can say that Athenaeus is referring to the arrangement of information, verbal and visual, into a treatise.

In any case, Athenaeus clearly asserts that technical writing, done properly, should glorify the author or anthologist at least as much as good technical design glorifies the original designer. He practices what he preaches, engaging enthusiastically with questions of what to include in his treatise, while maintaining such a distance from the technical subjects he describes that he prefers to cite whole passages from other authors rather than shape the descriptions himself. He advertises his text as sufficient to provide even a reader without practical experience with the knowledge he will need to put military machines to work in the field. Text is the central medium for knowledge transfer: Athenaeus exemplifies von Staden's first-person "author as reader."[150] His frequent references to the work of other authors mark him principally as a designer and craftsman of text rather than machinery, who has combed the landscape of texts on siege engines to find the best and most useful devices for his reader.[151]

Meißner argues that the emphasis in the early imperial period on textual rather than personal transmission of technical knowledge was due to a tendency to place members of the imperial family in positions of military leadership even if their prior experience was minimal or nonexistent.[152] Our knowledge of Athenaeus and his circumstances is too tenuous to make precise assertions about the immediate military needs of his patron. However, it can certainly be said that the instructional style and level of descriptive detail in this text are quite different from Philo's. Gone are the detailed, step-by-step instructions to manipulate components of a

[145] Hero, *De mensuris* 23.1. On the authorship of this text see Vitrac 2010, 6.
[146] Webb discusses the use of the term by rhetorical theorists like Hermogenes and Nicolaus of Myra as something equivalent to ekphrasis (Webb 2009, 51ff.).
[147] Tiberius Rhetor, *De figuris Demosthenicis* 43.
[148] Platt 2011, 192, 230. [149] Athenaeus 2004, 51. [150] von Staden 1994, 106.
[151] Meißner 1999, 193.
[152] Meißner 1999, 192. Athenaeus' treatise was probably addressed to M. Claudius Marcellus, nephew of Augustus.

complex device one by one. They are replaced by an overview of many different types of device, all presented as part of a landscape of textually based knowledge rather than inhabiting a "lived-in space" of material objects and hands-on manipulation.

Apollodorus, perhaps best known for his fall from grace under Hadrian, also composed a *kataskeuē* encompassing a broad range of structures and siege machines, which he affords considerably more descriptive detail than Athenaeus does.[153] The surviving text of the *Poliorcetica* is a remarkable combination of simple and practical devices, which the preface announces as the goal of the work, alongside impractically exotic contraptions like a flail-wielding catapult and a rock-breaking flamethrower. Indeed, Blyth argues that these devices, along with some others, are Byzantine interpolations, so care should be taken in ascribing any given part of the work to Apollodorus.[154] Overall, however, his text appears to be written by a practitioner to a patron, with the goal of satisfying immediate technological needs. Since the exact circumstances under which the machinery will be used are unknown, Apollodorus says he selected the designs most likely to be useful in a variety of situations, and quick and easy to build.[155]

Apollodorus fleshes out his authorial persona quite fully. He portrays himself as a technical rather than a textual practitioner; when he refers to his own authorial work it is largely to apologize for certain shortcomings. In this he falls quite in line with other Flavian technical authors.[156] He acknowledges the brevity of his text, saying that he is pressed for time,[157] and regrets that his work may be obscure for non-technicians because of his extensive use of technical vocabulary.[158] Even as his brevity may have inspired a broader audience to try tackling his technical subject matter, his compressed descriptive style and use of technical terminology might have been discouraging for a lay reader. Nor does he include anything like

[153] Cassius Dio, *Historiae romanae* 69.4. The *Poliorcetica* was probably addressed to Trajan (Apollodorus 2010, 20–4).

[154] Blyth identifies these anachronistic elements in the text based on a variety of factors including both the mechanical characteristics of the artifacts themselves and linguistic elements of their descriptions (Blyth 1992).

[155] Apollodorus, *Poliorcetica* 137.7–10.

[156] Meißner 1999, 249. Compare, for example, the preface to the text on surveying written by Apollodorus' likely near-contemporary Balbus, where Balbus represents himself applying his surveying skills to military service (Balbus 1848, 92–4). On this preface see Santini 1990, 137–42. Likewise, Frontinus explains the motivation for composing his work on the aqueducts of Rome by alluding to his professional responsibilities as *curator* of the water system (Frontinus, *De aquae ductu urbis Romae* 103).

[157] Apollodorus, *Poliorcetica* 138.5–7.

[158] Apollodorus, *Poliorcetica* 138.14–15.

the various historical or philosophical interludes that break up the more technically detailed parts of the works of Vitruvius, Hero, or Athenaeus, which are often viewed as refreshments for the non-technical reader.

Apollodorus uses the first-person singular only to describe in the preface how he chose appropriate exemplars of devices useful in siege warfare and drafted them in written form, then dispatched them along with an assistant trained to construct them through in-person observation of Apollodorus himself at work.[159] Along with the assistant, says Apollodorus, he sent "local craftsmen, and those otherwise capable of building and making" his designs.[160] He says this decision was motivated by his own experiences in battle, where he has seen first-hand "how the surprising demands of war demand versatile men and machines alike."[161] All these technicians are defined quite specifically in terms of their experiential qualifications for the work. A few additional remarks delivered to his addressee (whom he addresses in the vocative as δέσποτα) make the preface a lively epistolary space, full of detail about the processes of decision-making, training, and writing that went into producing the text that follows. However, the craftsmen and the imperial addressee are entirely creatures of the preface; they do not figure in the instructions given in the main body of the text.

In the preface, Apollodorus claims novelty (καινότης) for his work, but does not insist (as Philo does) on his own particular contributions to the devices whose construction is described in the text. The passages describing the construction of mechanical artifacts that appear most likely to have been Apollodorus' own work use constructions like Hero's "hands-off" passages: third-person imperatives, passive verbs, and δεῖ + infinitive constructions. For example, Apollodorus presents instructions for building a tortoise using passive verbs and third-person imperatives as follows:

> The construction of a tortoise is like this. Two crossbars are placed, separated from one another by no less than 12 feet, and beams are placed between each of the crossbars, 12 dactyls in thickness, one foot in width, and no less than 24 feet in length, 4 in number … Let other crossbars be riveted to these at the middles of the uprights, and let the side supports be laid under from the inside, holding up the middle crossbars and the one center ridge. (154.12–155.5)

[159] Apollodorus, *Poliorcetica* 137.3–6: Ποιήσας οὖν ὑποδείγματά τινα πρὸς πολιορκίαν εὔχρηστα ἔπεμψα διαγράψας, καὶ πᾶσιν ὑπελάλησα καὶ ὑπουργὸν ἀπέλυσα πάντα δείξας καὶ ἐπ᾽ αὐτοῦ ἐργασάμενος, ἵνα πρὸς τὰ ὑποδείγματα ὁμοίως, εἴ τις χρεία γένοιτο, ἐργάσηται.

[160] Apollodorus, *Poliorcetica* 138.7–9.

[161] Apollodorus, *Poliorcetica* 138.11–12.

The text also contains a few first-person plural verbs. These mostly refer to plans or preferences (βουλώμεθα, θέλομεν, θέλωμεν), along with a few generic phrases along the lines of "we'll build it (ποιήσομεν) thus." Sometimes, however, Apollodorus does use first-person plurals to refer to manual activities: drilling through a wall (148), devising a cover for a tower (173), or building a spy tower:

> In case we want to know the thickness of the walls or what is behind them (what the enemy is doing, or their number, or the opportune times and places), we will construct a spy tower like this for ourselves. (161.9–11)[162]

However, when he goes on to describe the actual process of constructing the tower, Apollodorus reverts to passive verbs:

> Two square beams of different widths are prepared, standing upright on beams that are as far apart from one another as the [square beams'] widths. Another beam is inserted between them on the ground, perpendicular to the one placed first. And from the ends of these beams, four stays support the upright beams from above at the ends. (161.11–162.3)

Instructions like these neither show Apollodorus at work, nor encourage the reader to imagine himself in that role; they contrast with the hands-on approach, which Philo uses consistently and which Hero dips into when he needs to explain a tricky manual operation. Whitehead labels this approach to issuing directives "deferential didacticism," saying that Apollodorus "needed to keep in mind the identity and status of his (notional) reader-in-chief."[163] This arm's length distance is perhaps to be made up by the technical practitioners, distinct from both author and addressee, who Apollodorus says will put his building instructions into action.

The descriptions in the main body of the text feature absolute dimensions rather than variable geometrical relationships, though the possibility of resizing the devices described is mentioned. Like Athenaeus, Apollodorus eschews geometrical abstraction in favor of well-defined, absolute dimensions. The diagrams in the surviving manuscripts of Athenaeus and Apollodorus reflect the lack of letter labels in the text; there are occasional word labels, but otherwise the devices are shown as wholes rather than sums of parts. By contrast, Hero hardly ever mentions the dimensions of the components he describes, marking them only with

162 It is worth noting that these instances fall in passages identified as likely interpolations in Apollodorus 2010. This of course does not mean that Apollodorus did not opt or would not have opted to use these forms where appropriate.

163 Apollodorus 2010, 33.

points on a lettered diagram as though they were as infinitely scalable as a purely mathematical construction.

The surviving texts on military machines suggest that artifacts were increasingly presented as distant from the author in imperial Roman literature as compared to Hellenistic literature.[164] Philo represents himself as a professional instructing a student colleague about the designs of another professional (Ctesibius) and the work he himself has done to improve on them. He emphasizes the chain of hands-on knowledge about those designs, which Philo claims to have obtained directly from those who had worked most closely with Ctesibius himself. In his own transmission of knowledge to Ariston, Philo invokes the possibility that textual transmission could form part of this chain, relying on the "simulated closeness" of his text. Biton presents his siege engines through historical exemplars, but still emphasizes his own expertise in a related technical field and evidently furnished considerable construction-oriented textual details in the form of the list of dimensions (*logothesia*).

Roman and Greco-Roman texts on military machines tend to downplay the direct connection between author and inventor, often emphasizing the texts and authors from whom they obtained their knowledge over the inventor of the device itself. Hero critiques the available texts on the material as inaccessibly technical, proposing to render the information into an introductory form suitable for propagation to a wide audience. Vitruvius does the same. Other authors of the imperial period like Athenaeus and Apollodorus represent themselves as writing for busy rulers, just as Biton had concisely dispensed information to his royal addressee, and as engineers today write "executive summaries" of their projects. Where engineering projects are controlled by technical laymen, status may come from generality and accessibility or from masterful verbal persuasion at least as much as from technological power.

The increased apparent distance between author and technology is accompanied by a partial fading-out of the first-person figure of the mentor who guides the reader in building the machine, and of his complement, the student addressee. This figure is everywhere in Philo and still strong in Biton, while in the later texts the author's first-person appearances are for the most part limited to remarks about the task of writing, rather than the tasks of building. In place of Philo's hands-on, personal instructive form

[164] On the transition in the first century CE from scientific treatises whose prefaces claim they are motivated by the dedicatee to treatises that claim they are motivated by the nature of the subject matter, see Alexander 1993, 75ff.

of address, the surviving Greco-Roman mechanical texts predominantly use an impersonal voice to introduce technological artifacts to the reader, effectively separating the reader from the object of description. Hero introduces direct, personal instructions to the reader within punctuated intervals of vividly "lived-in" construction scenes, separated by passages of impersonal or passive constructions. In the other Greco-Roman authors of this period, this mode of address is hardly present at all. Nor is this trend limited to military machines; these same authors handle the task of describing other kinds of machines using a similar dynamic of address and instruction.

The various assistants Apollodorus refers to would have been invaluable as carriers of the tacit knowledge learned by directly observing someone more skilled or experienced. Indeed, Meißner argues that these technicians embody experience, craft knowledge, and decision-making capacities that can only be transmitted in person.[165] Certainly this is at least partly true; there are aspects of any kind of manual work that cannot be completely encoded in text. Collins makes an important refinement to claims like Meißner's, distinguishing two types of such intermediaries. Some, who are "inscribed" with informational "strings," are bearers of knowledge imparted to them explicitly that they can parrot back to a recipient. Others have themselves been imbued through practice with tacit knowledge that they can in turn pass on through whatever non-linguistic activity they used to acquire it in the first place. Collins then makes the further point:

> One can say for sure that knowledge that can be passed on via intermediaries is explicit knowledge, but one cannot say that knowledge that is passed on without intermediaries is tacit knowledge. Close personal contact between teacher and learner enhances the transmission of explicit knowledge as well as tacit knowledge, so close contact proves nothing about the nature of the knowledge.[166]

What I want to call attention to here is not the acknowledged ubiquity of tacit knowledge in technical work, but the different ways ancient authors describing that work grapple with the challenge of including it in their texts.

The tacit knowledge possessed by such a practitioner would certainly also have been a vital supplement for anyone attempting to build the machines Philo describes, but his text is itself designed to facilitate

[165] Meißner 1999, 247. [166] Collins 2010, 87.

the process of coming to grips with the machine's material particulars. Philo creates a particular kind of *enargeia* in an attempt to help bridge this gap: the attentive and personal instructive environment, the insistent references to manual techniques and material properties, the "lived-in space" he creates. The instructions Apollodorus provides are much less detailed and multilayered than Philo's; he effectively outsources the task of bearing that tacit knowledge to his assistants, and does not even attempt to represent it in text. Vitruvius and Athenaeus Mechanicus reframe the problem, making the text itself the focus of their labors and relegating most of the material practicalities to the distant background. Hero's approach blends detached emphasis on authorial tasks with vignettes of manual activity where he, like Philo, attempts to create a scene sufficiently vivid to simulate the close and thorough knowledge transmission of personal collaboration.

Hero arguably does provide sufficiently detailed instructions for the reader to build the machines he describes, as has been done by Marsden and Schramm, among others.[167] However, this is not the main point of the text; Hero's objective is not to inspire an audience of civic-minded lay readers to build their own copies of long-outdated ballistic devices. He is instead working to situate mechanics in a new kind of competitive world of knowledge exchange, where innovative large-scale technology is not as highly regarded for its own sake as it reportedly was, for example, under technophiliac Hellenistic kings like Hiero of Syracuse and Demetrius Poliorcetes. In this world, as Meißner observes, systematization was seen as intrinsic to maintaining the continuity of the discipline and its contextualization as part of an established set of "sciences" (μαθήματα).[168]

Vitruvius, Athenaeus Mechanicus, and Apollodorus all likewise make explicit assertions about the civic value of the technical works they describe.[169] The brevity of their descriptions detracts from their ability to guide a reader in actually building the devices described, but is crucial to their claims to present mechanical technology as practical and straightforward, without wasting the patron's time. Athenaeus highlights the pride he takes in his work as editor and author, while Apollodorus modestly downplays his authorial talents, presenting himself first and foremost as a knowledgeable practitioner capable of getting just the right technology into his patron's hands. Apollodorus appears in his preface as an active

[167] See, for example, Baatz 1994, 216–17, 272–3.

[168] Meißner 1999, 183.

[169] On claims for the civic utility of military engines, see particularly Baldwin 1978; Meißner 1999, 95–9, 161–7, 192–4, 246–55, 277–89; Cuomo 2002.

figure, but in the instructions for building the devices themselves, he uses the same impersonal, hands-off instructional techniques as his immediate predecessors.

Rather than drawing the reader into a deep hands-on instructional space like Philo does, these authors show the reader many different types of artifacts, which by and large come together without the reader's being asked to imagine himself as a participant in their construction. This impersonal instruction technique moves away from guiding the reader to engage mentally and manually with the technology itself, and toward hands-off instructions, which instead guide him to think about mechanical technology as connected to relatively high-status disciplines like mathematics and philosophy, or about the civically valuable uses to which it can be put. All these texts are shaped by the need to integrate technical knowledge into a larger landscape of intellectual value, accessibility, and utility to the patron, the state, or the reading public, and it is this need that is served by their prologues and by the hands-off, impersonal way they address the reader.

Directive forms in other technical texts

The variations in personnel, forms of address, textual macrostructures, and so forth that these texts manifest are not at all limited to texts describing military machines. A full review of the forms of instruction taken in all surviving passages of technical ekphrasis is obviously not possible here, but a few examples will demonstrate the breadth of applicability of the descriptive-instructive techniques already seen earlier.

Ptolemy provides instructions for building several instruments used to collect harmonic and astronomical data. The harmonic *kanōn* was already encountered in the previous chapter; let us revisit the passage.[170] Ptolemy's linguistic patterns are reminiscent of Hero's. First, the device is described as a diagrammed object, defined geometrically through the use of mathematical terminology, letter labels, and third-person imperatives. Ptolemy, more than the authors on military machines, emphasizes the equivalency between the geometrical ideal of the design and the material device. This is not surprising, since the mathematical precision of the constructed *kanōn* is vital to its use as an instrument of mathematical proof.[171]

[170] Ptolemy, *Harmonica* 1.8.31–74.
[171] On this aspect of Ptolemy's language see Creese 2010, 70–1.

This is not to say that Ptolemy does not acknowledge the materiality of the instruments he describes. He gives them a physical presence by describing their materials, unpacking the tricky process of carefully dividing slabs of wood and tuning lengths of gut to control their sound as precisely as possible. The same is true for the astronomical instruments of his *Syntaxis mathematica*. Like Hero, when Ptolemy delves into the material minutiae of an instrument's construction, he shifts to the first person, creating the "lived-in space" that seems so essential for discussions of hands-on craft activity. Sometimes these are hortatory subjunctives; elsewhere he uses first-person aorists to personalize an account of discovery, construction, or refinement. Here he describes an instrument used to find the angle between the equator and ecliptic, which he praises for its simplicity of construction and use:

> We made this observation even simpler to execute by constructing, instead of the rings, a block of stone or wood, square and rigid, having one of its sides smooth and precisely sized up. Using a certain point toward one of its corners as a center we inscribed a quarter circle on it, joining from the point at the center to the inscribed circumference the lines containing the right angle subtending the quarter circle …
>
> Then we set up the inscribed side of the block along a meridian line drawn on the base plane so as to have a position parallel to the plane of the meridian, and with a plumb line hung between the cylinders, perpendicular and straight to the base of the horizon, we accurately found the line between them, again using small shims to correct any shortfall.[172]

Ptolemy frames his description not just as a finished device constructed by some unidentified figure, but as an artifact refined in design from an existing model, carefully crafted, and accurately positioned, all by Ptolemy himself. Many attributes of this device need to be manufactured to precise geometrical specifications: its edges must be exactly perpendicular, and they must be precisely lined up with the notional axes of the astronomical bodies that the instrument is built to observe. He notes that in order to attain the desired precision, shims may be required to counterbalance any accidental departures from the foursquare. He praises this device for its ease of manufacture: it is structurally simple, and only one side needs to be smoothed down.

The combination of material detail with first-person plural aorist verbs suggests to the reader that these findings resulted from an actual process of construction in which Ptolemy himself participated. Creese finds this

[172] Ptolemy, *Syntaxis mathematica* 1.1.66.5–67.8.

approach typical of Ptolemy's descriptions of astronomical instruments in the *Syntaxis*, where descriptions of past construction processes are more common than directives to perform new activities.[173] By contrast, in the *Harmonica*, instead of an account of a past process of construction the reader is usually exhorted to imagine an instrument coming together through imperatives and future indicatives, along with hortatory subjunctives.[174] For example, after reviewing the problems arising from the variable thickness of gut strings, Ptolemy proposes a method for assuring predictable performance from the *kanōn* string:

> Then, if the segment EK of the string is found to have an equal tone to KH, and also KΛ to ΛH, it is clear to us that it is indistinguishable in density. If not, let us change the test to another segment, or indeed another string, until the following is maintained, namely the identity of tone in [segments] that are similar and proportional and of equal length and which have the same density.[175]

The string is mapped onto a lettered diagram, but at the same time put into an experimental scene populated by Ptolemy and his reader. Its evenness or lack thereof will be "clear to us," and if it is found wanting in this respect "we" have work to do. The verb forms Ptolemy chooses determine whether he sets a scene of past experimental activities or proposes a new construction. The first approach offers the reader assurance that Ptolemy himself enjoyed success performing a certain set of procedures, with the implicit suggestion that the reader might do the same – the pattern Gibson describes as "I do this (and you can/should too)."[176] The second approach, whether delivered through imperatives or hortatory subjunctives, draws the reader closer to the action by making that suggestion explicit.

Ptolemy may not have been building military machines, but these passages are still about building instruments. Do the same patterns hold if we move further away from this subject matter? Galen makes frequent reference to the hands-on work of dissecting and otherwise exploring animal and human bodies. As in Philo and Hero, this manual work demands the presence of a human operator in the text, established by second-person verb forms and personal pronouns. Whether this operator is learning the

[173] Creese 2010, 56–7, 63–8.

[174] Creese 2010, 69. Ptolemy does occasionally give a first-person account of a past experimental process for his harmonic investigations as well; for example, in describing a method for making the *kanōn* easier to tune he says "we divided the remainder into twenty-five segments ..." (*Harmonica* II.16.20–2).

[175] Ptolemy, *Harmonica* II.8.51–6. [176] Gibson 1998, 71.

first principles of diagnosing from the pulse, or being guided through a dissection of the arm's musculature, this hands-on work requires actual hands, guided directly by the "voice" of the mentor.[177]

Like Hero, however, Galen tends to use these passages of guided manipulation sparingly within texts that largely steer clear of directly addressing the reader–anatomist, preferring an impersonal mode of address. The addressee is sometimes portrayed as engaging actively with the insights into the order of the universe that Galen's anatomical instructions claim to provide. In this vein, Galen promises in *On the usefulness of parts* that "you will find the same wisdom, power, and providence" in the motions of the cosmos as shapes human and animal bodies.[178] While Galen does not at this moment describe the anatomical manipulations that will yield such a discovery, it does fall under the heading of a personal act of observation, which, as we have seen, frequently involves first- or second-person verbs or pronouns. Nutton links these phenomena, suggesting that "sometimes the reader is addressed directly in some form of the second person, sometimes he or she is invited to join the writer in 'our' search or associated in matters that 'we' know or conclude together."[179] More often, however, the act of observing is less marked, and inducements to observe the connections between medicine and other bodies of knowledge are indicated through impersonal directive structures, like δεῖ + infinitive or -τέον constructions.

In Galen as in other Greco-Roman technical authors, personal verbs and direct imperatives are most common in passages offering guidance in hands-on interactions with material objects. In many cases, authors who elsewhere choose impersonal or third-person verb forms resort to the use of direct imperatives for these interludes of hands-on engagement. In Galen's case the body is the site of engagement, in dissection, pulse-taking, etc.; in Ptolemy's it is scientific instruments that need to be adjusted; in Hero's works certain mechanical components that have to be manipulated ad hoc. Technical work that involves this sort of hands-on engagement complicates the picture of objective, impersonal "distance" that Asper sketches in the case of mathematical texts. Other authorial

[177] For example: on the pulse, Ἁπτομένῳ δέ σοι φανεῖται διαστελλομένη κατὰ πᾶσαν διάστασιν ἡ ἀρτηρία (Galen, *De pulsibus libellus ad tirones*, Kühn v. 8.455.1); on the musculature of the arm, εἰ οὕτως πράξαις, εὑρήσεις ὅλον σαρκώδη διαμένοντα τὸν μῦν τοῦτον (Galen, *De anatomicis administrationibus*, Kühn v. 2.278.1). On the unusual predominance of second-person forms of address in the *De anatomicis administrationibus*, see Nutton 2009, 58.

[178] καὶ γὰρ σοφίαν καὶ δύναμιν καὶ πρόνοιαν ὁμοίαν εὑρήσεις ἐνταυθοῖ; Galen, *De usu partium*, Helmreich 1.175 (Kühn v. 3.238.11–13).

[179] Nutton 2009, 57.

tasks have their own typical verb forms, as when the author appears as an editorial or polemical *ego*. Galen makes particularly liberal use of the first person when he is staking argumentative claims. Even Galen, however, more often relies on impersonal directive forms; this is true as well, to a still greater degree, of other post-Hellenistic technical authors.

Conclusion

Technical authors in Greek and Latin had many different options for guiding the reader to imagine the objects – living creatures, works of art, technological artifacts, or whatever else – encoded in verbal descriptions. The reader is often explicitly directed to imagine or picture something; this is a commonplace of description, and does not necessarily draw the reader's attention to the link between description and instruction that is being forged. The directive work the text does to connect the reader to the object is more obvious when the object appears through instructions to "imagine" or "see" an element of its design, or even instructions for building it. As Hindle indicates, this kind of visualization has been of lasting importance to the process of mechanical design, which "requires mental arrangement, rearrangement, and manipulation of projected components and devices."[180]

Directive texts are particularly vivid when a particular viewer is established by the text through the use of direct imperatives or personal pronouns. The addressee, enjoined by the author to explore the objects of the text visually (and perhaps also manually), becomes a kind of surrogate for the reader, who vicariously performs the same visualizations. An author may also place himself in the role of an explorer or builder with prior experience, as Ptolemy does when he recounts his building activities in the first-person past tense. Establishing a personal presence contributes to the creation of a "lived-in space," where reader and author are established by the action of the text as working, perhaps collaborating, on discovery, experimentation, or construction. Even Hero, who typically delivers his instructions for designing machines via impersonal constructions, resorts to personal pronouns and verbal constructions when he shifts to the hands-on details of crafting components.

The alternative to creating a "lived-in space" is to depict the construction or use of an artifact through passive or impersonal verb forms. Many different types of verb forms can be deployed in this way, but in all of

[180] Hindle and Lubar 1986, 75.

them the encounter the text creates between reader and artifact is substantially changed from the "lived-in space" approach. Some texts use the distance imposed between reader and object to locate the artifact in a larger disciplinary or interdisciplinary context. Impersonal verb forms may also be used to address instructions to a patron or another audience where explicit directives might be inappropriate, to suggest a kind of objectivity, or to assimilate their material in some way to mathematical prose. Even the less personal styles of directive, however, still create a kind of virtual, textually mediated encounter between reader and artifact: the reader is still guided to engage with the object in his imagination, even if that does not mean imagining grappling manually with its material components.

CHAPTER 6

Knowledge and Artifact

Introduction

The layered bodies of knowledge that go into making technological artifacts and encoding them in text can be usefully differentiated into knowledge about their "inputs" and "outputs." Some texts focus on the knowledge that goes into designing and building an artifact: familiarity with the peculiar properties of materials, craft knowledge, experience deploying them in the field, and so forth. Others emphasize how careful attention to the structure and function of certain types of artifact can open up bodies of knowledge beyond the artifact itself, like astronomy, physics, biology, or mathematics. These two approaches to the knowledge embodied by technological artifacts can of course be combined in a single text, but even so they are usually separated formally as well as spatially – the former is more often accompanied by first-person accounts of the author's own experiences, for example, while the latter type is more likely to bear "mathematical" textual markers.

Useful as it is to distinguish between these two approaches, in doing so we run the risk of contributing to the attractive falsehood of a dichotomy between "scientific" and "technological" knowledge. In the discourse of the history of science, this dichotomy owes something to the fact, observed by de Solla Price, that "so many historians of science and virtually all of the philosophers of science are born-again theoreticians instead of bench scientists."[1] Many studies of ancient scientific work do indeed focus solely on the observational "outputs" that fed theoretical scientific knowledge, treating the instrumentation that practitioners used as an afterthought (if at all), subsuming them into a narrative that is decidedly "scientific" rather than "technological." Such studies discount the vast amount of practical craft knowledge that goes into the design and construction of

[1] de Solla Price 1980, 75.

those instruments. A thorough understanding of the scientific work that instruments enable requires understanding them on their own terms as crafted objects.

Gille characterizes as "abusif" the assimilation of technical knowledge to scientific knowledge; the two may well inform each other, but they each have their own peculiar character.[2] Studies of the epistemology of modern technology may focus on the "autonomy" of technological knowledge, as Vincenti does, and such studies have the positive effect of arguing that the epistemology of technology is a topic worthy of serious study in its own right.[3] However, arguments for the stronger claim that scientific and technological knowledge have nothing to do with each other have thus far failed to find purchase.[4] Moderate approaches that seek to articulate the distinctive qualities of technological knowledge while acknowledging that one of those distinctive qualities is in fact its interoperability with scientific knowledge have fared better. For example, Faulkner defines a spectrum of types of knowledge associated with various activities having to do with the production, use, and exchange of modern technology.[5] At the same time, however, she observes that technology contributes to science in a variety of ways, through the category of "knowledge relating to the natural world."[6] This chapter will attend to both the "input" and "output" knowledge that technical ekphraseis convey, with an eye for the factors that distinguish them from each other as well as the connections between them.

Both a "scientific" and a "technological" perspective are necessary to fully understand mechanical artifacts. Caws proposes, for both the practical and theoretical domains, "a mode of immediate interaction of man as knower or agent with the world, and another mode of activity that is independent of such immediate interaction and which in *both* cases I would call intellectual."[7] The "practical" side is the domain of *praxis*, which is continuous with intellectually mediated "technology" (the term here referring not to the artifact itself but to the knowledge it embodies). On the "theoretical" side, *empeiria* (gained from "an experimental acquaintance with things") is connected to intellectually mediated *theōria*.[8] This scheme

[2] Gille 1980, 199. [3] Vincenti 1990, 1.

[4] See Houkes 2009 for a detailed study of such recent approaches.

[5] Faulkner 1994, 442–7. [6] Faulkner 1994, 446–51.

[7] Caws 1979, 231.

[8] Caws cautions that this pairing should not be artificially neatened; there might, for example, be active aspects to *empeiria*, or the term *epistēmē* might be preferred to *theōria*, and so forth. The crucial parallel is between the complementary acts of encountering the world physically and theoretically, associated respectively with technological and scientific activity. Caws's use of Greek terms does

reconciles the "technological" and "scientific" as two different ways of intellectualizing one's encounters with things in the world, whether those encounters take place via the hand or the eye.

Kogge's treatment of the related problem of tacit knowledge refines and adjusts the terms Caws uses in some useful ways. Kogge interprets *empeiria* in its Aristotelian context as unifying *logos*, *epistēmē*, *technē*, and *sophia*, and suggests that as such *empeiria* is a good match for what we call "tacit knowledge."[9] While Kogge is primarily interested in the term's virtual disappearance from twentieth-century theoretical discourse, his analysis of *empeiria* in its Aristotelian context clarifies the enduring and multifaceted connection between practical and theoretical knowledge. *Empeiria* as Aristotle defines it in the *Metaphysics* comprises sensory perception (*aisthēsis*), memory (*mnēmē*), and a capacity for understanding and learning mediated by hearing.[10] In this view, a unified *empeiria* is produced by accumulating multiple memories about the same thing. Kogge points out in particular the relationship between *empeiria* and *technē*: the former produces the latter through another process of accumulation, when a single universalizing judgment about similar objects arises from many "concepts of experience" (ἐκ πολλῶν τῆς ἐμπειρίας ἐννοημάτων).[11]

As Kogge points out, this coheres closely with Aristotle's claim in the *Posterior Analytics* that *technē* and *epistēmē* emerge from *empeiria* of the "one beside the many" (τοῦ ἑνὸς παρὰ τὰ πολλά).[12] In both texts Aristotle links *technē* and *epistēmē* together, and both to *empeiria*, and *empeiria* in turn to a linguistically mediated process of learning.[13] The strong Aristotelian connection between *empeiria*, *technē*, and *epistēmē* is a reminder that "scientific" and "technical" knowledge ought not to be artificially divided. Behind neat scientific results and reliable engineering achievements lies a vast amount of experientially acquired craft knowledge and refinement, much of which may lend itself to textual encoding only partially or not at all.

not imply that he is particularly interested in Greek technology, or that he is striving to return discourse on technology to some imagined Aristotelian roots. For a more detailed look at Aristotelian and Platonic categories of knowledge in this context, see Owens 1991.

[9] Kogge 2012, 33–8.

[10] Aristotle, *Metaphysica* 980a27–981a1.

[11] Aristotle, *Metaphysica* 981a5–6; Kogge 2012, 34–5.

[12] Aristotle, *Analytica posteriora* 100a3–13.

[13] Kogge proceeds to track developing analyses of a complementary kind of knowledge outside the limits of what can be encoded in language; he localizes its first appearance to Bacon, whom he views not actually as opposing Aristotle (since Bacon's process of induction from sensory perception, through memory, to the formation of concepts precisely mirrors Aristotle's *empeiria*), but rather his Scholastic reception (Kogge 2012, 38–9).

Thinking about artifacts

Philo's *Belopoeica* is a text firmly focused on the knowledge "inputs" of the artifacts he describes. He consistently emphasizes the mechanical and craft knowledge built into his catapult designs, while keeping that knowledge tightly focused on the catapults themselves, rather than using it to motivate a treatise on materials science or ballistics. The structure of the treatise, which builds up the catapult's description by layering various bodies of knowledge on top of one another, contributes to the development of deep knowledge about the artifact itself. Philo alludes to different categories of catapults (torsion catapult, wedge-stretched arrow-shooter, bronze-spring, repeating catapult, and air-spring), but the text is not a catalogue of briefly described catapult types with historical exemplars, like Biton's. Nor is it an evolutionary history of catapults designed to locate mechanics in a civic and intellectual context, like Hero's.

Philo's approach highlights some complexities in the "oral" and "literary" structural markers that Asper defines for technical texts.[14] In many respects, Philo corresponds more closely to the "oral" indicators: he makes extensive use of references to the reader and to himself as a kind of collaborator; moreover, the text consists of only a few sections, which are not heralded individually with any kind of formulaic language. On the other hand, Philo does make use of some textual elements that fall nearer Asper's domain of the "literary": he uses a few tables and diagrams, issues a list of "headings" (κεφάλαια) corresponding to qualities desirable in a catapult and discusses each one in its own subsection of the work, and the text as a whole is framed both as a letter and as a part of a larger group of texts.

The work this text will do to develop and transmit technical knowledge is highlighted from the very beginning, where Philo tells Ariston that he is about to embark on the acquisition of knowledge that is "unintelligible and baffling to many people."[15] But Ariston will be able to learn it, with Philo's help, if he observes and follows the principles of design and calculation that Philo is going to teach him via the text. He begins by invoking the sculptor Polyclitus, reporting that "he said the good comes gradually through many numbers" (τὸ γὰρ εὖ παρὰ μικρὸν διὰ πολλῶν ἀριθμῶν ἔφη γίνεσθαι).[16]

[14] Asper 2007, 27–35. The former category includes irregularities like anacolouthon and references to a listener, while the latter includes regular substructures like enumerated lists, exact repetition of formulaic language, and schematized structure.

[15] Philo, *Belopoeica* 49.18–20.

[16] Philo, *Belopoeica* 50.7–8, tr. Cuomo. On Polyclitus' text as a precursor to post-classical "art history" written largely by non-artists, see Tanner 2005, 215–18. On the role played by the *Canon* in facilitating accurate copies of itself, see Tanner 2005, 288.

Philo is referring here to Polyclitus' *Canon*, a text prescribing the correct measurements and proportions for a sculpted human figure, supposedly embodied by the *Doryphoros* statue.[17] Tanner describes Polyclitus' emphasis on numerical precision as "an enthusiasm for measurement and number, which were preoccupations, even obsessions, of fifth-century intellectual culture."[18] In particular, Tanner finds in the detailed part-by-part prescriptions from the surviving fragments of the *Canon*, and in the anatomical detail of the surviving copies of the *Doryphoros*, evidence for the strong influence of medicine and mathematics on classical-era sculpture.

Philo, who had his own disciplinary priorities and worked within a Hellenistic cultural environment that differed significantly from that fifth-century world, repurposes Polyclitus' claims about numerical precision to make an argument about how technical knowledge must be developed and handled. To this end, he immediately reiterates Polyclitus' tag, repurposing it as a claim about his own discipline: "in the same way it happens that when works are completed through many numbers, those who make a small piecewise deviation end up distilling it into one enormous error."[19] If engineering is to proceed on the basis of anything other than laborious trial and error, all activities of design and construction have to be scrupulously correct. An error in any one calculation or measurement is compounded again and again as the design process goes on, and the ultimate result may be quite inaccurate.

Philo tells us a lot about how he believes technical knowledge is acquired and transmitted, both ideally and in practice. He provides a kind of speculative, teleological history of early catapult engineering, beginning on the threshold of the game-changing discovery of the relationship between the hole for the spring and the ballistic power of the machine:

> Some of the previous generation discovered the governing element, the foundation and measure of the construction of engines, namely the diameter of the hole. This had to be grasped neither by chance nor by guessing, but rather through an established method capable of yielding the

[17] On the type of numerical information the *Canon* might have provided, on other "canonical" sculptures, and on the controversy over whether the *Doryphoros* is in fact the target of the text, see Tanner 2005, 161–70. On the cultural preconditions for Polyclitus' text to have been produced in the first place, and its relationship to post-classical "art history" produced in general by non-artists, see Tanner 2005, 215–18.

[18] Tanner 2005, 164.

[19] Philo, *Belopoeica* 50.8–13. Likewise, Lucretius transitions in *De rerum natura* IV.500–21 from the optical illusion of roundness produced by viewing a faraway square building, to the problem of propagating and amplifying calculation errors through the process of constructing a building, where any slight error with the *libella* or *norma* is apt to bring the finished building to ruin.

> proportions of all sizes in the same way. It was not possible to determine this in any other way except expanding and reducing the circumference of the hole experimentally. I note that the previous generation did not bring this to its conclusion, nor did they establish the size, since their experience did not come from many trials, but anyway the matter was investigated.[20]

At some point, he says, these experiments yielded more methodical understanding of the factors that affect the performance of a catapult, specifically the role of the spring hole as the basis of the proportions of the rest of the catapult's components. Philo's connection of this discovery with the tag from Polyclitus is no accident, as the *Canon* was itself, at least in part, a text on proportion.[21] The next generation of engineers knew to search specifically for a factor that would create a predictable link between the size of the engine and its ballistic capacity; they carried out this search methodically, learning from errors and shaping the next series of experiments around them.

Philo regarded the professional opportunities and funding that were available to engineers in Alexandria as crucial to their progress from trial and error to analytical, reproducible knowledge. He does not explain what exactly enabled this methodological leap forward, but the model of experimental learning he describes depends so heavily on trial and error that he may simply be referring to the brute force such well-funded programs could make available to the problem of improving the catapult. He emphasizes that "not everything can be grasped through reason and mechanical methods, but many things are also found through experience."[22] That is, even this orderly experimental process never leads to a point where the problem can be handled completely by theory: hands-on experimentation, and the tacit knowledge it delivers, remain a vital part of the design process.

Other technical disciplines, for example architecture, likewise rely ultimately on a refining process of trial and error. The architects of the past, he claims, were "superlatively unskilled" (καθ' ὑπερβολὴν ἀτεχνῶν), having no knowledge of perspective and thus producing buildings that looked irregular and misshapen from the perspective of a viewer on the ground.[23] Heisel corroborates Philo's assessment of the shortcomings of Greek architectural drawings at this stage; reduced-scale drawings likely could not even have been produced for a long time because the details of

20 Philo, *Belopoeica* 50.20–32.
21 On the complex interaction of the practical and theoretical aspects of proportion in the *Canon*, see Tanner 2005, 166–9.
22 Philo, *Belopoeica* 50.21–32.
23 Philo, *Belopoeica* 50.49.

proportions or the corrections for optical illusions could not be properly represented.[24]

Gradually, Philo says, practitioners learned to enlarge or reduce components in such a way that the building would look symmetrical to the viewer. Developing these skills required exhaustive testing, but eventually produced a consistent methodology, which spared later architects the trouble their predecessors had had. For Philo, the goal is to work past the phase of blind experiment, to extract methodological knowledge from repetitious craft experience, and to work thereafter with that knowledge in order to get results more efficiently – a very close parallel to Aristotle's representation of the development of *empeiria*. This is the principal purpose that shapes the *Belopoeica*: to deliver this knowledge, structured in a framework that gradually assembles the different types of expertise required to design a catapult in the methodical fashion of which Philo approves. Cuomo connects Philo's reference to the sculptor Polyclitus at the beginning of his work to his concern for slow, careful work: learning from the mistakes of one's predecessors is what allows experience to triumph over chance.[25]

Philo's framework of streamlined knowledge about the catapult is built on a more or less arithmetical foundation. He commences with an explanation of the mathematical techniques that Ariston will need to calculate the size of the hole for the spring, which in turn determines the size of the missile the catapult will take:

> Take the weight of the stone for which the machine must be built; reduce it to units, and however many units are in the cube root of the number obtained, make the diameter of the hole that many fingerbreadths – with the addition of an extra tenth of the cube root found. If the weight does not have a rational root, take the closest number; if it is more, then try to lessen the tenth part proportionally; if it is less, then supplement the tenth.[26]

So, if the weight of the missile in drachmae is w, the hole diameter is found by rounding $\sqrt[3]{w}$ to the nearest integer, and introducing a multiplying factor of 1.1, plus or minus a "fudge factor" intended to offset the initial rounding of the cube root. This arithmetical operation is underdetermined in the text, and not mathematically exact – even in the initial calculations the reader is thrown into the craftsman's world of approximations and rules of thumb, regulated by the engineer's drive to get the most predictable results from such inexact methods.[27]

[24] Heisel 1993, 156. [25] Cuomo 2007, 51. [26] Philo, *Belopoeica* 51.23–34.
[27] Mitcham 1994, 193–9.

These rules of thumb are not deduced from mathematical first principles, but are rather generated iteratively through long practice and testing – the "many numbers" Philo invokes at the beginning of the treatise. Extracting even an approximate formula like this one through repeated testing would have been a grueling process, timed for no one's convenience and with no certain end. Philo's simple formula omits the details of its development; indeed, many of these details are probably elided out of necessity, developed as hands-on tacit knowledge that simply cannot be articulated verbally.

From this starting point in arithmetic, Philo proceeds to a description of the hole carrier. He maintains the elements of numbering and calculation familiar from the previous section, while also introducing a geometrical construction:

> The form of the hole carrier is rhomboid and not rectangular; moreover, since it has two legs that are not straight but rounded, it is not to be drawn by guesswork, but this too by a certain method. On a straight, precisely smooth plane, it is necessary that we take a compass and draw a circle of any size, not deviating at all; drawing a diameter across it, divide the circumference of one semicircle into eleven parts, and taking four parts from the diameter draw a line to the center, and this will be the acute angle of the hole carrier. Transferring the angle from the circle to the design, you will make the length, width, and height of the hole carrier by taking the measurement from the diameter of the hole, just as described in the arrangement.[28]

The reader maintains focus on the catapult's mathematical design principles, taking the diameter of the hole that was just arithmetically determined (within a certain margin for error) and imposing a geometrical superstructure on it. Once again, Philo emphasizes the importance of exactitude and of following the correct method; geometrical as well as arithmetical knowledge has to be handled precisely. This geometrical construction will later be translated into the material domain and used as a template. After providing an alternative geometrical construction method for shaping the hole carrier, Philo then introduces the mathematical process of sizing the rest of the device based on the size of the hole, including a variety of methods of scaling up from an arbitrarily small model to a full-scale catapult of any dimension.

The next part of the treatise (56.25–62.39) builds a layer of mechanical knowledge onto the mathematical framework, as Philo begins with

[28] Philo, *Belopoeica* 52.30–46.

a list of headings (kephalaia) denoting desirable mechanical features of the wedge-stretched catapult, which he claims to have developed himself from an older model. The reference to kephalaia might suggest that this text could qualify as a "Lehrbuch," and indeed Fuhrmann does briefly mention Philo's *Belopoeica*, on the grounds that the work forms part of the *Mēchanikē syntaxis* and that it contains these kephalaia.[29] It certainly is true that this work is but a part of the systematic-sounding *Syntaxis*, though since most of the other components do not survive we cannot know what parallel internal structures the other books might have had. Philo's *Parasceuastica et poliorcetica* does not have any *kephalaia*, and the *Belopoeica*'s references to his text on levers do not suggest that such headings were used there either. We also cannot know much about how the *Syntaxis* was put together as a whole, though as Fuhrmann notes the language at the beginning and end of the *Belopoeica* suggests some effort to link each book to the rest.[30] Philo does at least, as Fuhrmann observes, make use of consistent vocabulary between the initial list of *kephalaia* and the corresponding passages of text.[31] In any event, even if the work does not meet the precise criteria of a Lehrbuch, it is nevertheless carefully structured to unfurl the knowledge associated with the artifact so that each stage builds upon the last.

Philo's *kephalaia* include long-range firing, ease of construction and adjustment, quick assembly in the field, and impressive appearance. He explores each in turn, beginning with a more or less detailed and practical statement of the problem, allegedly based on his own experience. Each problem comes with a solution, making use where possible of mechanical explanations based on the principles of leverage, and of linear and circular motion more generally. This mechanical discussion is always grounded in the material properties and limitations of the catapult's components, rather than abstracting the mechanical principles from the material objects, but Philo does consistently make the effort to reach his final conclusions based on an unvarying set of mechanical laws.

Those laws are not systematically introduced in this treatise. Another part of the *Mēchanikē syntaxis*, such as the book on levers (*Mochlica*), might have included a systematic, proof-based exploration of the laws

[29] Fuhrmann 1960, 169–71. The *kephalaia* occupy only a relatively small part of the work, so that the work as a whole does not meet Furhmann's criterion of "Klassifikation" (18–22). Indeed, even this section probably does not qualify as an inset "Lehrbuch," as Philo does not go through the *kephalaia* in a way that remotely resembles Fuhrmann's approved processes of enumeration, definition, and discussion (26–7).

[30] Fuhrmann 1960, 169–70. [31] Fuhrmann 1960, 170.

of mechanics, but nothing of the kind survives in the remaining texts. However, the *Belopoeica* does allude to a system of mechanical principles described elsewhere in the *Syntaxis*. Philo begins his explanation of the catapult's throwing force by appealing to the principle of the lever, which he hopes will already be familiar to the reader from the *Mochlica*: "larger circles surpass smaller ones drawn about the same center, just as we showed in the *Mochlica*."[32] He then maps the parts of the catapult onto the theoretical construction: "the arm is an inverted lever, since the inner part of it is the fulcrum, and the weight is the bowstring, which as it is connected to the end of the arm discharges the load."[33] Spacing out the spring cords will move the "fulcrum" nearer the load and further from the force applied, increasing the mechanical advantage of the lever analogue. Thus the missile will be discharged with maximum force.

We do not know what form the demonstration (Philo uses the word ἀπεδείξαμεν) in the *Mochlica* might have taken. Perhaps it was mathematically rigorous, at least to the extent of showing the proportional relationship between a lever's length and its lifting power. Here, however, Philo does not present these laws with numerical or formulaic precision; instead, he explicates the physical principles of the catapult in a distinctly practical, hands-on fashion. The spring cords are not to be spaced a certain distance apart, nor even to occupy a certain fraction of the available space. We are told merely that they are to be spaced "as widely as possible," a measurement that is unique to each individual machine.

Philo's preference for experimental methods, refined at the scene of construction, over attempts at theoretical predictions is also evident in his instructions for building the spring itself and arranging the cord. Here, too, the technical details must be determined on a case-by-case basis at the time of construction. Philo cannot describe how much cord the spring should incorporate, only that it should have as much as possible without needing holes large enough to weaken the hole carrier:

> Since it is necessary for anyone who intends to shoot far to insert as much spring cord as possible (for it is not only we who consider power to consist mostly in this, but everyone else too), but the spring cord issues through the holes of the hole carrier; thus it will be necessary to make the holes of the hole carrier larger, for otherwise it will not have room for more spring cord.[34]

Philo once again locates his work in a universe of hands-on mechanical tweaking rather than abstract geometrical perfection. Here he explicitly

[32] Philo, *Belopoeica* 59.16–18. [33] Philo, *Belopoeica* 59.24–8. [34] Philo, *Belopoeica* 56.50–57.2.

abandons even the approximation of mathematical certainty; proof is replaced by belief, backed up by the assertion that "everyone else" (other designers? other builders?) believes the same thing. He then details the problems with this approach, which are similarly unquantifiable: enlarging the holes allows for more cord to be inserted, but "the surrounding edges are left extremely thin and are unnaturally weak." The particular balance between shooting power and component weakness must be found by experience.

In keeping with this hands-on approach, Philo attempts to develop in his reader a physical understanding of force and elasticity that is not precisely quantified or formulized. It is instead a kind of physical intuition, ultimately born from and refined through practical experience, but nevertheless presented as something that can be inculcated, at least in part, through the text. The reader's design intuition is developed, for example, in the passage where he is instructed to evaluate the strength of the hole carrier (57.24–30) by holding it up and (literally) eyeballing an estimate of its strength. This quick and informal approach to calculation relies more on hands-on experience and rules of thumb than rigorously proven mechanical principles. The same is true for the stone-throwing catapult: the required quantity of spring cord is "undetermined" (μὴ ὡρισμένον) (54.35), though Philo estimates it as weighing something in the neighborhood of twenty-five times the weight of the stone missile.[35] The non-quantifiability of the forces Philo describes here creates tension in the text, as their resistance to quantification is also a resistance to verbalization.

The tension becomes clearer still when Philo weighs in on the relationship between shooting force and the degree of twist in the spring. After the catapult is fired, the spring slackens and requires re-stretching. Philo critiques the common approach to this process as "absolutely useless, artless, and harmful to the shooting,"[36] as it involves the application of excessive (παρὰ φύσιν) force:

> It happens that those who want to tighten it can neither perform the re-stretching vertically or on a straight line, but instead by twisting, give a greater twist than is appropriate; they take this as helpful, when in fact they are seriously harming the tension and, I argue, slowing down the shooting and weakening the impacts. The device becomes slack because

[35] Marsden observes that as the same figure has just been given for the weight of the catapult's metal fittings, these quantities are "a little suspicious" (Marsden 1971, 161 n. 30).

[36] Philo, *Belopoeica* 58.5–6.

> the strands are bunched up into a spiral, and the spring becomes warped and is deprived of its natural force and tension because of the excessive twisting.[37]

Philo's approach here demonstrates the need for the mechanical engineer to engage with phenomena that lend themselves easily neither to formulaic nor textual encoding. The general process of tightening the spring to which Philo refers here is clear enough, but the particular phenomena to which he alludes (such as the bunching of the sinew strands) are easier to grasp by a quick hands-on experiment than from the text alone. It is difficult to verbalize precisely how much of a twist the "excessive twist" represents; Philo can only provide a sketchy recreation of an essentially craft-based process. Rather than presenting this as a type of proof (for example, with a form of *apodeiknumi*, as he uses for the claim about levers from the *Mochlica*), Philo makes a less rigorous claim: "I argue" (λέγω) that this is so.

Shortly afterward, Philo again addresses the question of the proper method for constructing the spring from stretched sinews. Here, rather than being concerned with exactly how far they should be twisted, he argues for the importance of stretching all the strands together and equally. Here he again appeals to his *Mochlica*, which apparently went beyond the principle of the simple lever:

> And drawing in and stretching the strands – not one by one, disorganized, with different tensions in the stretching, but encircling all the strands by hand and then stretching them equally, with the same tension as one another, using to stretch them the greatest force demonstrated in the *Mochlica*: I will give them the greatest twist they naturally allow, which will persist throughout and cannot in any way be lost.[38]

Once again, the spring is supposed to undergo a process that cannot be verbally encoded, but must be tried with the hand – the greatest natural twist in the spring cord cannot be predicted, but rather the line between insufficient and excessive twisting must be felt. Philo uses the theoretical knowledge of leverage that he had developed in the *Mochlica* as a jumping-off point for improvements in the practical effectiveness of the catapult. A problem in physics is made the basis of a problem in engineering; whatever the mathematical basis of the analysis may have been in the *Mochlica*, Philo does not explicitly carry it over into its application to the catapult spring. Knowledge of the lever continues to inform the design

[37] Philo, *Belopoeica* 58.11–22. [38] Philo, *Belopoeica* 61.16–24.

of the spring, but its representation here is more practical than analytical, more material than mathematical.

The significance of Philo's approach to mechanics may be underscored by a brief digression into a difference between how Hero and Vitruvius treat the problem of the lever. As Fleury observes, Vitruvius does not distinguish in vocabulary between the lever as a tool and the principle of the lever, although he does distinguish between the fulcrum (*hypomochlion*, *pressio*) and the lever's axis of rotation (*centrum*).[39] Hero, on the other hand, does distinguish between the lever's practical–material and theoretical aspects, and his subsequent discussion of the principle is mathematically precise, unlike Vitruvius'.[40] Fleury notes that Hero finishes his theoretical discussion of the lever with a note on the theoretical principle of the balance, while Vitruvius follows a description of a lever with a description of a balance. Although Vitruvius and Philo are writing two very different treatises, neither includes a rigorously quantified or formulized explanation of the mechanical principles at work. For Philo, this appears to be because he has explained it elsewhere and does not want to digress too far from the work at hand.[41] Vitruvius claims to be writing for an audience of elite laymen, so that mathematical rigor is less important than accessibility. Both authors make a conscious decision not to elaborate on mathematical content that they understand perfectly well.[42]

Philo later moves on to discuss the action of the springs as they relate to the catapult's projectile capacity. He recontextualizes his practical considerations about the design of the spring, now combining them with physical knowledge expressed in loosely quantitative terms. He establishes a chain of rough proportional relationships between the weapon's range, the force of its throwing arm, the force of the spring, and the velocity of the spring's movement. The firing range is determined by the force of the arms, which in turn depends on the springs. Since the arm is in the middle of the spring, half the spring is on the outside of the frame and drives the arm backwards and in, while the other drives it forward. Having sketched out the problem, Philo proceeds to make an assertion about the relationship between velocity and force:

> I claim therefore that whichever of the arms can most quickly recoil will send the missile farthest. For the swiftest impulse of the bowstring creates

[39] Vitr. *De arch.* 10.3.2; Fleury 1993, 67.

[40] Hero, *Mechanica* 2.2. [41] Philo, *Belopoeica* 72.2–6.

[42] For Philo's mathematical inclinations, see also the solution to the problem of finding two mean proportionals catalogued by Eutocius, at Archimedes and Eutocius 2004, 277–8.

> the quickest motion for the missile, so that in a given amount of time it is carried farther because of the continuity of the motion.[43]

The physical principles Philo invokes here are not proven in this text, and once again he "claims" (λέγω) the proportional relationship between spring velocity and projectile capacity rather than proving it. The approach in the *Mochlica* may have been more rigorously analytical, but we cannot know for sure. At any rate, the approach here privileges engineering solutions over theoretical or philosophical analysis.

Philo introduces this mechanical analysis in order to argue for a change in the catapult's design. While conventional catapults use two half-springs, he says, in fact only one is needed. For evidence of this claim, he draws on observation of the springs in action, combined with an assertion, apparently based in theory, about the relationship between force and spring velocity:

> Then is it true, as most have said, and we too just now said, that the power for the recoil of the arm comes from both half-springs, or only from one? We will clearly see this if we fix our attention here, for it will be very useful for the argument at hand. In stringing the bow, don't we clearly see both half-springs working against each other? Hence it happens that the arms are often worn out because they are moving two forces back and forth. In the release, it happens that both forces of the half-springs, coexisting, with speed equal to each other since they consist of forces equal and alike, move the bowstring at the same time. A single force would not contribute to the speed of the arm if it were not greater than the other.[44]

The initial stage of this argument is framed as a first-person plural process of discovery and "in the field" observation: "we" can see the principle in action through watchful attention to the bowstring at work, in the course of which the whole machine might break down. In the second stage, Philo reframes these observations impersonally, generalizing them to broadly applicable physical laws.

Philo's argumentative approach in this passage (and others like it) recalls the distinction Creese draws between "epideictic" and "apodeictic" harmonic demonstrations. "Epideictic" in this context refers to a kind of rhetorical performance of harmonic knowledge, in which an instrument might be used "rather like a stage prop" to demonstrate the topic of argument. "Apodeictic," by contrast, indicates a more formal process of deductive proof modeled ultimately on the Aristotelian *Analytics*. As such, the "apodeictic" demonstration is much better suited than the "epideictic"

[43] Philo, *Belopoeica* 68.37–43. [44] Philo, *Belopoeica* 68.43–69.9.

to participation in cutting-edge debate, as it can argue rigorously for controversial theses or prove assertions previously only assumed.[45] Creese, drawing on Barnes, argues that apodeixis in this Aristotelian sense was a teaching tool rather than a research method, connecting the distinction as well to the "heuristic and demonstrative methods" Archimedes refers to at the outset of the *Method*.[46]

Philo's circumstances are of course different than those Creese describes; there is nothing in the surviving texts on practical mechanics that corresponds perfectly to the arithmetically rigorous "apodeictic" demonstrations in the texts on harmonics.[47] However, this passage strongly resembles the "epideictic" passages that Creese refers to. Philo's frequent use of personal verbs and pronouns, as well as the rhetorical question with which he begins the passage, create a declamatory atmosphere. He then answers the question with a kind of virtual version of the "epideictic" use of the instrument Creese describes. Philo's description of the spring becomes the locus of demonstration, and the reader a virtual witness: "we clearly see both half-springs working against each other."[48] Thus, even though mechanics, unlike harmonics, had no institution of public demonstration of its principles, Philo manages to simulate a comparable environment of epideictic demonstration in the text. He creates, as it were, another kind of "lived-in space" – this one designed not for collaborative construction, but for demonstration.

The next passage is the closest Philo gets to an "apodeictic" demonstration. He links this argument to the claims derived from the previous "epideictic" passage: "what has been stated could also be demonstrated (δειχθείη) as manifestly true, just as we say, from the example about to be recounted."[49] The example claims (incorrectly) that a two-mina weight will fall more quickly than a one-mina weight, and more quickly than two one-mina weights joined together.[50] This assertion is framed as a notional

[45] Creese 2010, 136–7.

[46] Archimedes, *Method* 82.3–83.15; Barnes 1975, 80; Creese 2010, 152.

[47] However, for Hero's use of ἀπόδειξις and its cognates in the *Metrica* to indicate his debt to Archimedes, and in the *Dioptra* to refer to the calculations made with the eponymous instrument, see Tybjerg 2004, 33–4, 46–8.

[48] Philo, *Belopoeica* 68.49–50. The term "virtual witnessing" is of course owed to Shapin and Schaffer 1985, 60–5.

[49] δειχθείη γὰρ ἂν τὸ λεγόμενον σαφῶς ὑπάρχον, οἷον ἡμεῖς φαμεν, καὶ ἐκ τοῦ μέλλοντος ὑποδείγματος λέγεσθαι; Philo, *Belopoeica* 69.9–10.

[50] This long-standing misapprehension was not experimentally disproven until Simon Stevin's experiments, published in 1586, which anticipated Galileo's better-known experiments (Duhem 1991, ch. 1.12). Duhem traces the closely related idea that the force required to move a body of a given weight is proportional to the product of the object's weight and the velocity of motion to the Peripatetic

experiment, which anyone might carry out: "If someone (τις) takes two weights, the same in kind and form, one of one mina, the other of two, lifts them to a height and lets go, I argue that the two-mina weight will be carried downward more quickly."[51] There is nothing special about this particular set of numbers; Philo claims that the same theory is valid for any set of unequal weights. Various reasons for this have previously been adduced by certain scientists (τινες τῶν φυσικῶν): it might happen because of differences in air resistance, or because downward force (ῥοπή) is an attribute naturally proportional to mass (βάρος). The same result will obtain, he says, not just for a single two-mina weight, but for two one-mina weights tethered together, or indeed any combination of weights.

Philo does not take a stand on the explanatory principle; he is more interested in the result and its engineering applications:

> Since these things are thus established, it has clearly been demonstrated (ἐδείχθη σαφῶς) that there is nothing in the half-spring cooperating with the speed of the arm, since its speed is equal to the other. Therefore it was judged good to strip away and remove the one that could not aid the power of the machine.[52]

This change, says Philo, requires the replacement of the sinew-cord springs with bronze springs, as it is impossible to hold the arm in position with just one of the cord springs. All this analysis, couched in the language of mathematical proof, is brought in to explain the radical shift in design to the Ctesibian bronze-spring catapult. As it happens, this analysis, like Philo's earlier theorizing about weight and velocity, is technically incorrect. While it is true that the bronze-spring catapult operates by applying force to only one side of the arm, using just one half-spring would in fact move the arm forward at half the velocity of the traditional model, resulting in weaker projectile capacity.[53]

This passage more closely resembles "apodeictic" demonstration as Creese defines it (though the degree of mathematical rigor is obviously

Mechanica. Although his absolute assurance that the work is Aristotle's own is no longer taken for granted, Duhem goes on to argue persuasively in his second volume that the approach in the *Mechanica* can be derived from the ideas expressed at *Physica* 249b14ff., and perhaps also at *De caelo* 275a24ff. In any case, the formula Philo is working with appears to have Peripatetic origins.

51 Philo, *Belopoeica* 69.11–15. 52 Philo, *Belopoeica* 69.29–37.

53 Rihll offers a much more convincing account of this error and the likely evolutionary path of the bronze-spring catapult (Rihll 2007, 154). Based on the fact that the arrow-firing bronze-spring catapult shares with the stone-hurling air-spring catapult the design feature of the one-sided spring attachment to the arm, and that the air-spring catapult is itself likely an adaptation of the mechanism for the pumps that Ctesibius was famous for inventing, she suggests a lineage running through these devices, bypassing the sinew-cord spring mechanism completely.

lower). The "proof" is a deduction from the principle that a larger weight falls more quickly than a smaller one. This principle is placed in a context of contemporary physical debate between opposing schools of "scientists," as Creese suggests is more common for the apodeictic demonstrations. Here, as in the "epideictic"-style proof mentioned earlier, Philo once again creates a space within the text to stage a sort of "experimental proof," but this time he allows himself to describe that proof using forms of δείκνυμι rather than λέγω. First, from the principle that a larger weight falls more quickly than a smaller one, he derives the further conclusion that several smaller weights tethered together will fall more slowly than a single weight equal to their total weight; thus, that combining two equal forces will not result in greater velocity. This in turn allows him to reach an answer to the question in which he is really interested: does the payload of a catapult acquire additional velocity from the traditional design incorporating two half-springs? Philo concludes that "it has clearly been demonstrated" (ἐδείχθη σαφῶς) that it does not, so the design should be changed.

It should be noted that despite his claims of rigorous proof, Philo is not actually replicating any process of quantitative, experimentally based engineering design carried out either by himself or by Ctesibius. The connection he draws between quantitative physical laws and the design of the catapult is rather a rhetorical technique, a way of textually positioning these mechanical activities as connected with a theoretically robust, mathematicized understanding of physics. None of this is to say that Philo and Ctesibius did not genuinely engage with this body of knowledge; other passages in the *Belopoeica* provide evidence of engagement with mathematical methods. However, the design process of the catapult owes a lot to mechanical intuition gained from trial and error (a process to which Philo himself refers early in the treatise), and this particular attempt to retrofit the results of that process to a still-incomplete dynamical theory was a failure.

Yet, the very attempt clarifies the importance Philo places on linking his own engineering activities to such quantitative mechanical science as had been developed by his time. The union of these two bodies of knowledge shows up in Philo's language: the hypothetical pair of one-mina weights are to be "connected and swung as closely together as possible" (69.24) which is to say that even in the midst of this theoretical discussion Philo is imagining engineering challenges to be surmounted. Philo's physics is invariably bound to the material elements and devices in which that physics is displayed and put into practice. His conclusions are practical here as well: this whole text is focused on making better catapults, and obtaining a better understanding of the laws of physics is interesting only insofar as it furthers this goal.

Next he introduces the complex challenges of constructing the machine from materials of varying properties. Materials provoke a new set of scientific questions, such as how the particles (σώματα, 71.29) of metals change their arrangement after heating, cooling, and beating, as well as more practical matters of how to process metal so it will be stronger. The text is organized in such a way that the reader, having fully fleshed out his understanding of the wedge-stretched catapult (from the mathematical tools of design, to the mechanical principles of its ballistic action, to assembling all these parts into a complete machine), is now prepared to take in the design differences between entire machines from more summary descriptions. The descriptions in this part of the text demand that the reader comprehend these machines on every level at once, since Philo now makes claims based on mechanical, material, and functional principles.

Some knowledge of the science of the materials from which the spring is constructed is necessary for the enterprise of designing an efficient catapult, although it does not lend itself to neat theoretical explication. Understanding their properties is particularly important to the development of the bronze-spring catapult, where the resilience of the bronze used to make the spring is crucial for the effectiveness of the catapult. Philo argues that the bronze plates from which these springs are constructed have the peculiar property that they retain the curvature imposed after the metal's initial cooling, even after being stretched straight by the tension of the catapult. This assertion appears to have been somewhat controversial; Philo insists that it is indeed true, even though other resilient materials, like horn and certain woods, behave differently. These materials have a natural tendency to spring back to a straight form after being bent, so that they make good bows, but Philo argues that bronze of the type used for the springs maintains its bend once curved.[54]

Philo kindly notes that those who incorrectly suppose an analogy between the behavior of horn and wood on the one hand, and bronze and steel on the other, are to be forgiven. Their ignorance stems from their failure to observe the behavior of a more closely related material:

> Let the excuse be made for their being led into this belief, that they have not previously investigated this one step at a time. For the working of the plates mentioned previously has been seen in the swords called "Celtic" and "Spanish." Whenever they want to test if these are useful, they take the

[54] Philo, *Belopoeica* 70.48–71.10.

> grip in their right hand and the tip of the sword with the other; placing it sideways on their head they pull it down from both ends, until they touch their shoulders. After this they let go, quickly lifting off both hands.[55]

The result of this very informal "experiment" is that the sword is observed to straighten itself out when released, and it can do this any number of times. It should be noted here that Philo is playing a bit fast and loose with his materials; the actual subject of interest here is the bronze springs of the catapult, but Philo brings in the behavior of "iron" (actually steel) swords by way of proving his point about the resilience of metals. His argument that we should assume the same properties for beaten bronze as for this type of steel is largely implicit, and relies on the fact that the metals are worked in the same way.

The metal from which these swords are constructed has been the subject of informal experiments into the cause of their remarkable resilience: "Investigations have been made as to the cause (ἐξητάζετο οὖν, τίς ἐστιν ἡ αἰτία) of these swords turning out so springy" (71.24–6). What Philo is describing here is not a formal experimental approach to understanding the molecular structure of steel or beaten bronze. Rather, he brings the findings of expert craftsmen to bear on his argument that bronze behaves as he says it does. The test itself takes place not in some separate "laboratory" space dedicated to experimentation, but in the hands (and on the heads!) of the craftsmen who created the material. Evidently such behavior can only be expected from steel that is very pure, is worked extensively after smelting, and is "neither overly hard nor soft, but in between" (71.30–1). Philo here seems to be referring to the balance that must be struck between steel that is hard but brittle, and that which is too malleable.[56] Philo's subsequent remark, that the swords are "subjected to a severe beating when cold, and this is what produces resilience," confirms the relationship he understood between the processing of the steel and its later resilience.[57]

Thus far Philo has been building up the reader's understanding of a particular body of knowledge: the practices of smiths, carefully refined and developed over a long experimental period but not necessarily based on a theoretical understanding of why the processes work as they do. From

[55] Philo, *Belopoeica* 71.10–20.

[56] Steel is produced when iron ore absorbs carbon while being smelted with charcoal. Its hardness is enhanced by quenching the wrought iron quickly by immersion in water, as this prevents the loss of iron carbide (also known as cementite). The resulting brittleness can then be abated by repeating the process of hammering, heating, and quenching.

[57] On ancient metalworking processes, see White 1984, ch. 9; for more detail see Forbes 1950; Forbes 1955, vol. VIII.24.

here he moves on to a new level of explanation, linking the details of craft processes to a theory about the structure and behavior of the particles of the metal. This processing must be carried out with a delicate hand, as beating the material too forcefully alters its shape and over-compresses it. Swords manufactured in that way are inflexible and liable to snap in two, "because all the space is closed off, compressed by the beating."[58] Heating softens steel and bronze by rarefying their particles. Cooling and beating have the opposite effect; in the limit, these processes compress the interstitial voids between particles of the material.[59]

The desired effect of resilience in steel swords is derived from the moderate beating process developed by craftsmen. Philo connects that craft process to a theoretical explanation on the level of particle behavior: the more delicate, two-sided beating creates a three-layered plate of metal, whose particles are more densely packed on the outside and thus harder. Meanwhile, the middle layer remains relatively unaffected by the beating, its particles left with larger patches of void between them. Philo consciously refrains from engaging more deeply with the physics of all this: "Thus it happened that they became springy, as was shown earlier. Concerning bronze-spring engines and all that pertains to their construction, let enough have been said, lest we get carried away in expounding physical arguments further."[60] Here Philo is on the brink of a full-blown theoretical discussion of the nature of matter and void, but he pulls back at the last second, saying that he does not want to accidentally veer into a philosophical discussion.[61] Berryman accurately describes this as a "less than complete theory of elasticity," but then again advancing such a theory is, as Philo himself says, not his aim here.[62]

Philo concludes his work by reviewing a few other types of catapult: he lists their inventors, provides notes on their utility, and gives brief but detailed descriptions of how they function. Here he presents a variety of design possibilities for a catapult, although this section's postponement to the end of the work, and its relative brevity, prevent the work as a whole from being a catalogue of different catapults. Philo approaches this material as something more like a theme and variations, where each new variant, whether it is a welcome innovation or rejected as inferior technology, emerges from layering new knowledge onto the existing framework that

[58] Philo, *Belopoeica* 71.37–42. [59] Philo, *Belopoeica* 71.41–52.

[60] Philo, *Belopoeica* 71.52–72.5.

[61] However, the very assertion of interstices of void within apparently solid matter is itself a stance taken in an ongoing argument about matter and void; more on this follows.

[62] Berryman 2009, 127.

has previously been imparted to the reader. Philo's text is written in a manner suitable for someone who wants to learn how to build a catapult, or at least someone prepared to grapple with the mathematical and mechanical principles according to which Philo argues a catapult operates. Simply in order to comprehend the text, the reader must develop at least a passing understanding of practical mechanical principles, craft techniques, and calculation methods, even if he does not go on to put it into practice.

It is true that this text makes occasional excursions that connect mechanics to other bodies of knowledge, like his citation of Polyclitus in the prologue and his digressions on mathematics and the behavior of material particles. However, these excursions are modest in extent; they resemble neither the broad-ranging history of the catapult's evolution that Hero provides in his *Belopoeica*, nor the thorough history of the philosophical problem of the void presented in Hero's *Pneumatica*. Philo instructs the reader in designing and creating a certain kind of technological artifact, which is described in painstaking depth and detail. The bodies of knowledge that Philo presents to the reader locate the catapult squarely within an approach to mechanics that is partly quantitative and mathematicized, partly reliant on a deep familiarity with the properties of materials. Quoting Polyclitus at the beginning serves to expand and elevate the intellectual context of mechanics, yet Philo makes the quotation itself into a statement about mechanical technology, about the calculation processes used in designing and crafting the machine.

Meißner offers an alternative view of Philo's text, describing it as a general systematic instructional text for beginners, asserting that the work was written for an undefined addressee (as opposed to Biton's text, which is directed toward someone in a position of political power), and that the material is thus less concrete and more abstract than that in a text like Biton's.[63] He argues further that Philo's text represents an attempt by a technical specialist to make his subject matter accessible and attractive to laypeople.[64] It is certainly true that the *Belopoeica* does not appear to be directed toward a patron or other political authority; in this it is unlike the texts on the same topic by Biton, Athenaeus, and Apollodorus. The Ariston identified at the beginning of the text is not the only possible reader of the work, and one need not already be an expert in order to begin the text. Indeed, any reader can fit himself into the "lived-in space" the text creates, as long as he is willing to proceed through the tasks of

[63] Meißner 1999, 162–3. [64] Meißner 1999, 165.

calculation, measurement, imagination, and (if he chooses) construction that Philo guides him through.

However, this text gives no signs of being principally an introductory text for mechanical "tourists": it confronts the reader with a level of mathematical rigor and depth that is lacking in the other texts on artillery mentioned earlier, and indeed in any other surviving text on military devices. It is of course true that any reader could potentially pick up this text, like any text, with any set of expectations at all. However, if the work is read as it is written – that is, if the reader thinks through the instructions that make up so much of the text, if he follows Philo's several advisories on the vital importance of precise calculation and the development of mechanical intuition, and indeed, if he does anything more than let his eyes scroll idly down the pages – by the end of the work he will have had a transformative educational experience. He will not just have picked up the kind of overview of the subject that a curious layman might want to acquire, and could get from Biton's brisk catalogue or Hero's evolutionary history. Instead, he will have learned to think his way through the many "layers" of the catapult, to see the mathematical and mechanical principles that underlie the outer structure.

The knowledge of the catapult into which Philo inducts his reader is complex and deep, intensified by tasks of calculation and fleshed out by direct injunctions to imagine and build. Even in the likely event that the reader does not physically build the devices Philo describes, if he follows the text he will be personally implicated in the work of visualizing their construction and contemplating the principles that underlie their function. He will acquire a deep and multifaceted knowledge of these devices – even if the reader begins the text as a layman, he does not finish it so. Collins invokes the reader's role in actualizing instructions in the context of tacit knowledge: "the book is not the knowledge; the knowledge is the book and the person reading it – so long as they are the right kind of person."[65] Philo's work is structured so that its guidance, if heeded, *makes* the reader into the right kind of person – a person who will engage with catapults (real or notional) in a careful and disciplined way, with the fullest understanding obtainable from his level of acquaintance with them (hands-on or merely textual).

Philo does not promote understanding materials and artifacts as a way of accessing external truths about nature (such as the more general properties of the particles of a solid). When he hits a subject that ranges too

[65] Collins 2010, 45.

far from his primary interest of explaining the structure and operation of catapults, he stops himself. His interest is in unfolding the bodies of knowledge immediately connected to the engineering of catapults, in deep detail and with broad contextualization.

Before exploring some texts that do focus on those connected bodies of knowledge, it will be useful to look at another text that takes a deeply focused approach similar to that of Philo's *Belopoeica*. Hero's *Peri automatopoiētikēs*, which covers the design and construction of theatrical automata, likewise self-consciously weaves together different fields of knowledge into a unified approach to designing and constructing a particular class of artifacts. The automata Hero describes include mobile complexes whose outside frame moves to different locations while individual components within dance, dispense liquids, start fires, and so forth, as well as stationary theaters whose doors open and close to reveal miniature figures within performing the scenes of a play. The actions of both types of automaton are mostly the result of the motion of a counterweight, which is timed by the gradual flowing of small grains (Hero suggests millet or mustard seed) into a vessel, and which in turn controls the timing of the other components by pulling a cord wound around drums and axles in a complex pattern.

There is considerably more evidence of a bibliographically systematic approach in this text than in Philo's *Belopoeica*. Hero acknowledges that he draws on earlier texts on the same subject, and in particular that its second part is based on that of Philo, who wrote "in order and well-arranged" (ἐν τάξει καὶ εὐμεθόδως), though Hero criticizes the complexity of some of Philo's components.[66] He discusses the history of the technology as well as its associated discourse, beginning his own work by noting that automaton-making has long been valued because of its complex craftsmanship and the amazing spectacles it produces. Hero embraces both the technological and textual contexts within the claim that "each part (μέρος) of mechanics is included in automaton-making, because they are perfected one part at a time (κατὰ μέρος)."[67] That is to say, automaton-making is a kind of *summa* of the discipline of mechanics, encompassing all its diverse components in one complex genre of mechanical artifacts.

The explicit statement that there are "parts" (μέρη) of mechanics at all is remarkable in itself. The "parts" Hero refers to here are an important recurring element of his systematizing approach. In his *Belopoeica*, he

[66] Hero, *Peri automatopoiētikēs* 20.5.
[67] Hero, *Peri automatopoiētikēs* 1.1.

invokes *ataraxia* as "the greatest and most necessary part" (τὸ μέγιστον καὶ ἀναγκαιότατον μέρος) of philosophy, before proceeding to argue that it is most reliably produced by mastering ballistic weaponry.[68] Here it is the discipline of mechanics that is described as consisting of "parts," all of which are deployed in creating automata.

This is, of course, a bit of a rhetorical overstatement, but it is not entirely misleading. The technique of designing and building automata is presented as a kind of master class in mechanics, which takes mechanical knowledge of all kinds as its "inputs." The text that follows presumes that the reader is already able to think his way through numerous mechanical systems: axles with differential drums (which do the work of gears in this type of device), pulleys, springs, pivots, pneumatic systems, carpentry, and metalworking. He must likewise be prepared for the challenges of working with unpredictable material components such as screws, temperamental counterweights, and strings which must be stretched, oiled, and repeatedly tested to make certain their length will remain constant. This work thus requires some level of mastery of mechanical principles as well as material peculiarities, which can, in many cases, be acquired only through experience.

Constructing automata requires extremely precise craftsmanship, because the only human intervention occurs at the beginning of the sequence of actions, and everything thereafter depends on each action's happening with predictable timing. Hero suggests certain construction protocols, which take some of the guesswork and risk out of the operation of the automaton. The cords that transfer force throughout the automaton, triggering each event, are an obvious target for improvements. He recommends that these cords be repeatedly stretched for a long time, and then treated with a mixture of wax and pitch, to keep their length more or less constant. He likewise warns that whatever components are intended to turn must be perfectly rounded, and all pivot points perfectly smooth; all moving parts must be thoroughly oiled so there is no chance of sticking; "if not, none of the proposed effects will happen according to plan" (κατὰ λόγον).[69]

Like the descriptions of catapults in Philo's *Belopoeica*, Hero's guide to the design and construction of these two types of automaton presents the reader with a densely layered assembly of different bodies of knowledge. He treats questions of geometry and mechanics, consistency of

[68] Hero, *Belopoeica* 1.3; see also Fuhrmann 1960, 170–1.

[69] Hero, *Peri automatopoiētikēs* 2.4.

construction, and even stagecraft and aesthetics, while giving the reader a gradual sense of the variations that could be made on the mechanical themes he provides. All of these topics feature in Philo's text as well, even the question of the machine's visual impact. These two works, and others like them, share their detailed, methodical dispensation of the many different types of knowledge that must be collected and mastered in order to design, build, and improve machines. They are not concerned with using those machines, once built, to generate new knowledge about other subjects (perhaps ballistics, in the case of Philo's text, or, as here, the philosophy behind self-moving devices); the machine itself is the goal.

Thinking with artifacts

Another group of texts does use the knowledge unfolded in descriptions of artifacts as a jumping-off point to new types of knowledge – for example, about celestial bodies, harmonic theory, or the physics of material particles. There is not a hard-and-fast boundary between descriptions focused on the knowledge inherent in the artifact and those primarily concerned with the knowledge that can be gained through the artifact. Berryman, for example, speculates that the theory of elasticity and resilience elaborated by Hero in his *Pneumatica* might, like Philo's in his *Belopoeica*, have derived from observing the compression of different components of Ctesibius' machines.[70] Hero's discussion of this phenomenon is quite different from Philo's: Hero is far more enthusiastic about pursuing physical knowledge about the behavior of particles. Yet both texts undeniably participate in the give and take between science and technology. The difference is in the weight each text gives to these components rather than a strict division of kind.

Ptolemy's *Harmonica* includes an elaborate explanation of the effects of material density on the tones emitted by a stringed instrument, couched within a text whose interest in artifacts is principally focused on their knowledge "outputs." How does this differ from Philo's discussion of the effects of material density on the resilience of the catapult's spring? The difference, again, is one of emphasis. Ptolemy is interested in using what he knows about material densities to approximate a theoretical, quantified system of harmonics as closely as is possible through instruments. That is to say, he wants to use one body of knowledge (the material properties of the components of a musical instrument) to access another (the

[70] Berryman 2009, 173.

quantitative structure of harmonic relationships). Philo, on the other hand, deploys his knowledge of material density within an investigation about material behavior done for its own sake, not as a way of accessing a connected body of knowledge.

The use of artifacts to discipline natural phenomena and render them interpretable is predicated on what Baird calls "working knowledge."[71] Working knowledge can be the (subjective) possession of a person who "has knowledge sufficient to do something." Pappus' characterization of the various types of mechanical practitioner and their activities is a reminder of the continuity between people who possess "working knowledge" and the devices that embody it.[72] The *mēchanikos* possesses mathematical, physical, material, and craft knowledge, which can be deployed in the design and production of an artifact that will then function in a controlled, predictable way. By contrast, what Baird calls the "objective" form of working knowledge is borne by "devices made by humans that exhibit particular phenomena over which we have substantial material – if not linguistic – control," and by the techniques that go into building and refining those devices.

"Working knowledge" is thus closely linked to the nuanced distinction Kogge makes between "Können" and "Wissen," which he connects loosely to Polanyi's "knowing how" and "knowing that."[73] Kogge argues that while the two pairs are not to be identified, it is not unreasonable to think that in all knowledge, even the type of knowledge typically labeled "knowing that" or "Wissen," components that are not linguistically explicable inevitably remain. Passages of technical ekphrasis are part of the effort to add linguistic control to the material control of the craft process, and so to render that process more easily transferable and predictable. High predictability is especially valuable where the artifacts concerned are designed to provide reliable knowledge about phenomena external to the artifact.

Gaining this kind of control, even over the instruments themselves, is a non-trivial proposition. The processes for calibrating mechanical instruments designed to deliver data about natural phenomena must be carefully verified and regulated. This is a constant challenge in communicating scientific and technological knowledge, as Tybjerg's fascinating study of the early modern Danish astronomer Ole Rømer shows. A letter

[71] Baird 2004, 45–6. [72] Pappus, *Synagoge* 8.1024–6.
[73] Kogge 2012, 32; Polanyi 2009, 6–7.

sent by Rømer to Leibniz indicates the patience, humility, and commitment to precision required:

> Since furthermore I have the most mechanical of all the disciplines in the literary world (this is how I conceive of astronomy) that requires senses and hands to the same degree as thought, then there will always be something lacking or needing to be examined or adjusted, which cannot be added on the basis of speculations (this would be conceit) or achieved at a convenient moment: as a rule it requires tedious waiting.[74]

Rømer was deeply concerned with designing and regulating instruments so that they would yield precise and accurate astronomical knowledge – indeed, his concern ran so deep that he was possessed by the idea that the instruments themselves ought to be made in such a way that they would themselves reveal observational errors.[75]

Among the ancient scientists and engineers who created comparable opportunities for observation, none was more explicit about his search for material control over the means of observation as Ptolemy. In the *Harmonica*, he writes about the vital role played by sensory observation (specifically sight and hearing) in allowing the rational part of the mind access to observable phenomena. The interaction of these two senses is crucial to the development of rational knowledge from sensory input:

> Therefore not only through each one grasping its own material, but also joining in one another's actions in learning and theorizing about those things completed through their proper rationale, these senses and the most rational bodies of knowledge based on them pervade further into the beautiful and useful.[76]

He selects astronomy and harmonics in particular as bodies of knowledge that are exclusively accessed through one sense or the other: astronomy is related to sight, as its objects are only accessible through that sense, whereas harmonics is related to hearing for the same reason. In both cases, the sensory data is then analyzed and rationalized through the application of mathematical theory: "Using indisputable instruments, arithmetic and geometry, to get at the quantity and quality of the primary movements; these are, as it were, cousins, descended from the sisters sight and sound, and reared as close relatives by arithmetic and geometry."[77] The work of the harmonic scientist, like that of the astronomer, is to connect the rational faculty of the soul with the data available in the natural world.

[74] Tybjerg 2012, 77. [75] Tybjerg 2012, 97. [76] Ptolemy, *Harmonica* III.3.71–4.
[77] Ptolemy, *Harmonica* III.3.79–84.

This data is acquired originally through the sensory organs, and subsequently rationalized and controlled using observational instruments like the *kanōn*.[78]

Barker observes that instruments narrow the gap between a rationally defined theoretical object and a manifestation of that object that is accessible to the senses.[79] A circle, defined as the plane figure made up of all points equidistant from a given point, can be effectively constructed in the sensory domain using a compass.[80] Likewise, the *kanōn* is an instrument that, when properly handled, gives our senses access to a close representation of the theoretical system of mathematically constructed harmonic relationships. Ptolemy acknowledges the need for finely calibrated instruments to make quantitative harmonic relationships comprehensible to the ears:

> Therefore, for analogous occurrences concerning sounds and hearing, just as for the eyes, there is need of some logical criterion for them using proper instruments, as there is need of the ruler for the straight line, and the compass for the circle and its subdivisions. Similarly, for hearing (which together with sight is the servant of theory and the part of the soul that possesses reason), there is need of some method derived from reason, concerning the things they were not born able to judge precisely; in accordance with this method they will not refute [one another] but agree how things are. The instrument of such a method is called the harmonic *kanōn*, from its common predication and its straightening [*kanonizein*[81]] what is lacking in the senses for grasping the truth.[82]

78 On this see also Creese on the Duris fragment, where he argues that the *kanōn* referred to there is a monochord; "The context is also explicitly Pythagorean: this is the right place to hear about the monochord. Finally *logoi* (ratios) are exactly, and only, what a *harmonikos* demonstrates by means of a *kanōn*" (Creese 2010, 100). Ptolemy's claims about the utility of instruments to aid the senses in the exploration of nature would be echoed centuries later by Robert Hooke, who likewise notes the deficiencies of the senses, and proposes that they might be remedied by the use of artificial instruments like microscopes and telescopes. The result of this application of properly designed instruments is a brand-new appreciation for the richness and order of the world: "The next care to be taken, in respect of the Senses, is a supplying of their infirmities with *Instruments*, and, as it were, the adding of *artificial Organs* to the *natural*; this in one of them has been of late years accomplisht with prodigious benefit to all sorts of useful knowledge, by the invention of Optical Glasses" (Robert Hooke, *Micrographia* preface).

79 Barker 2006, 22.

80 Cf. Ptolemy *Harmonica* 1.1.21–8 on the difference between a figure that looks to the eye like a circle and one that has been "properly constructed" according to mathematical principles, using a compass.

81 The verb *kanonizein* can refer to either of the uses of a ruler, which can serve either as a straightedge or a measuring tool. Barker translates the verb as "straightening," in the sense of disciplining sensory phenomena so that they yield useful theoretical results. Solomon, by contrast, prefers the latter, saying that this kind of *kanōn*, like any other, measures something, specifically "truthful harmonic intervals" (Ptolemy 2000, 7 n. 30).

82 Ptolemy, *Harmonica* 1.1.52–2.3.

Geometers are sometimes said to work with mathematical objects, but as Netz observes they actually operate on a diagram. In order to maximize the precision and accuracy of their work, they use compass and straightedge to bring the sensible diagrammed objects as close as possible to the geometrical ideals they represent. Likewise, the harmonic scientist must produce a sensible analogue of the harmonic relationships he explores, and he too requires an instrument to bring these sensible representations as close as possible to the arithmetic ideal. While the instruments themselves are often elided in scientific discussions of the phenomena they make available to the senses, a deep understanding of the instrument and its limitations is absolutely vital for studying the data received through them, and Ptolemy is emphatic about the importance of refining his instruments.

Ptolemy describes various types of *kanōn*, differing in the number of strings and types of bridges used in their construction, in the *Harmonica*. Let us begin, as he does, with the single-stringed *kanōn*. As seen earlier, Ptolemy initially assimilates the instrument to idealized geometrical elements, underscoring his commitment to accessing the mathematical structure of harmonic ratios. As the description continues, however, he introduces more material attributes, which he gradually insinuates into the mathematical language of the earlier part of the description:

> Now, joining a *kanonion*[83] to the string and dividing off the length EH on it, so as to make the comparisons more handily, first at a bisection made of the whole length (as K) and then at the bisection of the half (as Λ) we will place narrow bridges (ὑπαγώγια), quite slender and smooth, or indeed other bridges (μαγάδια), slightly taller than the others, but quite similar to them as far as position, equality, or similarity over the line in the middle of their convexity, which will be below the bisection of the *kanonion* or that of the bisection of the half. If the part of the string EK is found to have pitch equal to KH, and moreover KΛ to ΛH, [the string's] homogeneous composition will be clear to us. If not, let us apply our scrutiny to another part, or indeed another string, until the following is preserved, that is consistent pitch in similar, proportional, and same-sized parts that have the same tension.[84]

The markers of geometrical perfection – the doubly bisected string, the letters that label perfectly symmetrical points – are still present here. However, at the same time material factors creep in, with all the approximations they require. The bridges are to be "quite slender and

[83] Here this word means "ruler" or "measuring rod."

[84] Ptolemy, *Harmonica* 1.8.43–56.

smooth," and "slightly taller than the others": they have real thickness and roughness that can only be minimized, and the difference between the heights of the two sets of bridges is imperfectly defined. Meanwhile, the real possibility arises that the string will turn out to be too inconsistent to carry out the experiment correctly. Ptolemy moves gradually from mathematical generalities to material particularities. It is a *particular* string that may be useless, as opposed to the completely general and idealized ΑΕΗΔ; it is a particular set of bridges, with particular height relationships to one another, that divide up the string, and so on. Any imperfections and peculiarities will accumulate, and as they add up they complicate Ptolemy's stated goal of finding an instrument that can provide sensory access to an accurate representation of harmonic relationships.

This representation is, of course, never perfect. Creese perceives the "collision of the mathematical and physical construction of the instrument"[85] that occurs in trying to use the verbal tools of the mathematical *ekthesis* to represent a material artifact. Ptolemy wrestles with the theoretical and practical problems of constructing a stringed instrument that will provide uniform and predictable resonant frequencies when plucked. Ultimately what he presents to the reader is merely the best approximation of such an ideal instrument, and he is thorough in identifying the instrumental flaws that interfere with this approximation. The single-stringed *kanōn* is particularly plagued by structural inconsistencies. Both the consistency and the position of the string, he says, are often not checked with sufficient precision, and this in turn makes the quantitative ratios of its segments imprecise. Ptolemy further complains of a kind of user error in which people mark out the divisions on the string by stretching out the string and adjusting the positions of the bridges until they hear the note they are seeking. This defeats the whole purpose of the *kanōn*, which is designed to correct the ear's deficiencies by providing a more precise alternative.[86]

Other practical problems also affect the *kanōn*'s ability to mediate between reason and perception. The note played cannot be changed quickly or easily, and friction between strings and bridges causes the tones to shift while the bridge's position is being changed. Other instruments used for discovering harmonic ratios, including wind instruments, strings with weights suspended from them, hammers of varying weights, and filled or empty bowls, have their own sets of problems stemming from

[85] Creese 2010, 70–1. [86] Ptolemy, *Harmonica* II.12.22–8.

the difficulty of ensuring material consistency between them.[87] Ptolemy's effort to reconcile his quantitative system with the unpredictable material reality of the instruments he uses to explore the system will require an ongoing interplay between theory and practice that constantly threatens the accuracy of his method.

Ultimately, Ptolemy identifies the fifteen-stringed *kanōn* as a superior instrument for harmonic demonstrations.[88] This instrument allows the harmonicist to realize a double diapason system, rather than the single diapason system accessible from the instruments he previously described.[89] Even this instrument, however, is still susceptible to departures from theoretical perfection. Ptolemy recommends making the movable bridge taller than the fixed bridges, in order to ensure that the string will contact it firmly. But even if the bridges are made according to his recommendation with circular cross-sections, the string will not meet the movable bridge at a single point: a small extended length of string will contact the bridge. Ptolemy addresses this difficulty by recommending the use of fixed and movable bridges with equal circles as cross-sections, so that even if this happens the extended lengths of string will be equal (Fig. 7). He offers a complete geometrical proof of his claim:

> Let the base of the *kanōn* be imagined on the straight line AB, and let the lines ΑΓΔ and BEZ be drawn up on it at right angles, and on the centers Γ and E let arcs of circles HΔ and ΘZ be drawn as the convex peripheries of the bridges, so that BZ is greater than AΔ. Let the straight line be drawn touching the peripheries at H and Θ, ΘH, and let the lines HΓ and ΘE be linked; let HΘ be cut at K by an extension of ΓΔ, and similarly at Λ by an extension of EZ. Since the lines marked through the middle of the breadth of the bridges (μαγάδια) fall on the points Δ and Z, if AΔ and BZ are extended to these, then the contact points of the strings and the bridges (ἀποψάλματα) coincide at Δ and Z. It is clear that the attachment of the *kanonion* to the ends K and Λ will indicate the length KΛ, while its attachment on the line between the real contact points and the bridges (ἀποψάλματα) will form HΘ. So the triangle ΓHK will have equal angles to EΘΛ, since the line ΓK is parallel to EΛ and ΓH to EΘ. Because of this, as the line EΘ is to ΓH, so ΘΛ will be to HK. If ΓH and EΘ are equal, that is if segments of equal circles form the arcs ZΘ and HΔ, the line ΘΛ will be

[87] Ptolemy, *Harmonica* I.8.1–24. On the dubious stories about the origins of harmonic science such as Pythagoras' fortuitous encounter with anvils of differing sizes that produced proportionally differing sounds, Plato's account of Hippasus and the metal discs, etc., see Creese 2010, 94–7.

[88] Ptolemy, *Harmonica* III.1–2.

[89] Solomon notes that in Ptolemy's time a fifteen-stringed lyre had become quite common, so the fifteen-stringed *kanōn* probably owes much to this contemporary instrument (Ptolemy 2000, 128 n. 6).

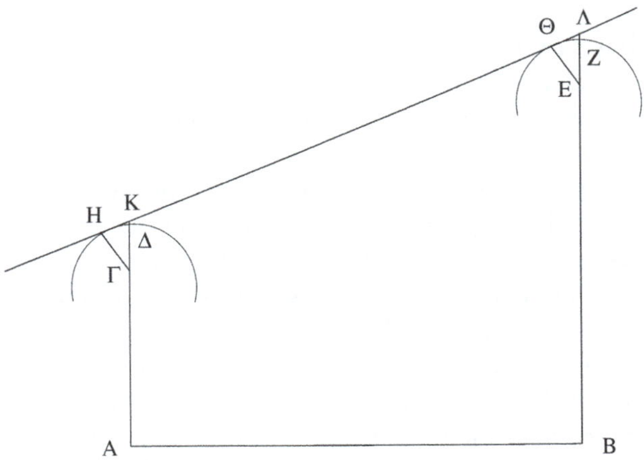

Figure 7. Kanōn *with circular bridges.*

> equal to HK, and KΛ to the whole of HΘ, so that the distance measured by the *kanonion* will not differ in any way from the real one.[90]

Ptolemy thus answers a problem arising from material complexity (the string sounds differently if it is firmly pressed by the bridge) by representing and reworking the instrument geometrically, and providing a geometrical argument that his improved structure is theoretically unimpeachable. Far from ignoring material realities in favor of an idealized view of the instrument, he builds a mathematical model of the structural problem and uses it to find a solution. Ptolemy is very comfortable operating with mathematical objects, and so this model of the angled string serves to move a material problem into his preferred domain.

Not every material aspect of the instrument lends itself well to such a mathematical translation. Even in these less tractable situations, Ptolemy tries to work toward roughly quantitative physical explanations, even if he is not able to transform the problem into mathematical terms so elegantly as in the case above. His approach to harmonics aims to find a happy medium between the Pythagoreans, who relied too heavily on mathematics, to the exclusion of sensory information, and the Aristoxenians, who erred in the other direction.[91] He therefore attempts a quantitative physical explanation for sensible attributes of sound such as pitch and volume.

[90] Ptolemy, *Harmonica* III.2.114–32. The diagram is after Paris suppl. gr. 59, fol. 210v.

[91] Ptolemy, *Harmonica* I.2. On Ptolemy's approach as a variety of Pythagoreanism, see Barker 1989, vol. II, 270–3; Barker 2006, ch. 4.

This explanation begins with a more qualitative one, where he connects these attributes to structural attributes of the instrument whose vibrations move the air. Depending on whether an instrument's material is dense or porous, thick or thin, rough or smooth, and so forth, its vibration will be different.[92]

He eventually finds a way of quantifying (however approximately) the parameters defining the constitutions of the materials used in harmonic investigations, using "amount of substance" as a scalar proxy for all of them. One material is of a denser (πυκνότερος) constitution than another when, for a given bulk (ὄγκος), it possesses a greater amount of substance (οὐσία); a piece of material is considered thicker (παχύτερος) than another of the same constitution when a given length of it possesses more substance than the same length of the second. Denser and thinner (λεπτότερος) materials will produce higher sounds; sparser (μανότερος) and thicker materials lower ones.[93] This occurs, Ptolemy says, because thinner things allow quicker penetration of the material and thus beat more quickly, whereas thicker things allow more extensive penetration.

Now he works his way back to less abstract territory, explaining the macroscopic musical properties of different materials, and differently shaped pieces of the same material, in terms of these microscopic differences of consistency. Bronze and gut are denser than wood and linen respectively, Ptolemy argues, so they make a higher sound. He does not spell out here that the "amount of substance" should remain the same; the reader is presumably meant to have internalized this from the previous discussion of the general case. Ptolemy then goes on to subdivide his materials even further. Comparing two pieces of bronze with the same density and bulk, but different thicknesses, he asserts that the thinner (λεπτότερος) will produce a higher tone; for two gut strings with the same density and bulk, the same is true of the thinner (ἰσχνότερος); likewise for hollow things (τὰ κοῖλα) compared to solid ones (τὰ ναστά), and finally for denser, thinner pipes.[94]

Ptolemy then makes a surprising revelation; after all this talk of density and thickness, it turns out that these are not the real governing factors after all; they are only proxies for higher tension (*tonos*). Why then use them as explanatory tools? Because they are available to the senses, whereas *tonos* in the material physics sense Ptolemy is using it here is not. Ptolemy's goal here is to reduce the many possible forms of experimental

[92] Ptolemy, *Harmonica* 1.3.1–64. For a thorough explanation of Ptolemy's views on how pitch and volume are governed by these factors, see Barker 2006, 33–53.

[93] Ptolemy, *Harmonica* 1.3.35–42. [94] Ptolemy, *Harmonica* 1.3.44–51.

harmonic instruments to a single variable. The first part of this approach is already familiar from Philo's *Belopoeica*, where the variations in elasticity of materials were explained in terms of their constitution from particles of different density. Ptolemy adds another layer of explanation to the argument, describing not just the low-level structure of these materials but the variable *tonos* that keeps them together. Sound, he explains, is "a continuous tuning of air," which circulates within vibrating objects before making its way to the outside, where it causes the air to vibrate in turn. In the process, the sound undergoes a transformation that depends on the particular tension in the vibrating object.

As we have already seen, Ptolemy later goes on to apply these theories to refining the design of instruments so as to avoid problems such as unequal stretching of the strings.[95] Ptolemy never denies the material and functional realities of the devices he describes; to do this would be to limit the precision of the instrument, and so to damage its very ability to represent harmonic or astronomical truth. Barker notes that Ptolemy does not here analyze or defend the physical theory behind his recommendations for construction: "Ptolemy is concerned only with the techniques by which an apparatus can satisfactorily be set up, given that the theories in 1.3 are sound."[96] In other words, from this point on in the text these are treated as practical rather than theoretical problems. He does not return to the physical theories he has previously explained, about how the sensible qualities of sound are determined by the density, and thickness, and shape of the materials from which an instrument or striker is constructed. Going forward, the emphasis will be very much on the artifacts' "outputs."

However, it is important that this discussion did take place, as it signals Ptolemy's transition from one kind of theoretical knowledge, to an embodiment of that knowledge in material instruments, and then to the exploitation of those instruments to produce a new kind of theoretical knowledge. One must agree with Ptolemy's claim that *tonos* is the governing force that determines the density, thickness, and smoothness of materials in order to follow his arguments about how sound can be manipulated by material objects. Using this knowledge, a special class of material objects (harmonic instruments) is designed and constructed for optimal predictability of *tonos* and thus of frequency production. Physical "experiments"[97] are carried out using these instruments, which turn

[95] Ptolemy, *Harmonica* 1.8.51–6. [96] Barker 2006, 195.

[97] On whether Ptolemy did actually perform the "experimental" activities he describes in the *Harmonica*, see Barker 2006, 192–258.

sensible qualities derived from the *tonos* of the instrument into quantifiable measurements. Then, the quantitative results of the "experimental" work with these instruments are applied to harmonics, which indeed does not end with music in this scenario but, idiosyncratically, involves applications to psychology and cosmology as well.

It is in this final stage that the way the *Harmonica* elucidates the knowledge encoded by artifacts differs most profoundly from that in a work like Philo's *Belopoeica*. Both authors explain in detail the constitution of the materials from which a certain type of artifact is constructed. Both kinds of artifacts mediate between visible macroscopic mechanical properties and the invisible microscopic world of material properties. However, the ends of the investigation are quite different. Philo investigates the materials used in a system that performs a mechanical action for its own sake, trying to make that action as predictable as possible. Ptolemy, on the other hand, investigates the materials used in a system that performs a mechanical action in order to model some property of the world, trying to make the model as perfect as possible. Ptolemy's instruments are built to reveal the structure of something in the world, and described in such a way as to help the instrument's user discover this structure for himself.

Even phenomena that are not perceptible by the senses under normal conditions can be illuminated through correctly deployed knowledge of mechanical instruments. The question of the void, for example, had vexed philosophers at least since Parmenides.[98] In his *Pneumatica*, Hero takes an instrument-based approach to examining this physical problem; he says he is following (unnamed) predecessors for his physical views, not inventing them anew. Based on what is known about the theory of the void propounded by Strato of Lampsacus, as transmitted by Simplicius and others, it has been argued that Hero is working with Strato's theory or one that closely resembles it, perhaps mediated through Erasistratus.[99] This theory, which Furley labels the "microvoid" theory, states that all materials are riddled with interstices of void, though these are imperceptible to the senses.[100] An alternative view holds that Simplicius misinterpreted Strato, who actually argued that both extended void and microvoid potentially exist, but are actually filled with something; the microvoid is in this view

[98] On ancient theories about the existence and nature of the void, see particularly Furley 1989, ch. 13; for their application in practical mechanics, see von Staden 1996.

[99] This "maximalist" view (to use Furley's term) of Strato's influence on Hero is defended principally in Gottschalk 1965.

[100] Strato fr. 56A; Furley 1989, 150–8.

the potential of some kinds of matter to have channels opened up in them.[101]

Berryman questions whether it is even proper to talk about a "theory of matter" underlying Hero's pneumatic work. Berryman argues that Diels "inaugurated an unfortunate tradition" of regarding ancient allusions to the void, Hero's among others, as attempts to prove a single homogeneously expressed theory "and not merely showing that there is *an effect to be explained*."[102] Where Diels suggested that Philo and Hero had essentially the same pneumatic view as that of Strato, and that that view depended on acceptance of a "theory" of the *horror vacui*, Berryman argues that the demonstrations Hero makes do not imply that he was trying to prove that a matter-attracting void existed and was the cause of the phenomena he observed.

Hero fittingly begins his *Pneumatica* with a preface in which he laments that theoretical philosophical argument on the topic of the void may proliferate endlessly without producing any definite result. Hero's task here differs crucially from Ptolemy's in the *Harmonica* in the degree to which the phenomena being discussed are subject to debate. In the *Harmonica*, the instrument gives precise order to phenomena that are already perceptible by means of the senses; here it provides access to a phenomenon that is ordinarily unavailable to the senses. The instrumental approach is Hero's solution to the apparent problem of unproven (yet deceptively convincing) theories about the void, where "it is possible for those who say that there is no void at all to produce many attempts at arguments besides these, and perhaps to appear quite persuasive in words, even though there is no corroborating perceptible proof (ἀπόδειξις)."[103] Hero proposes to add a new set of tools to the debate, appealing to the power of the senses to solve problems in natural philosophy.

The problem, of course, is that no extended vacuum exists under natural conditions, so that the relevant phenomena had thus far been

[101] This "minimalist" view is defended primarily in Gatzemeier 1970. Furley finds, however, that in the part of Hero's preface that is most often supposed to derive directly from Strato (1.pr.65–91), an extra assumption is needed to explain why the microvoid in the air does not expand indefinitely when the air is sucked through a tube out of a vessel containing water, so that there is in turn no need for the water to rise into the tube. This assumption resembles a claim made by Epicurus (*Letter to Herodotus* 46), and Furley identifies Hero as more likely than Strato to draw on both Stoic and Epicurean sources for a theory of matter. For this and other reasons, Furley gravitates toward the "maximalist" rather than the "minimalist" interpretation of Hero's degree of dependence on Strato.

[102] Berryman 2009, 166–7; Diels 1893.

[103] Hero, *Pneumatica* 1.pr.190–3.

inaccessible to the senses. In the *Pneumatica*, however, Hero proposes to produce an extended vacuum by artificial means, so that the void's effects will at last be available to sense perception, and the debate put to rest. If sensible phenomena can be engaged to show that a continuous vacuum may be artificially produced, and that tiny interstices of void are distributed in substances naturally, then even the most plausible-sounding argument from those who believe there is no void will be invalidated.[104] Hero's hope that the demonstrations his pneumatic apparatus makes possible will silence the ongoing verbal back-and-forth of the philosophers recalls Rømer's struggle to produce mechanical astronomical instruments in order that, as Tybjerg says, "they demonstrate astronomical knowledge mechanically and they can be used with no further argument."[105]

Since no extended vacuum is available under normal conditions, these "sensible phenomena" can only be accessed artificially, through the instruments Hero will go on to describe. These instruments are not nearly so complex or rigorously designed as the air pump designed by Robert Boyle in the seventeenth century, which became an iconic figure of the history of science with Shapin and Schaffer's highly influential *Leviathan and the Air-Pump*. Hero's instruments are of a more rough-and-ready kind, but there are nevertheless some key similarities between his textual presentation of the experiments and other activities he suggests and the rhetorical features of later discourse on pneumatics identified by Shapin and Schaffer.

In his preface Hero addresses the microscopic physical basis of macroscopic material properties. He asserts that continuous void does not exist in nature, but an interstitial void (Furley's "microvoid") does. The microvoid is responsible for certain remarkable properties of some materials, notably "adamant," which is so hard that it might be believed to possess no interstitial spaces at all:

> since it does not allow burning or breaking, and when struck on anvils and hammers it enters them whole. Now, this happens to it because of its uninterrupted density. For since the particles of fire are broader than the voids in the stone they do not enter into it, but only touch the outside surface.[106]

Thus far, Hero's description of the material properties of adamant is similar to Philo's discussion of the effects of cooling and beating on the flexibility of steel. In both texts, an account at the level of particles serves

[104] Hero, *Pneumatica* 1.pr.194–200: ἐπὶ τῶν φαινομένων καὶ ὑπὸ τὴν αἴσθησιν πιπτόντων. On the epistemological powers of philosophy and mechanics in Hero's time, see Tybjerg 2003, 452–3.

[105] Tybjerg 2012, 89.

[106] Hero, *Pneumatica* 1.pr.58–65.

to explain macroscopic properties of some particular known material. The known flexibility of steel prepared in a certain way emerges from the "sandwiching" of more and less dense layers of particles; the known hardness of adamant is a result of its void spaces being too small to allow softening fire particles into it. The appeal to interstitial void as an explanation for the hardness and density of materials is, however, the end of the resemblance between the way physics is handled in Philo's *Belopoeica* and in Hero's *Pneumatica*. Hero, unlike Philo, is very much interested in describing his artifacts in a way that highlights the access they provide to knowledge about the natural world.

The two authors' works on pneumatics may also be compared, somewhat more cautiously. While Philo's is no longer extant in Greek, it does survive in Arabic and Latin. The Latin manuscript known as N[107] contains in its first chapter this statement on the relationship between his work and that of philosophers interested in the void:

> Hence also philosophers who gave consideration to natural things, disagreeing with the erroneous opinion of the majority, said that no vessel was empty, as far as they believed, but rather full either of air or of some other bodies.[108] I, however, desiring to serve the interest of brevity in my present concerns, wish neither to protract their words on my tongue nor the opinions of the debaters, lest my discussion acquire an unwieldy wordiness (*ne dispensiosam sermo meus prolixitatem incurrat*).[109]

Philo's text, as the preface promises, describes pneumatic artifacts with an exclusive focus on their structure and function, just as his *Belopoeica* focuses exclusively on the practical properties of catapults' materials. Hero, unlike Philo but quite like Ptolemy in the *Harmonica*, goes on to build these observed phenomena into a larger system of physical principles.

Constructing such a system demands more than attaching a physical explanation to familiar phenomena; its principles must be broadly generalizable, ideally even to phenomena not usually observable under natural conditions. Hero describes the behavior of particles in a pneumatic device that artificially imposes low or high pressures on fluids and gases:

> The particles of the air press against one another; however, they do not coincide in every part, but have some void spaces in between, just like sand at the beach. (Here the grains of sand must be understood to be analogous to the particles of the air, while the air in between the grains of sand is

[107] Munich, Bayerische Staatsbibliothek, CLM 534 fols. 26v–29r.

[108] The text is difficult here; see Philo 1974, 80 n. 204.

[109] Philo, *Pneumatica* 1.

> analogous to the voids of the air.) Therefore when any force is applied, the air is subject to compression and collapse into the spaces of the voids, against the nature of the particles squeezed against one another (παρὰ φύσιν τῶν σωμάτων πρὸς ἄλληλα θλιβομένων).[110] Upon relaxation, the air is restored to its original arrangement by the tension of its particles, just as happens to shavings of horn and dried sponges: when compressed and released, they are restored to the same space and return to their same bulk.[111]

Similarly, he says, an extended vacuum can be created, but when the force that creates it is removed, the separated particles of air will rush back together quickly because there is nothing to get in their way.

Hero makes a number of significant physical claims here.[112] There is no extended void in nature, he says, although there are tiny pockets of void within all sorts of substances, including the air. Materials are assembled from particles of varying sizes and configurations, which explains variable macroscopic material properties like compressibility and elasticity. Within a vacuum, nothing interferes with the movement of particles. His analogy between grains of beach sand and particles of matter is meticulously constructed; he takes pains to remind the reader that, while the grains of sand stand in for the air particles, the air between the grains is actually void. He follows this up with another analogy: he asks the reader to imagine that the particles of the air work like an extremely sparse sponge – so, they can be compressed temporarily, and then spring back to occupy their natural volume. They can also be expanded temporarily if the ambient pressure is decreased, and then return to their natural volume when normal pressure is restored. Berryman argues that for the particles to fill in the interstitial voids and then spring back into their original shape, the particles themselves must be taken to be deformed under pressure, a remarkable departure from the atomist view of the attributes of particles.[113]

On the macroscopic side of things, Hero argues that it is possible to create a temporary extended vacuum artificially through various pneumatic devices. Several of these devices appear to have been adapted from vessels used in the practice of medicine, including cupping glasses and siphons.

110 On the history of this phrase, and its eventual attachment to the idea that nature and mechanics were seen as opposed in antiquity, see particularly Schiefsky 2007, 68; Berryman 2009, 44–8. Here, as in the examples Berryman cites, it makes sense to interpret the phrase as "artificially," rather than "against nature" in a strong sense.

111 Hero, *Pneumatica* 1.pr.65–85.

112 For the context of these claims in Greek physical theory, see von Staden 1996, 87–92; Berryman 2009, 167.

113 Berryman argues convincingly that Hero's interest in the laws underlying this material behavior was likely sparked by Ctesibius' mechanical innovations, which served to integrate compression effects into the discourse on matter (Berryman 2009, 169–70).

They were not purpose-built precision equipment based on pneumatic theory, and so differ profoundly from dedicated scientific apparatus like Boyle's air pump. Yet these devices nevertheless provide a theater for witnessing physical phenomena not found in nature, even though they were not originally developed for that purpose. For example, Hero notes that if one takes a narrow-necked vessel and sucks the air out of it, it will be possible to suspend the vessel from one's lips.[114] He argues that this is because the newly created vacuum draws the lip towards it to fill the empty space, proving that there was previously an extended vacuum in the vessel.

Nor is this of only theoretical interest; he points out that medical practitioners using small mouth-evacuated vessels to transport small quantities of liquids rely, consciously or unconsciously, on this same principle. When the air has been sucked out of the vessel, its mouth covered, and the inverted vessel placed into the liquid, when the mouth is uncovered, "the water is drawn up into the emptied space, even though the motion is against the nature (παρὰ φύσιν) of the water."[115] Likewise, cupping glasses, another tool of the medical trade, adhere to the skin with suction when heated, rather than falling down under their own weight, and in fact draw additional matter through the body's pores (ἀραιώματα), since

> the fire integrated into them destroys and rarefies the air contained in them, just as other bodies are destroyed by the fire and change into lighter substances (I am speaking of water, air, and earth). That they are destroyed is clear from the remains of coals. For these are demonstrated to maintain the same size as they had from the beginning, before the burning, or a little less, but they are much changed in weight from that at the beginning.[116]

The informal experiments that Hero says may be carried out using these devices use the extended vacuum to reveal physical principles, including the rarefaction of the elemental forms of water, air, and earth into lighter versions of themselves through the application of fire. These experiments provide access to phenomena that are unobservable under natural conditions; for example, by using force to create a vacuum, which in turn drives a liquid to behave in a way that is "against its nature" – as Berryman suggests, in the sense of an artificially imposed condition rather than a subversion of natural principles. Hero defines the "natural" behavior of liquids later on, citing the Archimedean principle that liquids at rest have a surface that is spherical and concentric with the earth, and that liquids

[114] Hero, *Pneumatica* 1.pr.85–90. [115] Hero, *Pneumatica* 1.pr.96–8.
[116] Hero, *Pneumatica* 1.pr.104–12.

not at rest will move until they reach this state; he then uses this principle to explain the behavior of liquids in a siphon.[117]

Observations of the behavior of fluids at pressures higher or lower than atmospheric pressure are predicated on the provision of this experimental apparatus. The dimensions and structure of these devices do not have to be as carefully tuned as those of the instruments Ptolemy describes. However, the artifacts constructed for pneumatic experiments are essential to the observations that Hero calls upon the reader to make, and so the descriptions of artifacts that make up most of the *Pneumatica* are essential to the work of scientific explanation threaded throughout that text.

One of these, a sphere with a siphon projecting from it, is described as follows: it should be spherical, in the neighborhood of a certain volume, too thick to be crushed, carefully sealed up with solder and then perforated to allow the insertion of a bronze siphon (which is further defined as a "narrow tube"), then sealed up again around the edges of the perforation.[118] Even the distance the siphon is to be allowed to project past the sphere is specified. Note that the reader is not instructed in the process of *making* the vessel, which comes into the text ready-made. But Hero does engage his reader in the observational experiment itself, using the first-person plural exhortation, "let us watch what happens" (σκοπῶμεν δὴ τὰ συμβαίνοντα).[119] Indeed, the same inclusive language is used throughout the discussion that follows, covering the experimental apparatus, its use, and the physical conclusions that can be reached through it. These intimations of inclusivity and universality suggest that the *Pneumatica* is a conduit through which anyone can gain knowledge.

Hero asserts that the globe is not empty (as it might be casually labeled), but contains air, and that this air behaves in a certain way. This experiment is structured as a kind of proof by contradiction: he considers the possibility that, as his detractors argue, there is no such thing as void, even interstitial void, and that therefore an attempt to blow more air into the globe would cause it to burst. Hero notes that this does not happen, and that it is in fact rather easy to blow more air into the vessel; from this he concludes that the air particles inside have been packed more tightly together:

> Yet if someone (τις) wishes to put the siphon to his mouth and blow into the sphere, he will introduce a lot of extra air, without the air that was already in there making its way out. Since this always happens, it is clearly

[117] Hero, *Pneumatica* 1.2. [118] Hero, *Pneumatica* 1.pr.200–13.
[119] Hero, *Pneumatica* 1.pr.213.

> demonstrated (δείκνυται) that there is some compression of the particles already present in the sphere into the interstitial void. However, the compression occurs artificially (παρὰ φύσιν) because of the force of the aggregation.[120]

This passage marks the conscious creation of an artificial (παρὰ φύσιν) environment, which is central to the value of this experiment. Hero then suggests a further observational experiment, which anyone who has such a globe in hand can carry out:

> Then if someone (τις) blows in, places his hand at the mouth, and immediately covers the siphon with a finger, the air will remain trapped in the sphere. If anyone uncovers it, the aggregated air will rush outside again with a blast and a roar in the expulsion, as we established before, because of the expansion of the air that was previously there, and this happens on account of its tension.[121]

Hero is, of course, implicitly making a general assertion about the behavior of matter and void everywhere. Yet the centrality of the artifact, the sphere, is absolutely crucial to his argumentative process. The apparatus is where all the work happens:

> In turn, if someone wishes to draw out the air present in the sphere through the siphon with his mouth, a lot of it will follow, even though no other substance balances it out in the sphere, just as was said earlier about the cupping glass. So through this it is completely proved (δείκνυσθαι) that a large continuous void is produced in the sphere, for it is not possible for the remaining particles of air to be produced at this moment, so as to refill the space left by the expelled particles.[122]

This experiment is tightly tethered to its apparatus, both by physical necessity and by Hero's rhetorical approach. The vacuum is produced *in the sphere*, he emphasizes; the sphere is the locus of proof. Hero does briefly return to the microvoid that exists in water, air, and wine at the end of the preface. However, this discussion is structured so as to set up the remainder of the work, which will explore the ramifications of the principles just established in new types of apparatus using those three substances.

The physical assertions made in the *Pneumatica* are predicated on a manual operation, carried out on a well-defined material artifact, which produces a result that is perceptible to the senses. This task can be carried out by anyone (τις), while the overall process is framed as a collaborative observational experiment: "Let us watch what happens." Hero's

120 Hero, *Pneumatica* 1.pr.235–41.
121 Hero, *Pneumatica* 1.pr.241–9.
122 Hero, *Pneumatica* 1.pr.250–6.

use of indefinite pronouns and hortatory constructions suggests that the physical knowledge gained from this process is (on the one hand) universally applicable, and (on the other) the result of a particular set of hands-on actions, which the reader is encouraged to replicate. Whether the universality of principle or the particularity of experience is emphasized in these experimental accounts, the apparatus is always the focus of discovery.

The text itself is equally essential to this process of discovery, despite Hero's disparaging words about the dangerous deception of plausible-sounding philosophical arguments. Purely theoretical claims cannot be considered sufficient to settle questions like whether void exists, but the elemental theory and analysis provided in the text are a crucial complement to the sensible phenomena to which Hero appeals. The argument the text conveys is necessary to turn those raw phenomena into proof of the theoretical claim. Hero's reliance on the explanatory powers of the text is reminiscent of Latour's claim that texts exert an important persuasive power in the modern scientific environment:

> What is so important in the images and in the inscriptions scientists and engineers are busy obtaining, drawing, inspecting, calculating and discussing? It is, first of all, the unique advantage they give in the rhetorical or polemical situation. "You doubt what I say? I'll show you."[123]

The inscriptions upon which modern laboratory science relies are, as Latour points out, a highly mobile distillation of laboratory work, a transformation of masses of data (derived from painstaking experiments based on complex theories) into a more easily digestible and transmissible kind of knowledge. This very process of distillation, of shaping raw data into useful inscriptions, is crucial for supporting theories with experimental evidence. The mobility and efficiency of the inscriptions themselves furthermore make them a useful rhetorical tool, marshalling experimental results with a great density of information, via textual technologies that can be widely reproduced. These inscriptions are essential to an intellectual enterprise that relies on the ability of its practitioners to make their ideas convincing in a competitive and polemical environment. While its environment of production was of course quite different from that of the modern scientific texts with which Latour is here concerned, Hero's text makes a comparable rhetorical effort to convince the reader of his claims to truth about physical phenomena.

[123] Latour 1986, 14.

Shapin and Schaffer emphasize that Boyle's pneumatic experiments were a way of staking argumentative claims, not only that it was possible to create an extended vacuum, but also that experimentally derived evidence was a valid source of natural philosophical knowledge. Boyle's views required the gradual aggregation of "witnesses" who would assent to his claims; once a certain critical mass of witnesses had been accumulated, the results of the air pump experiments could be considered to be what Shapin and Schaffer call "matters of fact."[124] People could be turned into witnesses in three basic ways. They could observe Boyle's experiments, or alternatively they could witness replications of these experiments, performed according to predetermined experimental protocols. The third, most widely influential method was what Shapin and Schaffer label *virtual witnessing*. This method

> involves the production in a *reader's* mind of such an image of an experimental scene as obviates the necessity for either direct witness or replication … The technology of virtual witnessing was not different in kind to that used to facilitate actual replication. One could deploy the same linguistic resources in order to encourage the physical replication of experiments or to trigger in the reader's mind a naturalistic image of the experimental scene.[125]

The "linguistic resources" used to enable "virtual witnessing" strikingly resemble the descriptive techniques used in technical ekphraseis. Virtual witnessing replaces the experimental apparatus and procedure with a textual representation, designed and detailed so that the reader of the text assents to the physical claims being made just as though he had witnessed the experiment in person. As Shapin and Schaffer observe, the textual technology required to instruct a reader in carrying out an experiment for himself is the same as that required to create an image in the reader's mind of the experiment being carried out without his intervention. This is a reminder of the close kinship between instructive and descriptive texts; they both, in their own way, turn the mind of a reader into the mind of an actor. This holds true whether or not the reader actually goes so far as to perform the actions of construction or experiment that the text encourages him to visualize.

Hero implicates the reader of the *Pneumatica* in a comparable experience of "virtual witnessing." Of course he does so from a technical and rhetorical perspective far removed from the highly systematized world of modern experimental science (and even from Boyle's experiments at the

[124] Shapin and Schaffer 1985, 80–9. [125] Shapin and Schaffer 1985, 60–1.

Royal Society). However, Hero's basic strategy bears an important resemblance to Boyle's, and to that of the experimental scientists whose practices Latour describes. He stakes claims about physical phenomena, and about the validity of the experimental methods he proposes, through a text that is self-consciously rhetorical, and which takes pains to turn the reader into a witness of the experiments described. Creating observational instruments is vital to the work of this text, and so Hero works to provide a description such that the reader can recreate the instrument and experiment, at least in his imagination. Hero's use of universalizing constructions with τις and hortatory first-person plurals serves to emphasize the possibility that the reader can be included in this experimental enterprise.

The way the rest of the text is structured further develops Hero's argument that experiments carried out by means of mechanical artifacts can unfold new knowledge about a theoretically based physics. The *Pneumatica* has been characterized as a catalogue of toys, with a theoretical preface tacked on.[126] The "toy box" view of the *Pneumatica* is quite misleading, however; Tybjerg offers particular insight into how ancient mechanical wonders functioned as objects of serious philosophical interest in their own right.[127] The artifacts Hero describes are meant to expand upon the theoretical principles introduced in the preface. He threads these principles throughout the entire text in such a way that the dozens of pneumatic devices he describes are implicated in an increasingly complex science of pneumatics, which unfolds throughout the work. It is not incorrect to identify the *Pneumatica* as a book about astounding mechanical marvels. Yet these marvels are framed and systematized by the text so as to bring the reader to understand a particular facet of the behavior of the natural world: the effects of pressure on the behavior and motion of fluids.

Hero imposes this systematic approach with a light but steady hand. He does not label each device in the main body of the work with the lesson it teaches about the behavior of fluids. However, he does immediately make it clear that systematization is at the heart of what he sees as his original contribution to the discipline:

> Since the study of pneumatics has been found worthy of attention by ancient philosophers as well as mechanicians – the former explaining its properties logically, the latter through the mechanisms of perceptible bodies – we also think it is necessary to bring into order (εἰς τάξιν ἀγαγεῖν)

[126] See, for example, Brumbaugh 1966, 94; White 1984, 181–2.
[127] Tybjerg 2003.

what we have inherited from our predecessors, and add what else we have discovered.[128]

Hero acknowledges the work done by his predecessors both on the practical and the theoretical sides of the discipline, and announces his plan to expand on this work and to put the whole complex in order. This organizational scheme will, furthermore, help to integrate the discipline of pneumatics with other bodies of knowledge:

> For in this way it will turn out to help those who afterward wish to engage with the sciences (ἐν τοῖς μαθήμασιν ἀναστρέφεσθαι). Believing this to follow after the system of water clocks already described by us in four books, we are writing this one immediately in order, and about the same system, as previously mentioned. For through the interweaving of air, fire, water, and earth, and through three elements or four entwined strands, varied compositions (ποικίλαι διαθέσεις) are effected, some serving life's most necessary requirements, others making a display of shock and wonder.[129]

Hero's priority here is to establish a consistent investigative and discursive method, which can be applied to different disciplines. Order is important; Hero specifies that mastery of pneumatics will precede the reader's study of the exact sciences (μαήθματα). The reference to his own prior work on water clocks, and the need to harmonize the approaches of the two, likewise suggests that he intends readers to proceed through these two works in the same order in which he wrote them. Hints about the mechanical system he envisions add to the clues to his mechanical system that Hero gives in the *Peri automatopoiētikēs*, which apparently stipulates that automaton-making should be the last subdiscipline learned. He invokes his "varied compositions" (ποικίλαι διαθέσεις) in direct connection with his own project of constructing a system of multidisciplinary knowledge, suggesting that he is referring to textual as well as mechanical superstructures.

This passage introduces another concept that will be important to the organizational scheme of the *Pneumatica*: the elements or building blocks from which both matter and knowledge are generated. He returns to the elements at the end of the preface, promising that these building blocks can be combined in such a way that they generate something new: "Now that these things have been clarified, we will describe the θεωρήματα that arise from the interweaving of the elements mentioned. For through them it is possible to discover quite varied and marvelous motions."[130] His use

[128] Hero, *Pneumatica* 1.pr.1–7. [129] Hero, *Pneumatica* 1.pr.8–17.
[130] Hero, *Pneumatica* 1.pr.343–7.

here of the word θεωρήματα is intriguingly ambiguous. Is he referring to spectacular wonders, to objects meant for contemplation, or to abstract theorems?[131] In the course of the *Pneumatica*, Hero moves nimbly between theoretical principles and the mechanical devices that make them visible. The devices themselves are both spectacular wonders, which conceal their working principles from the casual observer, and instruments that reveal physical truths to those who possess an explanatory key like the text of the *Pneumatica* itself. It is difficult, perhaps intentionally so, to know which of these Hero means to emphasize here. The common ground is the foundation of all the complexities of the work, material and conceptual, in a few simple elements.

Hero makes it clear from the outset that he intends to proceed through combinations of these elements into more and more complex constructions. He begins in the preface with an elemental theory, focusing on how the physical elements transform one another. These include the transformation of water into air by fire, of dense vapors into rarer ones by heat, and of water into earth by absorption. In the main body of the work, he continues in a similar combinatorial spirit, now applied to individual mechanical devices. He explicitly announces his intention to organize the devices themselves according to a comparable set of principles: "Now let us begin to create designs of devices for action, beginning from the lesser in elementary fashion" (τῶν δὲ εἰς ἐνέργειαν κατασκευαζομένων νῦν ἀρξώμεθα κατασκευὰς ποιεῖσθαι ἀπὸ τῶν μικροτέρων ἀρξάμενοι στοιχείου χάριν).[132] Hero uses the expression στοιχείου χάριν here, as he does στοιχείου ἕνεκα earlier, to indicate that his work will progress, like Euclid's *Elements*, from the simple to the more complex.[133] Tybjerg describes this as a claim to create a "geometry of machines" by progressively combining simpler devices into more complex ones.[134]

A look at the overarching structure of the first few chapters will give an idea of Hero's pattern. He begins with the bent siphon (1.1 and 1.2), which illustrates a familiar set of principles: once the air has been sucked out of a siphon, the liquid will continue flowing out of the vessel in which the siphon stands.

Next he introduces a new kind of siphon, distinguished as "concentric" or "enclosed." This device uses the same means of operation (ἐνέργεια) as

[131] Hero elsewhere uses this same word to refer to a diagram at *Cheiroballistra* 2.3, 2.6, and 2.7 – an attractive middle ground between theoretical principle and working device, but not a sensible translation here. Compare Tybjerg on Hero's exploitation of the ambiguity of the verb κατασκευάζειν "to geometrize mechanics" (Tybjerg 2004, 48).

[132] Hero, *Pneumatica* 1.7.1–3. [133] Hero 1997, 42 n. 1. [134] Tybjerg 2004, 50.

the bent siphon, but keeps the siphon enclosed, a concentric pipe within the larger vessel, rather than allowing it to protrude from the side.

Hero points out, however, that the stream of fluid through all these siphons has an irregular (ἀνώμαλος) velocity while the siphon is fixed in place. This happens for the same reason that water flowing out of a pierced vessel has irregular velocity: because a greater weight pushes it out at the beginning than at the end. In these siphons the velocity shift is proportional to the excess length of the outer leg of the siphon. Now Hero implicitly asks the question: *how can a siphon be constructed that will yield uniform velocity?* This brings us to the next device (1.4): a siphon that will move liquid with uniform velocity. This device relies on an inner basin that sinks along with the siphon, maintaining a constant excess length of the outer leg even as the fluid level changes.

The next complication emerges from the question: *how can a siphon be constructed that provides uniform velocity for an arbitrary amount of time, and then changes to a new uniform velocity?* This brings us to the next device (1.5), which allows the user to turn a screw in order to choose the velocity.

But all these siphon designs only allow drawing off water through the mouth if the siphon is quite small, which is inconvenient. So now Hero needs to engineer a vessel to draw air (and hence water) out of the siphon (1.6). The next section is where Hero places his injunction to proceed from the lesser devices to the more complex. The first element of this progression turns out to be a sphere (1.7) that, dipped in wine, will collect the wine and then dispense it when a hole is unstopped.

Next, he uses the same principle to devise a sphere (1.8) that will dispense either hot or cold water (or, using the same principle, either water or wine).

After this, he introduces a vessel (1.9) that will again discharge two liquids, but can also dispense them mixed.

And so the work continues, gradually introducing more advanced principles of fluid dynamics and more complex devices. Their complexity does not increase monotonically; no one could argue that the syringe (2.18) is more mechanically complex than the windmill-driven water organ that concludes the first book. The placement of the syringe is perhaps chosen so as to juxtapose it with another device of medical application, a complex cupping glass designed to work without heat. Hero's organizational scheme, in other words, is not without flexibility and complexities of its own.

However, the basic principles upon which the *Pneumatica* is organized reveal a great deal about the work Hero intends the text to do. His careful taxonomy is important for explicating the construction and use of pneumatic devices, not just for their own sake but as a way of unfolding more general knowledge about the behavior of fluids. Hero's text is thus structured according to a logic different from that of Philo's *Belopoeica*. Where Hero constructs his text artifact by artifact in order of increasing complexity, Philo structures his so as to introduce a series of types of skills needed to design and construct catapults. Philo's structural organization leads the reader more and more deeply into the complex skill set needed to master a single type of artifact, avoiding digressions that would bring the discussion too far afield from the center of focus. Hero's brings the reader into contact with a whole constellation of artifacts organized by their theoretical complexity, so gradually deepening his understanding of a single "scientific" problem, the physical principles of compressible fluids.

The structure Hero imposes upon the discipline of pneumatics through the structure of his text is crucial to the work of interdisciplinary systematization that the preface promises. The text is shaped to give didactic structure to its material: if the reader proceeds from beginning to end he receives an education in the principles of pneumatics and their combinations, even though that education is framed as a series of encounters with artifacts rather than a list of rules or abstract principles. The "toy box" view of the *Pneumatica* ignores the text's structure, which is easy to do given that the descriptive passages are discrete and the principles that connect them largely implicit. But to claim that the text *can* be read this way is not the same as claiming that it *ought to* be read this way. Of course one can just reach into the text anywhere, just as one could with *Tristram Shandy* or any other text whose structure conceals its own organizational principles. But to cherry-pick elements out of Hero's *Pneumatica* in this way is to miss the work the text does to show its reader just how those elements are meant to fit together into a larger picture.

The organizational principles behind the *Pneumatica* have been observed from a different perspective by Meißner, who compares the structure of Hero's text to that of the *Elements* of Euclid. The *Pneumatica*, like Euclid's text, begins with the elementary components of the discipline, the first objects with which the reader's eye or hand will engage. Meißner argues that the process of aggregating these increasingly complicated combinations of relatively simple elements helps to highlight the physical questions that underlie this artifact-based investigation.[135] The systematic recombination of familiar mechanical elements can generate

[135] Meißner 1999, 60.

new knowledge in much the same way as the recombinations of Euclid's mathematical *Elements*. Such a high-level perspective on the structural principles of the *Pneumatica* illuminates what it means for a text like Hero's to be part of a technical discipline, and for that discipline in turn to be part of a larger system of technical disciplines. These disciplines are complexly interconnected by overlaps in content, but also in the variety of forms adopted to communicate that content.

Euclid's text of course differs from Hero's in some important ways, notably the verb forms the two authors choose. Asper points out Euclid's work as an exemplar of impersonality. In the *Pneumatica*, by contrast, the universal yet embodied τις invoked in the text's "experimental" passages creates an effect very like Shapin and Schaffer's "virtual witnessing." The agents here, witnesses who balance specific experience with universalizable knowledge, are quite different from the detached "Helping Hand" of mathematics.

What does this have to do with technical ekphrasis? Mechanical artifact texts do a peculiar kind of work, which is in some ways very different from that of mathematical texts (as well as texts from other disciplines that frequently fall under the rubric of the "technical," such as handbooks of rhetoric or law). Texts like the *Pneumatica* draw on certain defining qualities of texts like the *Elements* (such as its organizational structure), but put that structure into a very different kind of context by rejecting other defining qualities (such as its impersonal, "objective" presentation of objects detached from the reader). Mechanical artifact texts have the opportunity to create engagement, cognitive if not manual, between the reader and a particular set of material objects. Hero's project relies on bringing artifacts into virtual contact with the reader, a kind of material proof and, he alleges, the way to get to the kind of truth the philosophers can only squabble about endlessly. This kind of contact lies at the core of the textual connection that technical ekphrasis makes between a reader and a technological artifact.

Conclusion

The predominance of works by Hero and Ptolemy in the field of texts that primarily focus on how mechanical artifacts can be used to access new knowledge is striking. Of course, this has something to do with the great productivity of both of these authors. It is also partially due to the happy fact that many of their works were evidently well used and popular, and survived with relatively strong manuscript traditions, while the works of their predecessors (including those they mention by name) largely perished. Yet the accidents of survival are not the only force at work here. Both Hero and Ptolemy had a particular interest in building up bodies of theoretical knowledge based on observations of physical phenomena carried out through instruments. There was, of course, a long-standing Alexandrian heritage of mechanical experimentation, which had flourished at least since the time of Ctesibius.

What Hero and Ptolemy add to this, however, is the formal integration of mechanical apparatus into the experimental practice of a kind of system of "natural philosophy." Hero foregrounds the mechanisms he describes, systematizing them along with the theoretical principles he uses them to demonstrate; however, he does not produce anything like systematic data. Ptolemy, on the other hand, produces highly systematized data, often delivered in elaborate data tables, but is more reticent on the subject of the experimental apparatus that produces it. Both authors, however, are agreed on the necessity of refining their instruments for scientific practice.

Structural variations like this, as well as the different degrees of logical or mathematical rigor involved in the textual presentation of mechanical artifacts, recall some of the questions of genre raised earlier. The distinction Creese raises between "epideictic" and "apodeictic" demonstrations in harmonics illustrates the role artifacts can play in these demonstrations. Creese notes of the *Sectio canonis* that because of its Platonic background it makes use of more rigorous forms of argument than the "epideictic" demonstration more common in contemporary harmonic discourse. It is here, Creese argues, that harmonics first adopts the apodeictic argumentative strategies and standards already presented by Aristotle and "adopted in the conventions of Euclidean mathematical discourse."[136]

This shift reflects a change in the bodies of knowledge to which harmonic science is assimilated. Rather than associating harmonic demonstrations with rhetoric and other technical disciplines (like medicine and architecture) that seem to have made use of public demonstrations of craft skill, harmonicists make claims on the territory of mathematics and philosophy, where demonstrations happen largely through word and diagram, whether in text or in person. The same shift may be reflected in the form of the text (for example, the distinction between a standardized repetition of formulas and local structures and a more discursive and heterogeneous format), as well as in the linguistic markers the author chooses (like

[136] Creese 2010, 151; these questions are addressed as well at Lloyd 1991, ch. 3; Netz 1999, 292–3.

a change from direct address and rhetorical interjections to the impersonal verb forms characteristic of mathematical prose).

The way an author builds his text into the larger framework of knowledge gained *about* and *by means of* mechanical artifacts, and the various forms of address that dominate them, suggests some further answers to the questions of genre with which we began. The same kind of distinction Creese brings up between "teaching tools" and "research methods"[137] is at work, for example, when Philo shifts from using ἀποδείκνυμι to indicate a comparatively rigorous theoretical claim about levers, to using λέγω for an unquantifiable claim about a material artifact that really has to be seen, and perhaps manually tested, to be believed. These are different types of knowledge, and they call for different grammar, vocabulary, and other textual techniques. For example, even if Philo had wanted to enlist the help of a diagram in his assertion about the "bunching" of the spring cord, it is not clear that a visual representation of the phenomenon would have clarified the spatial arrangement of the materials better than his experience-based account of the twisting process. This part of his problem fades into the territory of tacit knowledge, standing at least in part outside the reach of the text. By contrast, Ptolemy's argument about the string's contact with the bridges is framed in such a way that it pairs very easily with the diagram and the accompanying geometrical proof.

Standards of proof, large-scale textual structures, and linguistic markers like forms of address and imperatives may furnish an indicator of the type of work an author casts himself as doing – though probably not, in the context of mechanical artifacts, of anything that might properly be called "genre."[138] Still, they offer something resembling the interplay Asper cites between title, formal attributes, and the context of communication as a kind of precursor to genre.[139] All the different descriptive techniques we have been tracking – from small-scale features like the author's choice of forms of address or technical and rhetorical vocabulary, to book-level features like the overall structure of the text, to supra-textual systematization and outreach to other bodies of knowledge – can contribute to this analysis.

Because they offer so many variations at all these different degrees of textual localization, mechanical artifact texts provide an opportunity to approach the classification of technical texts in a new way. Rather than focusing on a text's conformity to rigid structural patterns or the authorial

[137] Creese 2010, 152. [138] On the problems here see Asper 2007, 50–1.
[139] Asper 2007, 53.

intentions that might be presumed from the author's own programmatic statements, this approach analyzes how these texts actually deploy the knowledge they contain. The very heterogeneity of the corpus of mechanical artifact texts illuminates consistent variations in patterns, suggesting how to fill out a "schema" system of attributes and their interrelations along lines such as Sinding suggests. Studying these patterns yields not a set of generic labels for texts, but a pragmatic approach to discovering the different ways they disseminate such a variety of types of knowledge, in a variety of ways, to a variety of readers. All the variations analyzed throughout these chapters add up to this: the "written machine," the textual encounter with the artifact created by an author's particular way of decoding the knowledge it embodies.

CHAPTER 7

Conclusion: The Limits of Description

Despite the wealth of highly refined descriptive strategies available to ancient authors of mechanical artifact texts, nevertheless there are occasions when they acknowledge their descriptive efforts have failed. Sometimes this failure is a matter of "information overload"; these authors frequently find themselves describing complex and unfamiliar machines, and they stop short of revealing every detail in text. Sometimes it is a matter of ineffability: mechanical craft, like artistic creation, has facets that simply do not lend themselves to verbalization. Some gaps can be filled with the use of diagrams, but some aspects of machines elude textual representation entirely. The tension between what can and cannot be encoded in text recalls Galen's remark in *De usu partium* that he cannot take the time to list all the joints of the animal in the text, but the student must proceed to examine all the parts in his hands-on practice.[1]

Failure to capture a subject adequately in text is of course something of an authorial trope. We can think, for example, of the limits Polybius sets on his description of Hamilcar's assaults on Italy. He recuses himself from describing the battles in detail, on the grounds that he is faced with a situation like a boxing match; there the two combatants fight in a flurry of blows, and the audience cannot anticipate or mark every one (indeed, he says, neither can the fighters themselves). The best one can do, says Polybius, is to get a general idea of the skill and strength of each fighter, and so it is with Hamilcar's attacks and counterattacks: "no writer could succeed in enumerating them, and moreover the material would ultimately be endless as well as useless for the audience."[2] A really complete description of such complicated things is unattainable, and even if it were possible the reader would be stupefied by the assault of information.

The problem of tedium is, for ancient authors as well as modern ones, a particular challenge for writers of technical material. Vitruvius ruminates

[1] Galen, *De usu partium* II.182 Helmreich. [2] Polybius, *Historiae* 1.57.3–4.

in the prologue to his fifth book on the difficulty of making his material readable. Writing about architecture, he says, is not like writing history or poetry: history offers the reader the promise of novel surprises, while poetry's elegance of form and content hold the reader's attention up to the very end. Architecture, by contrast, requires the use of technical vocabulary, which a reader may find obscure, while the didactic material may expand beyond what the reader can handle. Vitruvius promises that he will do his best to keep his discourse concise and explain technical terms as clearly as possible, the better to make them memorable and comprehensible.[3] He reiterates this plan elsewhere in the work, as when he promises to discuss the principles of water organs as concisely and precisely as possible (*quam brevissime proximeque attingere potero*).[4] Nor was the pressure to write briefly for an audience with limited patience for technical detail limited to Roman authors: think, for example, of the concise mechanical prescriptions Biton makes for Attalus.

Vitruvius elsewhere reflects on a more serious problem of description. Even if the reading audience had an unlimited appetite for technical vocabulary and mechanical facts, some of the details of the processes of designing and constructing machines are simply impossible to capture fully in text. This constraint applies particularly to the details of craftsmanship. For example, he qualifies his promise to explain the hands-on process of stringing and tightening catapults by saying that he will describe the application of tension to the hair or sinew spring "to the extent that I can cover it in writing" (*quantum comprehendere scriptis potuero*).[5] The inexpressible aspects of the process are not the materials the machine is constructed from, nor the way in which the strings are attached to the arm of the catapult, nor the mechanism used to tune the tension. The missing element is the hands-on perfection of the tuning; there is something about this process, as with many processes of mechanical craft, that cannot be adequately captured in writing.

The problem Vitruvius recognizes might be compared to Galen's discussion of the "unsayable" (ἄρρητα) attributes of the pulse in his *De dignoscendis pulsibus* and *De locis affectis*.[6] They cannot be adequately conveyed by the text alone; Galen can hint at their qualities, but only enough that the reader can go out and learn them himself through manual acquaintance with the thing itself. Galen compares these limitations to the problem of verbally describing a person's likeness; no one can recreate an

[3] Vitr. *De arch.* 5.pr.2. [4] Vitr. *De arch.* 10.8.1. [5] Vitr. *De arch.* 10.11.8.
[6] On this issue see Reinhardt 2011.

individual's form (ἰδέα) accurately, but one can, for example, give verbal "indications" (γνωρίσματα) sufficient to recognize runaway slaves from heralds' descriptions.[7] Knowledge of this type simply cannot be obtained in its entirety except by personal experience; an author can only hope to provide enough textual "indications" for the reader to go on until he has a direct encounter with the object. These "unsayable" attributes of objects and processes are part of the territory of tacit knowledge, which always lurks beyond the boundaries of the text.

Mechanical artifacts may likewise include elements deemed too rich and complex, or simply "unsayable" for other reasons, to be fully encoded in a text, so the text must again be seeded with "indications" that will allow the reader to comprehend an indescribable phenomenon if he sees it in person. Sizing an arrow-shooting catapult according to the length of the desired missile is a problem whose mathematical outline (sizing the components of the catapult proportionately to the diameter of the spring hole) had long been settled by Vitruvius' time. However, Vitruvius cautions that in order to perfect the catapult's range, practical testing must complement the theoretical design. If the frame turns out to be rather tall, or "upward-stretched" (*anatonus*), Vitruvius says that the tension of the spring will be a bit lessened, so compensation has to be made by filing down the length of the shooting arms. If, on the other hand, it is "downward-stretched" (*catatonus*), then the arms should be made a bit longer so they can be more easily pulled down.[8] Sizing a catapult like this, through tentative additions and subtractions, is unpredictably dependent on materials, and hence indescribable; Vitruvius can only offer vague quantifications such as *paulo longiora*. The mathematical formulations that he has just described are necessary in order to accelerate the design process beyond pure trial and error, but practical experience and experiment are necessary to finalize that process.

A similar acknowledgment that textual representation fails at some point, where hands-on interaction or at least a visual encounter with the artifact becomes necessary, follows Vitruvius' description of the water organ:

> As far as I could manage, I have tried to relate this difficult material quite clearly in writing; however, this is not an easy matter, nor accessible to everyone's understanding, except for those who have familiar experience (*exercitationem*) with this type of thing. But if anyone understands too little

[7] Galen, *De dignoscendis pulsibus*, K v. 8.774.
[8] Vitr. *De arch.* 10.10.6.

> from my writing, when he gets to know the machine itself, he will immediately find everything carefully and precisely arranged (*cum ipsam rem cognoscet, profecto inveniet curiose et subtiliter omnia ordinata*).[9]

To be sure, there is an element of the standard *captatio benevolentiae* here, as Vitruvius begs pardon for the specific failures of his text. At the same time, however, a larger concern is at work: the machine is on some level obscure except to those who have had an encounter with the thing itself, preferably often enough to constitute *exercitatio.*

As we have seen, ancient authors use a wide range of inventive descriptive strategies to suggest through text the kinds of knowledge variously labeled "practical," "tacit," "concrete," or "know-how." Some passages augment visual aspects of the description with references to attributes of the artifact that appeal to other senses, so that the reader acquires the impression of immediacy implied by appeals to the imagined sense of touch or hearing. Others use techniques of "directive reading," including direct address to the reader and instructions for imagining or manipulating the components of a machine in certain ways, to communicate this kind of knowledge even through the distance of the text. Even authors armed with these techniques, however, still sometimes struggle with the task of describing technological artifacts.

Mechanical artifacts were of course not the only kind of artifact in antiquity to provoke critical inquiry into the challenges of textual representation. The *Canon* of Polyclitus, regarded as a pairing of text and artifact, was particularly provocative. Pliny presents the *Canon* as such a successful model for later sculptors that Polyclitus "is thought to have exhausted all the resources of art in one work."[10] In particular, Pliny reports Varro's description of such statues as "*quadrata* and almost copies," suggesting that the *Canon* might have transmitted the dimensions of the statue in enough detail to allow for quite exact reproduction.[11] Such reproductions were, however, not universally lauded. As Perry points out, Quintilian does not vilify imitation, but advises his reader that it is not enough by itself, as this leads to work analogous to that of painters who

[9] Vitr. *De arch.* 10.8.6.

[10] *solusque hominum artem ipsam fecisse artis opere iudicatur*; Pliny, *NH* 34.55.

[11] *quadrata tamen esse ea ait Varro et paene ad exemplum*; Pliny, *NH* 34.56. On the sculptural significance of the term, see McCartney 1932; McEwen 2003, 268–70. Stewart interprets Polyclitus' reference to "precision" (ἀκρίβεια) as a refusal to accept anything other than exact repetition (Stewart 1998, 274). Stewart's strong assertion seems rather problematic, particularly as he follows it up with a claim that Polyclitus himself may have indulged in some variation, as well as conceding that craftsmen are subject to the limitations of their materials and skills. For a much more detailed treatment of the mechanisms and products of ancient sculptural copying, see Bartman 1992.

copy paintings by measuring and replicating their features mechanically.[12] However, Quintilian argues that copies (visual or verbal) are not as meaningful as new creations: something crucial to their cultural significance is lost even in a perfect act of copying.[13] Similarly, Pausanias refuses to describe the measurements of Pheidias' statue of Zeus at Olympia: its dimensions have been measured, but repeating them cannot capture the experience of seeing it.[14]

Dionysius of Halicarnassus is more explicit about what exactly does get lost in strictly textual transmission of the details of such an artifact. He distinguishes two types of mimesis in sculpture, which he then analogizes to rhetorical imitation. The first is a "natural" kind, which is attained through instruction (κατήχησις) and familiarity (συντροφία); the second is learned only through the precepts (παραγγέλματα) of the craft.[15] Works created by this second kind of mimesis, he says, lose the "natural grace" that comes from the first kind, and have a "deliberate and unnatural" effect. Perry observes that Dionysius does not describe the first kind.[16] Indeed, how could he, since this superior type of mimesis is defined by its inaccessibility to purely verbal instructions? There is evidently some aspect of certain works of art that it seems cannot be captured in even a detailed report of their physical characteristics. They are to some extent "black-boxed," inaccessible to analysis and textual encoding.

We do, of course, often manage to get inside this "black box," to judge and analyze artworks and machines. Yet sometimes an element of an artwork remains beyond the reach of textual encoding, and this may be true of technological artifacts as well. In mechanical artifact texts, the gap between the object and its textual encoding usually has less to do with extraordinary artistry, and more to do with the messy unpredictability of handling objects in the real world. The "natural grace" Dionysius finds lacking in the artistic products of instruction by precepts has its mechanical analogue in the complications of craftsmanship that likewise cannot

[12] Quintilian, *Institutiones oratoriae* 10.2.4–7: *ante omnia igitur imitatio per se ipsa non sufficit ... turpe etiam illud est, contentum esse id consequi quod imiteris*. On this passage see Perry 2002, 158–63; Perry 2005, 95–6. Exact duplicates of paintings and (especially) sculptures were clearly in some demand in the Roman world in particular, as several references to such copies survive in text: see e.g. Cicero, *Academica* 2.85; Pliny the Younger, *Epistle* 4.28; Lucian, *Zeuxippus* 3–7.

[13] This still allows for what Tanner describes as "eclectic copying," which involved the composition of new works as harmonious recombinations of individual elements from a variety of models. See Tanner 2005, 288–95; on eclecticism see also Perry 2005, 111–49. On the connotations of the "exact copy" and the "free copy" in Roman art, see Perry 2005, 90–9.

[14] Pausanias, *Graeciae descriptio* 5.11.9. On this topic see also Platt 2011, 224–7.

[15] Dionysius of Halicarnassus, *De Dinarcho* 7.27–36.

[16] Perry 2002, 158.

be satisfactorily transmitted in text. The craft process has certain hands-on aspects that must allow for human error, human grace, and material variation.

So we saw for Hero's odometer: its gears are turned by screws, and while it is theoretically simple to calculate the gear ratio of this mechanism, in practice screws (particularly in the days before precision machining) do not operate perfectly. We may recall that it was this device Hero labeled the "written machine" (*graphomenon organon*), and indeed the other parts of the machine, being less difficult to manufacture precisely, do lend themselves well to textual encoding. Yet when it comes to the screws, the problems of material reality creep into the design, as Hero concedes that the calculation of gear sizes might not be as simple and geometrically perfect as his initial description would suggest. Baird comments on the particular difficulties of manufacturing predictably functioning screws, which persisted even into the modern era:

> We can imagine a perfect screw, but we cannot make a perfect screw. And if we don't recognize that all material reality departs from the ideal of perfection – and design and build to compensate for this – we won't get very far. Ideas may be perfect, but reality is more interesting.[17]

Baird is here commenting on Henry Augustus Rowland's article "Screw" for the ninth edition of the *Encyclopedia Britannica*, in which he devotes extensive attention to the errors that affect a screw's performance. Rowland argues that "as the mountings of a screw introduce many errors, the final and exact test of its accuracy can only be made when it is finished and set up for use," a predicament strikingly similar to Hero's.[18] The problems of design and craft that emerge from this complicated reality have an interest of their own.

A similar descriptive challenge arises for Eratosthenes' mesolabe. This device especially demands precision, because it must convince the reader that it can solve a mathematical problem with the same accuracy as more cumbersome geometrical methods. Eratosthenes acknowledges the difficulty, but cannot offer precise written advice for how to surmount it:

> But so as we may also take the two means by a machine, a box is fixed (made of wood, or ivory, or bronze), holding three equal tablets, as thin as possible. Of these, the middle is fitted in its place, while the other two are moveable along grooves (the sizes and the proportions may be as anyone wishes them; for the arguments of the proof will yield the conclusions in

[17] Baird 2004, 152. [18] Rowland 1902, 506.

> the same way). And, for taking the lines in the most precise way, it must be done with great art, so that when the tablets are simultaneously moved they all remain parallel and firm, touching each other throughout.[19]

He gives hints – the thinness of the panels, the warning that the operation "must be done with great art" (φιλοτεχνητέον). But these hints hardly solve the problem; this is a task that must be done manually, working carefully and gradually to achieve perfection by hand.

Pappus, proposing a different mechanical solution to the same problem, runs into the same descriptive difficulties. Here, as in Eratosthenes' solution, a method is available that is promised to be easier than the competing geometrical solutions to the problem. Here too, however, the method is almost impossible to verbalize: one just has to try it for oneself, over and over, until success happens:

> Let the ruler be brought around between B and Γ, until the part of it that was moved away between the straight lines ZE and EB becomes equal to the part between the straight line BE and the circumference of BKΓ. Trying this repeatedly, and shuffling the ruler around, we will easily achieve this (τοῦτο γὰρ πειράζοντες αἰεὶ καὶ μετάγοντες τὸ κανόνιον ῥᾳδίως ποιήσομεν).[20]

Vignettes of craftsmanship and calculation like this mark the borderline between the kinds of knowledge the machine embodies that lend themselves to textual transmission (dimensions, materials, step-by-step assembly instructions) and those that are more difficult to put into text, which correspond roughly to Kogge's *empeiria*, Polanyi's "tacit knowledge," or Feibleman's "concrete knowledge." Their operation in the machine has to be figured out by hand, through a manual process of perfection like fiddling with the screw by hand, or shaping and fitting the sliders so that the tablets will slip smoothly across one another. Particularly given the limited standardization and predictability of much of the means of production available in the ancient world, some aspects of the construction process defy description; some kinds of knowledge simply have to be acquired through manual practice.

Authorial declarations of failed attempts at perfect description prompt a similar question about this project: what still remains undescribed? A great deal, certainly. Although chronological and generic breadth was one goal of this investigation, for detailed textual analysis I have concentrated principally on a few key texts, including Philo's *Belopoeica*; Biton's

[19] Eutocius, *Commentarii in libros de sphaera et cylindro* 94.15–96.2, tr. Netz. On the mesolabe's "three-dimensionality," see Archimedes and Eutocius 2004, 295–6 n. 167.

[20] Pappus, *Synagoge* 8.1072.

Kataskeuai; Vitruvius' *De architectura*; Hero's *Belopoeica*, *Pneumatica*, and *Peri automatopoiētikēs*; and Ptolemy's *Harmonica* and *Syntaxis mathematica*. This approach emerged from considerations of both space and sense. With so many different artifacts, texts, and genres to consider, to give all of them full consideration would have required an immensely lengthy treatment, in which the analytical themes would have been at risk of getting lost.

So, I do not pretend that this work provides a complete guide to the descriptive techniques used in ancient literature on technological artifacts. The advantage of this chronologically and culturally extended approach, which seemed to outweigh the disadvantages for an initial study, is that it allows us to see how technological artifacts were treated in literature among the varied cultures of the ancient Mediterranean. The intersection between "technical" works and contemporary literature of other kinds promises to be a particularly rich vein, which I have only been able to touch on here. A chronologically localized study would permit a more detailed treatment of these issues.

Technical artifact texts emerged as a class, for all practical purposes, in Hellenistic Greece. This is not to say, of course, that the artifacts themselves appeared out of nowhere during this time; they developed (as technologies do) more or less gradually, sometimes in Greece, sometimes elsewhere. But the Hellenistic era was undeniably a period of extraordinary mechanical productivity and innovation. Every major branch of the mechanical family tree that would develop in the Mediterranean over the next few centuries had already budded in this period, though of course they would all continue to adapt and expand, flourishing to different degrees in different cultural environments. It is possible, then, to look at the developments and innovations made in these artifacts by later cultures as a kind of "reception" of Hellenistic technology, and the literary reflection of this technological culture as a kind of reception in its own right. Indeed, the low survival rates of Hellenistic prose literature mean that Hellenistic mechanical technology lives on most of all in the literature (technical and otherwise) of later periods. Understanding the dynamics that shaped this multifaceted literary reception can therefore help us understand the astonishing technological innovations of this era.

As these machines found their way into the literature of other cultures, the way they were described and presented to the reader changed. Roman authors writing in Latin had to confront problems of translating the terminology of their Greek predecessors and contemporaries. They likewise worked against the backdrop of a prevailing literary taste that left comparatively little room for highly technical accounts of the mechanical features of the artifacts that had begun to spread throughout the Roman world. None of this is to say that interest in the delights of mechanical innovation vanishes from Roman literature; from Cicero's philosophical work to the poetry of Ovid, Statius, and Claudian, technological achievements remain a subject of lasting cultural interest. Prose works like Vitruvius' *De architectura*, Frontinus' *De aquae ductu urbis Romae*, and the anonymous *De rebus bellicis* demonstrate an enduring demand for more detailed explorations.

Greco-Roman authors like Athenaeus Mechanicus, Hero, and Apollodorus draw, much more often than do authors of technical works in Latin, on the textual tools of Greek geometry used by their Hellenistic predecessors. However, they seldom create a sustained, technically demanding encounter with any given artifact, but more often produce broad-ranging, comparatively superficial guides to their use. Like their Latin-language counterparts, Greco-Roman authors largely eschew the "directive reading" techniques that address the reader directly to command him to participate in flights of imagination or feats of construction. Changes in the style and presentation of technical knowledge in texts between Hellenistic and Roman authors do not represent the obliteration of Hellenistic technical knowledge, but rather the appearance of a new potential audience and a new route to status for texts containing technical knowledge.

Hero and Ptolemy privilege the systematization of disciplines over specialized knowledge of a single subject, a knowledge-management strategy that increasingly gained traction in the Roman imperial period. "Systematizing" texts like these approach the task of describing mechanical artifacts, whether as aids to the exploration of natural phenomena or as the focus of study in their own right, in a way that emphasizes the continuity between the different bodies of knowledge involved. This approach need not come at the cost of technical detail, of course; Ptolemy's work is precisely quantified, deeply informed, and often not easy reading. Nor is any given author bound by some authority to systematize at all times: the surviving part of Hero's *Cheiroballistra* provides deep technical detail, and as far as is possible to tell makes no particular attempt to contextualize that device as part of a wider interdisciplinary landscape.

Such diachronic changes do not of course tell the whole story, nor are they universally true of every text. Probably it would indeed be impossible to construct a robust causal explanation for the various changes observed here on the basis of the surviving evidence.[21] My aim here, however, has been rather to outline the variations, both subtle and drastic, in the strategies ancient authors devised for describing technological artifacts in text, against a rough chronological and cultural backdrop. We have seen mechanical artifacts described from inside and outside, through the eyes of designers and spectators, positioned at the reader's very fingertips or at an insurmountable conceptual distance. We have witnessed them in action and seen them broken down; in either case their working principles might be explained in considerable detail or effaced in favor of telling an interesting story about the artifact's original invention. While some technical ekphraseis depend entirely upon verbal description, others adduce additional tools like illustrations (schematic or veristic) or tables of dimensions. Some strongly emphasize the many and varied layers of material, mechanical, or mathematical knowledge available about the artifact itself, while others highlight the artifact's utility for exploring external bodies of knowledge, like astronomy, harmonics, or pneumatics.

This broad spectrum of descriptive approaches corresponds to the many possible ways of seeing, understanding, and valuing the objects that surround us. These possibilities are particularly rich in the case of technological artifacts, which together with the texts that bring readers into vicarious contact with them create a mutually reinforcing system of knowledge transfer. I hope that this particular text has put the existing body of knowledge about ancient mechanical technology to good use, and that it will perhaps serve to open up some new questions for study.

[21] One final acknowledgment of a book to which this one owes so much: for a more detailed account of the cautions appropriate to addressing a similar problem, see Netz 2009, 230–5.

Bibliography

Ager, S. L., and Faber, R. (eds.) (2013) *Belonging and isolation in the Hellenistic world*. Toronto.

Albrecht, M. von, and Schmeling, G. L. (1997) *A history of Roman literature: from Livius Andronicus to Boethius: with special regard to its influence on world literature*. Leiden; New York.

Alexander, L. (1993) *The preface to Luke's Gospel: literary convention and social context in Luke 1.1–4 and Acts 1.1*. Cambridge; New York.

Apollodorus (2010) *Apollodorus Mechanicus, Siege-matters = Poliorketika*. Whitehead, D., ed. Stuttgart.

Archimedes and Eutocius (2004) *The works of Archimedes: translated into English, together with Eutocius' commentaries, with commentary and critical edition of the diagrams*. Netz, R., ed. Cambridge.

Aristotle (2000) *Problemi meccanici*. Bottecchia Dehò, M. E., ed. Soveria Mannelli (Catanzaro).

Asper, M. (2001) "Dionysios (Heron, Def. 14. 3) und die Datierung Herons von Alexandria." *Hermes* 129(1): 135–7.

(2007) *Griechische Wissenschaftstexte: Formen, Funktionen, Differenzierungsgeschichten*. Stuttgart.

(2009) "Science and fiction in Callimachus." In Harder, A., Wakker, G. C., and Regtuit, R. F. (eds.), *Nature and science in Hellenistic poetry*. Leuven: 1–18.

Athenaeus (2004) *On machines = Peri mēchanēmatōn*. Whitehead, D. and Blyth, P. H., trans. Stuttgart.

Aujac, G. (1993) *La sphère: instrument au service de la découverte du monde: d'Autolycos de Pitanè à Jean de Sacrobosco*. Caen.

Baatz, D. (1994) *Bauten und Katapulte des römischen Heeres*. Stuttgart.

Baird, D. (2004) *Thing knowledge: a philosophy of scientific instruments*. Berkeley.

Bal, M. (1991) *Reading "Rembrandt": beyond the word-image opposition: the Northrop Frye lectures in literary theory*. Cambridge; New York.

Balbus (1848) "Expositio et ratio omnium mensurarum." In Lachmann, K. (ed.), *Die Schriften der römischen Feldmesser* (vol. 1). Berlin: 91–108.

(1996) *Présentation systematique de toutes les figures*. Guillaumin, J.-Y., trans. Naples.

Baldwin, B. (1978) "*The De rebus bellicis*." (vol. XVI). Prague.

Barker, A. (1989) *Greek musical writings* (vol. II). Cambridge; New York.
(2006) *Scientific method in Ptolemy's Harmonics*. Cambridge.
Barnes, J. (1975) "Aristotle's theory of demonstration." In Barnes, J., Schofield, M., and Sorabji, R. (eds.), *Articles on Aristotle* (vol. I). London: 65–87.
Bartman, E. (1992) *Ancient sculptural copies in miniature*. Leiden; New York.
Basalla, G. (1988) *The evolution of technology*. Cambridge; New York.
Bauer, J. M., and Herder, P. M. (2009) "Designing socio-technical systems." In Meijers, A. (ed.), *Philosophy of technology and engineering sciences*. Amsterdam; London; Boston: 601–30.
Baumann, M. (2011) *Bilder schreiben: virtuose ekphrasis in Philostrats Eikones*. Berlin; New York.
Baxandall, M. (1985) *Patterns of intention: on the historical explanation of pictures*. New Haven.
Beagon, M. (1992) *Roman nature: the thought of Pliny the Elder*. Oxford; New York.
Bergmann, B. (2007) "A painted garland: weaving words and images in the House of the Epigrams in Pompeii." In Newby, Z. and Leader-Newby, R. E. (eds.), *Art and inscriptions in the ancient world*. Cambridge; New York: 60–101.
Berryman, S. (2003) "Ancient automata and mechanical explanation." *Phronesis* 48(4): 344–69.
(2007) "The imitation of life in ancient Greek philosophy." In Riskin, J. (ed.), *Genesis redux: essays in the history and philosophy of artificial life*. Chicago: 35–45.
(2009) *The mechanical hypothesis in ancient Greek natural philosophy*. Cambridge; New York.
Berthelot, M., and Ruelle, C.-E. (eds.) (1888) *Collection des anciens alchimistes grecs*. Paris.
Bing, P. (2002) "The un-read Muse? Inscribed epigram and its readers in antiquity." In Regtuit, R. F., Harder, M. A., and Wakker, G. C. (eds.), *Hellenistic epigrams*. Sterling, VA: 39–66.
Blyth, P. H. (1992) "Apollodorus of Damascus and the *Poliorcetica*." *Greek, Roman, and Byzantine Studies* 33(2): 127–58.
Boehm, G., and Pfotenhauer, H. (eds.) (1995) *Beschreibungskunst, Kunstbeschreibung: Ekphrasis von der Antike bis zur Gegenwart*. Munich.
Bömer, F. (1953) "Der Commentarius: Zur Vorgeschichte und literarischen Form der Schriften Caesars." *Hermes* 81(2): 210–50.
Boudon-Millot, V. (2012) *Galien de Pergame: un médecin grec à Rome*.
Bowersock, G. W. (1989) "Philostratus and the Second Sophistic." In Easterling, P. E., and Knox, B. M. W. (eds.), *The Cambridge history of classical literature* (vol. I). Cambridge: 95–8.
Brodersen, K. (1992) *Reiseführer zu den sieben Weltwundern: Philon von Byzanz und andere antike Texte*. Frankfurt am Main.
Brown, F. E. (1963) *Vitruvius and the liberal art of architecture*. Lewisburg, PA.
Brumbaugh, R. S. (1966) *Ancient Greek gadgets and machines*. New York.
Bruun, C. (1991) *The water supply of ancient Rome: a study of Roman imperial administration*. Helsinki.

Buchner, E. (1982) *Die Sonnenuhr des Augustus*. Mainz am Rhein.
Bunge, M. (1979) "Philosophical inputs and outputs of technology." In Doner, D. B., and Bugliarello, G. (eds.), *The history and philosophy of technology*. Chicago: 262–81.
Burford, A. (1972) *Craftsmen in Greek and Roman society*. Ithaca, NY.
Büssing, H. (1984) "Optische Korrekturen: Vitruvs Empfehlungen verglichen mit der attischen Architektur des 5. Jhs. v. Chr." In Wesenberg, B. (ed.), *Vitruv-Kolloquium des Deutschen Archäologen-Verbandes e.V., durchgeführt an der Technischen Hochschule Darmstadt, 17. bis 18. Juni 1982*. Darmstadt: 27–40.
Callebat, L. (1994) "Rhétorique et architecture dans le 'De Architectura' de Vitruve." In Gros, Pierre (ed.), *Le projet de Vitruve: objet, destinataires et réception du De architectura*. Rome: 31–46.
Campbell, D. B. (2011) "Ancient catapults: some hypotheses reexamined." *Hesperia* 80(4): 677–700.
Campbell, J. B. (2000) *The writings of the Roman land surveyors: introduction, text, translation and commentary*. London.
Cantor, M. (1968) *Die römischen Agrimensoren und ihre Stellung in der Geschichte der Feldmesskunst: eine historisch-mathematische Untersuchung*. Wiesbaden.
Carettoni, G. (1960) *La pianta marmorea di Roma antica: forma urbis Romae*. Rome.
Catamo, M., Lanciano, N., Locher, K., Lombardero, M., and Valdes, M. (2000) "Fifteen further Greco-Roman sundials from the Mediterranean area and Sudan." *Journal for the History of Astronomy* 31: 203–22.
Catullus (1990) *Catullus: a commentary*. Fordyce, C. J., ed. Oxford; New York.
Caws, P. (1979) "Praxis and techne." In Doner, D. B., and Bugliarello, G. (eds.), *The history and philosophy of technology*. Chicago: 227–38.
Cech, B. (2012) *Technik in der Antike*. Darmstadt.
Chevallier, R. (1976) *Roman roads*. Berkeley.
Chouquer, G., and Favory, F. (2001) *L'arpentage romain: histoire des textes, droit, techniques*. Paris.
Cicero (1984) *De re publica*. Büchner, K., ed. Heidelberg.
Clarke, K. (1999) *Between geography and history: Hellenistic constructions of the Roman world*. Oxford.
Coarelli, F. (1992) "Le plan de via Anicia: un nouveau fragment de la Forma Marmorea de Rome." In Hinard, F., and Royo, M. (eds.), *Rome: l'espace urbain et ses représentations*. Paris; Tours: 429–56.
Collins, H. M. (2010) *Tacit and explicit knowledge*. Chicago; London.
Conso, D., Gonzalès, A., and Guillaumin, J.-Y. (2006) *Les vocabulaires techniques des arpenteurs romains: actes du colloque international, Besançon, 19–21 septembre 2002*. Besançon.
Coulton, J. J. (1977) *Ancient Greek architects at work: problems of structure and design*. Ithaca, NY.
Coxhead, M. A. (2012) "A close examination of the pseudo-Aristotelian Mechanical Problems: the homology between mechanics and poetry as technē." *Studies in History and Philosophy of Science Part A* 43(2): 300–6.

Creese, D. E. (2010) *The monochord in ancient Greek harmonic science*. Cambridge; New York.

Cuomo, S. (2000) *Pappus of Alexandria and the mathematics of late antiquity*. Cambridge; New York.

(2002) "The machine and the city: Hero of Alexandria's Belopoeica." In Tuplin, C., and Rihll, T. E. (eds.), *Science and mathematics in ancient Greek culture*. Oxford; New York: 165–77.

(2007) *Technology and culture in Greek and Roman antiquity*. Cambridge; New York.

(2011) "A Roman engineer's tales." *The Journal of Roman Studies* 101: 143–65.

Damerow, P., Renn, J., Rieger, S., and Weinig, P. (2002) "Mechanical knowledge and Pompeian balances." In Castagnetti, G. (ed.), *Homo faber: studies on nature, technology, and science at the time of Pompeii*. Rome: 93–108.

Dean-Jones, L. (2003) "Literacy and the charlatan in ancient Greek medicine." In Yunis, H. (ed.), *Written texts and the rise of literate culture in ancient Greece*. Cambridge; New York: 97–121.

Deinlein, B. (1975) *Das römische Sachbuch*. Friedrich-Alexander-Universität, Erlangen-Nürnberg.

De Jong, J. J. (1989) "Greek mathematics, Hellenistic architecture and Vitruvius' *De architectura*." In Geertman, H., and de Jong, J. J. (eds.), *Munus non ingratum: proceedings of the international symposium on Vitruvius' De architectura and the Hellenistic and Republican architecture, Leiden, 20–23 January 1987*. Leiden: 100–13.

DeLaine, J. (1996) "De aquis suis? The 'Commentarius' of Frontinus." In Nicolet, C. (ed.), *Les littératures techniques dans l'antiquité romaine*. Geneva: 117–45.

de Solla Price, D. J. (1974) *Gears from the Greeks: the Antikythera mechanism, a calendar computer from ca. 80 B.C.* Philadelphia.

(1980) "Philosophical mechanism and mechanical philosophy: some notes toward a philosophy of scientific instruments." *Annali dell'istituto e museo di storia della scienza de Firenze: 75–85.*

Diederich, S. (2007) *Römische Agrarhandbücher zwischen Fachwissenschaft, Literatur und Ideologie*. Berlin; New York.

Diels, H. (1893) "Über das physikalische System des Straton." In *Sitzungsberichte der Preussischen Akademie der Wissenschaften* (vol. XII). Berlin: 101–27.

Dilke, O. A. W. (1971) *The Roman land surveyors: an introduction to the agrimensores*. Newton Abbot.

Dipert, R. R. (1993) *Artifacts, art works, and agency*. Philadelphia.

Drachmann, A. G. (1954) "The plane astrolabe and the anaphoric clock." *Centaurus* 3: 183–9.

(1963) *The mechanical technology of Greek and Roman antiquity: a study of the literary sources*. Copenhagen; Madison.

(1973) "The crank in Graeco-Roman antiquity." In Teich, M., and Young, R. (eds.), *Changing perspectives in the history of science: essays in honour of Joseph Needham*. London: 33–51.

Duhem, P. M. M. (1991) *Les origines de la statique*. Paris.
Dunsch, B. (2012) "Arte rates reguntur: nautical handbooks in antiquity?" *Studies in History and Philosophy of Science* 43(2): 270–83.
Elder, C. (2004) *Real natures and familiar objects*. Cambridge, MA.
Ellul, J. (1990) *La technique, ou, l'enjeu du siècle* (2nd edn. rev.). Paris.
Elsner, J. (1995) *Art and the Roman viewer: the transformation of art from the pagan world to Christianity*. Cambridge; New York.
(2001) "Introduction: the genres of ekphrasis." *Ramus: critical studies in Greek and Roman literature*. 31(1/2): 1–18.
(2009) "A Protean corpus." In Bowie, E., and Elsner, J. (eds.), *Philostratus*. Cambridge; New York: 3–18.
Evans, H. B. (1994) *Water distribution in ancient Rome: the evidence of Frontinus*. Ann Arbor.
Evans, J. (1998) *The history and practice of ancient astronomy*. New York.
(1999) "The material culture of Greek astronomy." *Journal for the History of Astronomy* 30: 237–307.
(2004) "The astrologer's apparatus: a picture of professional practice in Greco-Roman Egypt." *Journal for the History of Astronomy* 35: 1–44.
Faber, R. (1995) "Vergil Eclogue 3.37, Theocritus 1 and Hellenistic ekphrasis." *The American Journal of Philology* 116(3): 411–17.
Fantuzzi, M., and Hunter, R. L. (2004) *Tradition and innovation in Hellenistic poetry*. Cambridge, England; New York.
Faulkner, W. (1994) "Conceptualizing knowledge used in innovation: a second look at the science–technology distinction and industrial innovation." *Science, Technology, & Human Values* 19(4): 425–58.
Federspiel, M. (1991) "Sur la definition euclidienne de la droite." In Vuillemin, J., and Rāshid, R. (eds.), *Mathématiques et philosophie de l'antiquité à l'age classique: hommage à Jules Vuillemin*. Paris: 115–30.
Feke, J. (2014) "Meta-mathematical rhetoric: Hero and Ptolemy against the philosophers." *Historia Mathematica* 41(3): 261–76.
Fleury, P. (1990) "Les textes techniques de l'antiquité: sources, études et perspectives." *Euphrosyne* 18: 359–94.
(1993) *La mécanique de Vitruve*. Caen.
(1994) "Le *De architectura* et les traités de mécanique ancienne." In Gros, Pierre (ed.), *Le projet de Vitruve: objet, destinataires et réception du De architectura*. Rome: 187–212.
(1996) "Traités de mécanique et textes sur les machines." In Nicolet, C. (ed.), *Les littératures techniques dans l'antiquité romaine*. Geneva: 45–76.
Fögen, T. (2000) *Patrii sermonis egestas: Einstellungen lateinischer Autoren zu ihrer Muttersprache: ein Beitrag zum Sprachbewusstsein in der römischen Antike*. Munich.
(2003) "Metasprachliche Reflexionen antiker Autoren zu den Charakteristika von Fachtexten und Fachsprachen." In Horster, M., and Reitz, C. (eds.), *Antike Fachschriftsteller: literarischer Diskurs und sozialer Kontext*. Wiesbaden: 31–60.

(2004) "Zur Transformation griechischer Wissensbestände durch römische Fachschriftsteller." In Hassler, G., and Volkmann, G. (eds.), *History of linguistics in texts and concepts*. Münster: 433–54.

Fögen, T. (2009) *Wissen, Kommunikation und Selbstdarstellung: zur Struktur und Charakteristik römischer Fachtexte der frühen Kaiserzeit*. Munich.

Folkerts, M. (1992) "Mathematische Probleme im Corpus agrimensorum." In Behrends, O., and Capogrossi Colognesi, L. (eds.), *Die römische Feldmesskunst: interdisziplinäre Beiträge zu ihrer Bedeutung für die zivilisationsgeschichte Roms*. Göttingen: 311–36.

Forbes, R. J. (1950) *Metallurgy in antiquity: a notebook for archaeologists and technologists*. Leiden.

(1955) *Studies in ancient technology*. Leiden.

Fowler, B. H. (1989) *The Hellenistic aesthetic*. Madison, WI.

Fowler, D. H. (1995) "Further arithmetical tables." *Zeitschrift für Papyrologie und Epigraphik* 105: 225–8.

Fowler, D. P. (1991) "Narrate and describe: the problem of ekphrasis." *The Journal of Roman Studies* 81: 25–35.

(1999) *The mathematics of Plato's Academy: a new reconstruction*. (2nd edn.). Oxford; New York.

Fowler, D., and Taisbak, C. (1999) "Did Euclid's circles have two kinds of radius?" *Historia Mathematica* 26(4): 361–4.

Freeth, T., Bitsakis, Y., Jones, A., and Steele, J. M. (2008) "Calendars with Olympiad display and eclipse prediction on the Antikythera Mechanism." *Nature* 454(7204): 614–17.

Freeth, T., Bitsakis, Y., Moussas, X., Seiradakis, J. H., Tselikas, A., Mangou, H., Zafeiropoulou, M., Hadland, R., Bate, D., Ramsey, A., Allen, M., Crawley, A., Hockley, P., Malzbender, T., Gelb, D., Ambrisco, W., and Edmunds, M. G. (2006) "Decoding the ancient Greek astronomical calculator known as the Antikythera Mechanism." *Nature* 444(7119): 587–91.

Freeth, T., and Jones, A. (2012) "The cosmos in the Antikythera mechanism." *ISAW Papers* 4.

Fuhrmann, M. (1960) *Das systematische Lehrbuch: ein Beitrag zur Geschichte der Wissenschaften in der Antike*. Göttingen.

Furley, D. J. (1989) *Cosmic problems: essays on Greek and Roman philosophy of nature*. Cambridge; New York.

Gale, M. (2004) "The story of us: a narratological analysis of Lucretius' *De rerum natura*." In Gale, M. (ed.), *Latin epic and didactic poetry: genre, tradition and individuality*. Swansea; Oakville, CN: 49–71.

Gallagher, R. L. (2001) "Metaphor in Cicero's 'De re publica.'" *The Classical Quarterly* 51(2): 509–19.

Gara, A. (1994) *Tecnica e tecnologia nelle società antiche*. Rome.

Gatzemeier, M. (1970) *Die Naturphilosophie des Straton von Lampsakos*. Meisenheim am Glan.

Gee, E. (2000) *Ovid, Aratus, and Augustus: astronomy in Ovid's Fasti*. Cambridge; New York.

Genette, G. (1980) *Narrative discourse: an essay in method.* Ithaca, NY.
(1982) *Figures of literary discourse.* Sheridan, A., trans. New York.
(1988) *Narrative discourse revisited.* Ithaca, NY.
Gibbs, R. (2003) "Prototypes in dynamic meaning construal." In Gavins, J., and Steen, G. (eds.), *Cognitive poetics in practice.* London; New York: 27–40.
Gibbs, S. L. (1976) *Greek and Roman sundials.* New Haven.
Gibson, R. K. (1998) "Didactic poetry as 'popular' form: a study of imperatival expressions in Latin didactic verse and prose." In Atherton, C. (ed.), *Form and content in didactic poetry.* Bari: 67–98.
Gille, B. (1980) *Les mécaniciens grecs: la naissance de la technologie.* Paris.
Goldhill, S. (1994) "The naïve and knowing eye: ekphrasis and the culture of viewing in the Hellenistic world." In Goldhill, S., and Osborne, R. (eds.), *Art and text in ancient Greek culture.* Cambridge; New York: 197–223.
(2001) "The erotic eye: visual stimulation and cultural conflict." In Goldhill, S. (ed.), *Being Greek under Rome: cultural identity, the Second Sophistic and the development of empire.* Cambridge; New York: 154–94.
(2007) "What is ekphrasis for?" *Classical Philology* 102(1): 1–19.
Goldhill, S., and Osborne, R. (eds.) (1994) *Art and text in ancient Greek culture.* Cambridge; New York.
Gombrich, E. H. (1960) *Art and illusion: a study in the psychology of pictorial representation.* New York.
Gottschalk, H. B. (1965) *Strato of Lampsacus: some texts.* Leeds.
Gow, A. S. F., and Page, D. L. (eds.) (1965) *The Greek anthology.* London.
Graf, F. (1995) "Ekphrasis: die Entstehung der Gattung in der Antike." In Boehm, G., and Pfotenhauer, H. (eds.), *Beschreibungskunst, Kunstbeschreibung: Ekphrasis von der Antike bis zur Gegenwart.* Munich: 143–55.
Gros, P. (1994) "Munus non ingratum: le traité vitruvien et la notion de service." In Gros, P. (ed.), *Le projet de Vitruve: objet, destinataires et réception du De architectura.* Rome: 75–90.
(1996) "Les illustrations du *De architectura* de Vitruve: histoire d'un malentendu." In Nicolet, C., and Gros, P. (eds.), *Les littératures techniques dans l'antiquité romaine.* Geneva: 19–44.
(2006) "Introduction générale." In *Vitruve et la tradition des traités d'architecture: fabrica et ratiocinatio: recueil d'études.* Rome: 399–435.
Guillaumin, J.-Y. (1994) "Géométrie grecque et agrimensorique romaine: la science comme justification d'une idéologie." *Dialogues d'histoire ancienne* 20(2): 279–95.
Gutzwiller, K. (2002) "Art's echo: the tradition of Hellenistic ecphrastic epigram." In Regtuit, R. F., Wakker, G. C., and Harder, M. A. (eds.), *Hellenistic epigrams.* Sterling, VA: 85–112.
Hall, J. J. (1983) "Was rapid scientific and technical progress possible in antiquity?" *Apeiron* 17: 1–13.
Hannah, R. (2008) "Timekeeping." In Oleson, J. P. (ed.), *Oxford handbook of engineering and technology in the classical world.* Oxford; New York: 740–58.
Hanson, N. R. (1958) *Patterns of discovery: an inquiry into the conceptual foundations of science.* Cambridge.

Haselberger, L. (1999) "Curvature: the evidence of Didyma." In Haselberger, L. (ed.), *Appearance and essence: refinements of classical architecture – curvature: Proceedings of the Second Williams Symposium on Classical Architecture Held at the University of Pennsylvania, Philadelphia, April 2–4, 1993*. Philadelphia: 173–84.

Hassall, M. (1979) "The inventions." *British Archaeological Reports*: 77–95.

Healy, J. F. (1999) *Pliny the Elder on science and technology*. Oxford; New York.

Heath, M. (1987) *The poetics of Greek tragedy*. London.

Heath, T. L. (1963) *A manual of Greek mathematics*. New York.

Heffernan, J. A. W. (1991) "Ekphrasis and representation." *New Literary History* 22(2): 297–316.

(1993) *Museum of words: the poetics of ekphrasis from Homer to Ashbery*. Chicago.

(1999) "Speaking for pictures: the rhetoric of art criticism." *Word & Image* 15(1): 119.

Heisel, J. P. (1993) *Antike Bauzeichnungen*. Darmstadt.

Heliodorus, Damianus, and Geminus (1897) *Damianos Schrift über Optik: mit Auszügen aus Geminos: Griechisch und Deutsch*. Schöne, R., ed. Berlin.

Herbert, A. J. (1965) *The structure of technical English*. London.

Hermogenes (1997) *L'art rhétorique*. Patillon, M., trans. Paris.

(2005) *Invention and method: two rhetorical treatises from the Hermogenic corpus*. Kennedy, G. A., trans., Rabe, H., ed. Atlanta.

Hero (1995) "Heron of Alexandria's On automaton-making." Murphy, S., trans. *History of Technology* 17: 1–44.

(1997) *Les Pneumatiques d'Héron d'Alexandrie*. Argoud, G., and Guillaumin, J.-Y., eds. Saint-Étienne.

(2003) *Erone di Alessandria: le radici filosofico-matematiche della tecnologia applicata: definitiones: testo, traduzione e commento*. Giardina, G. R., ed. Catania.

Heron Byzantius (1909) "De strategematibus." In Schneider, R. (ed.), *Griechische Poliorketiker* (vol. 1). Berlin: 4–80.

Herrmann, A., and Breidenbach, W. (1952) *Das delische Problem (die Verdoppelung des Würfels)*. Leipzig.

Hesberg, H. (1984) "Römische Grundrißpläne auf Marmor." In Deutsches Archäologisches Institut (ed.), *Bauplanung und Bautheorie der Antike*. Berlin: 120–33.

Hindle, B. (1981) *Emulation and invention*. New York.

Hindle, B., and Lubar, S. D. (1986) *Engines of change: the American industrial revolution, 1790–1860*. Washington, D.C.

Hine, H. M. (2009) "Subjectivity and objectivity in Latin scientific and technical literature." In Taub, L. C., and Doody, A. (eds.), *Authorial voices in Greco-Roman technical writing*. Trier: 13–30.

(2011) " 'Discite … Agricolae': modes of instruction in Latin prose agricultural writing from Cato to Pliny the Elder." *Classical Quarterly* 61(2): 624–54.

Hodder, I. (2012) *Entangled: an archaeology of the relationships between humans and things*. Malden, MA.

Hodge, A. T. (1991) *Future currents in aqueduct studies*. Leeds.

Horster, M. (2003) "Literarische Elite? Der soziale Kontext römischer Fachschriftsteller." In Horster, M., and Reitz, C. (eds.), *Antike Fachschriftsteller: literarischer Diskurs und sozialer Kontext*. Wiesbaden: 176–97.

Houkes, W. (2009) "The nature of technological knowledge." In Meijers, A. (ed.), *Philosophy of technology and engineering sciences*. Amsterdam; London; Boston: 309–50.

Hovgaard, W. (1926) "The arsenal in Piraeus and the ancient building rules." *Isis* 8(1): 12–20.

Høyrup, J. (1997) "A note on water-clocks and on the authority of texts." *Archiv für Orientforschung* 44/45: 192–4.

Hyde, W. W. (1938) "The recent discovery of an inscribed water-organ at Budapest." *Transactions and Proceedings of the American Philological Association* 69: 392–410.

Hyginus and Frontinus (2005) *Les arpenteurs romains, vol. i: Hygin le Gromatique; Frontin*. Guillaumin, J.-Y., trans. Paris.

Ihde, D. (1990) *Technology and the lifeworld: from garden to earth*. Bloomington.

Ivins, W. M. (1969) *Prints and visual communication*. New York.

Jacob, C. (1998) "La bibliothèque, la carte et le traité: les formes de l'accumulation du savoir à Alexandrie." In Argoud, G., and Guillaumin, J.-Y. (eds.), *Sciences exactes et sciences appliquées à Alexandrie*. Saint-Étienne: 19–37.

Jaeger, M. (2008) *Archimedes and the Roman imagination*. Ann Arbor.

Janson, T. (1964) *Latin prose prefaces: studies in literary conventions*. Stockholm.

Jauss, H. R. (1982) *Toward an aesthetic of reception*. Bahti, T., trans. Minneapolis.

Jones, A. (2009) "Mathematics, science, and medicine in the papyri." In Bagnall, R. S. (ed.), *The Oxford handbook of papyrology*. Oxford; New York: 338–57.

Kenney, E. J., and Clausen, W. V. (eds.) (1996) *The Cambridge history of classical literature* (vol. ii). Cambridge.

Keyser, P. T. (1988) "Suetonius *Nero* 41.2 and the date of Heron Mechanicus of Alexandria." *Classical Philology* 83: 218–20.

(1998) "Orreries, the date of [Plato] Letter ii, and Eudoros of Alexandria." *Archiv für Geschichte der Philosophie* 80(3): 241–67.

Keyser, P. T., and Irby-Massie, G. L. (2008) *The encyclopedia of ancient natural scientists: the Greek tradition and its many heirs*. London; New York.

Kittay, J. (1981) "Descriptive limits." *Yale French Studies* (61): 225–43.

Koelb, J. H. (2007) *The poetics of description: imagined places in European literature*. New York.

Kogge, W. (2012) "Empeiría: vom Verlust der Erfahrungshaltigkeit des 'Wissens' und vom Versuch, sie als 'implizites Wissen' wieder zu gewinnen." In Loenhoff, J. (ed.), *Implizites Wissen: epistemologische und handlungstheoretische Perspektiven*. Weilerswist: 31–48.

Koller, D., Trimble, J., Najbjerg, T., Gelfand, N., and Levoy, M. (2006) "Fragments of the city: Stanford's digital Forma Urbis Romae project." In Haselberger, L., and Humphrey, J. W. (eds.), *Imaging ancient Rome: documentation, visualization, imagination: Proceedings of the Third Williams*

Symposium on Classical Architecture, Held at the American Academy in Rome, the British School at Rome, and the Deutsches Archäologisches Institut, Rome, on May 20–23, 2004. Portsmouth, RI: 237–52.

König, A. (2007) "Knowledge and power in Frontinus' On aqueducts." In König, J., and Whitmarsh, T. (eds.), *Ordering knowledge in the Roman empire*. Cambridge; New York: 177–205.

König, J., and Whitmarsh, T. (2007a) *Ordering knowledge in the Roman empire*. Cambridge; New York.

(2007b) "Ordering knowledge." In König, J., and Whitmarsh, T. (eds.), *Ordering knowledge in the Roman empire*. Cambridge; New York: 3–41.

Konstan, D. (1994) "Foreword: to the reader." In Schiesaro, A., Mitsis, P., and Clay, J. S. (eds.), *Mega nepios: il destinatario nell'epos didascalico*. Pisa: 11–22.

Krenkel, W. (2003) "Sprache und Fachsprache." In Horster, M., and Reitz, C. (eds.), *Antike Fachschriftsteller: literarischer Diskurs und sozialer Kontext*. Wiesbaden: 11–30.

Krieger, M. (1992) *Ekphrasis: the illusion of the natural sign*. Baltimore.

Kronenberg, L. (2009) *Allegories of farming from Greece and Rome: philosophical satire in Xenophon, Varro and Virgil*. Cambridge; New York.

Kusukawa, S. (2011) *Picturing the book of nature: image, text, and argument in sixteenth-century human anatomy and medical botany*. Chicago; London.

Lachmann, K. (ed.) (1848) *Die Schriften der römischen Feldmesser* (vol. 1). Berlin.

Laird, A. (1999) *Powers of expression, expressions of power: speech presentation and Latin literature*. Oxford.

Lakoff, G., and Johnson, M. (1981) *Metaphors we live by*. Chicago.

Langacker, R. W. (2008) *Cognitive grammar: a basic introduction*. Oxford; New York; Auckland.

Langslow, D. R. (2000) *Medical Latin in the Roman empire*. Oxford; New York.

Latour, B. (1986) "Visualisation and cognition: drawing things together." *Knowledge and Society: Studies in the Sociology of Culture Past and Present* 6: 1–40.

Lavan, L. (2007) "Explaining technological change: innovation, stagnation, recession and replacement." In Lavan, L., Zanini, E., and Sarantis, A. C. (eds.), *Technology in transition: A.D. 300–650*. Leiden; Boston: xv–xl.

Leach, E. W. (1988) *The rhetoric of space: literary and artistic representations of landscape in Republican and Augustan Rome*. Princeton, NJ.

Leader-Newby, R. E. (2007) "Inscribed mosaics in the late Roman Empire: perspectives from east and west." In Newby, Z., and Leader-Newby, R. E. (eds.), *Art and inscriptions in the ancient world*. Cambridge; New York: 179–99.

Lehoux, D. (2007) *Astronomy, weather, and calendars in the ancient world: parapegmata and related texts in classical and Near Eastern societies*. Cambridge; New York.

(2012) *What did the Romans know? An inquiry into science and worldmaking*. Chicago.

(2013) "Seeing and unseeing, seen and unseen." In Lehoux, D., Morrison, A. D., and Sharrock, A. (eds.), *Lucretius: poetry, philosophy, science*. Oxford; New York: 131–51.

Lelgemann, D. (2010) *Die Erfindung der Messkunst: angewandte Mathematik im antiken Griechenland*. Darmstadt.

Lenel, O. (1920) *Zum sogenannt Gnomon des Idios Logos*. Heidelberg.

Leng, R. (2004) "Social character and pictorial style, and the grammar of technical illustration in craftsmen's manuscripts in the late Middle Ages." In Lefèvre, W. (ed.), *Picturing machines 1400–1700*. Cambridge, MA: 85–111.

Lessing, G. E. (1965) *Laokoon*. Reich, D., ed. London.

Lewis, M. (1999) "When was Biton?" *Mnemosyne* 52(2): 159–68.

Lewis, M. J. T. (2001) *Surveying instruments of Greece and Rome*. Cambridge; New York.

(2007) "Antique engineering in the Byzantine world." In Lavan, L., Zanini, E., and Sarantis, A. C. (eds.), *Technology in transition: A.D. 300–650*. Leiden; Boston: 367–78.

Lloyd, G. E. R. (1987) *The revolutions of wisdom: studies in the claims and practice of ancient Greek science*. Berkeley.

(1991) *Methods and problems in Greek science*. Cambridge; New York.

Long, P. O. (2001) *Openness, secrecy, authorship: technical arts and the culture of knowledge from antiquity to the Renaissance*. Baltimore.

Lukács, G. (1971) "Narrate or describe?" In Kahn, A. (trans.), *Writer and critic, and other essays*. New York: 110–48.

McCartney, E. S. (1932) "Signa quadrata." *The Classical Weekly* 25(10): 80.

McEwen, I. K. (2003) *Vitruvius: writing the body of architecture*. Cambridge, MA.

Manieri, A. (1999) "Colori, suoni e profumi nelle 'Imagines': principi dell'estetica filostratea." *Quaderni Urbinati di Cultura Classica* 63(3): 111–21.

Männlein-Robert, I. (2007) "Epigrams on art: voice and voicelessness in Hellenistic epigram." In Bing, P., and Bruss, J. S. (eds.), *Brill's companion to Hellenistic epigram*. Leiden; Boston: 251–71.

Marsden, E. W. (1969) *Greek and Roman artillery: historical development*. Oxford.

(1971) *Greek and Roman artillery: technical treatises*. Oxford.

Martin, A., and Wiggs, C. L. (1996) "Neural correlates of category-specific knowledge." *Nature* 379(6566): 649–52.

Masterson, M. (2004) "Status, pay, and pleasure in the 'De architectura' of Vitruvius." *The American Journal of Philology* 125(3): 387–416.

Mattern, S. P. (2013) *The prince of medicine: Galen in the Roman empire*. New York.

Meißner, B. (1999) *Die technologische Fachliteratur der Antike: Struktur, Uberlieferung und Wirkung technischen Wissens in der Antike (ca. 400 v. Chr.–ca. 500 n. Chr.)*. Berlin.

(2003) "Mündliche Vermittlung und schriftliche Unterweisung in der antiken Berufsausbildung." In Horster, M., and Reitz, C. (eds.), *Antike Fachschriftsteller: literarischer Diskurs und sozialer Kontext*. Wiesbaden: 153–75.

Mendelsohn, E., and Elkana, Y. (eds.) (1981) *Sciences and cultures: anthropological and historical studies of the sciences*. Dordrecht; Boston.

Meskell, L. (ed.) (2005) *Archaeologies of materiality*. Malden, MA.

Mitcham, C. (1994) *Thinking through technology: the path between engineering and philosophy*. Chicago.

Mitchell, W. J. T. (1994) *Picture theory: essays on verbal and visual representation.* Chicago.

(2005) *What do pictures want?: the lives and loves of images.* Chicago.

Mokyr, J. (1996) "Evolution and technological change: a new metaphor for economic history?" In Fox, R. (ed.), *Technological change: methods and themes in the history of technology.* Amsterdam: 63–83.

Netz, R. (1999) *The shaping of deduction in Greek mathematics: a study in cognitive history.* New York.

(2004) *The transformation of mathematics in the early Mediterranean world: from problems to equations.* New York.

(2009) *Ludic proof: Greek mathematics and the Alexandrian aesthetic.* New York.

Neugebauer, O. (1947) "The water clock in Babylonian astronomy." *Isis* 37(1/2): 37–43.

Newby, Z. (2009) "Absorption and erudition in Philostratus' *Imagines.*" In Bowie, E., and Elsner, J. (eds.), *Philostratus.* Cambridge; New York: 322–42.

Newby, Z., and Leader-Newby, R. E. (eds.) (2007) *Art and inscriptions in the ancient world.* Cambridge; New York.

Nicolet, C. (1991) *Space, geography, and politics in the early Roman empire.* Ann Arbor.

(1996) "Introduction." In Nicolet, C., and Gros, P. (eds.), *Les littératures techniques dans l'antiquité romaine.* Geneva: 1–17.

Nicolet, C., and Gros, P. (eds.) (1996) *Les littératures techniques dans l'antiquité romaine.* Geneva.

Nightingale, P. (2009) "Tacit knowledge and engineering design." In Meijers, A. (ed.), *Philosophy of technology and engineering sciences.* Amsterdam; London; Boston: 351–74.

Nutton, V. (1978) "The beneficial ideology." In Garnsey, P., and Whittaker, C. R. (eds.), *Imperialism in the ancient world: the Cambridge University Research Seminar in Ancient History.* Cambridge; New York: 209–22.

(2009) "Galen's authorial voice: a preliminary enquiry." In Taub, L. C., and Doody, A. (eds.), *Authorial voices in Greco-Roman technical writing.* Trier: 53–62.

Oleson, J. P. (1984) *Greek and Roman mechanical water-lifting devices: the history of a technology.* Toronto; Buffalo.

(1986) *Bronze Age, Greek, and Roman technology: a select, annotated bibliography.* New York.

(2004) "Well pumps for dummies: was there a Roman tradition of popular, sub-literary engineering manuals?" In Minonzio, F. (ed.), *Problemi di macchinismo in ambito Romano macchine idrauliche nella letteratura tecnica, nelle fonti storiografiche e nelle evidenze archeologiche di età imperiale.* Como: 65–86.

(2005) "Design, materials, and the process of innovation for Roman force pumps." In Pollini, J. (ed.), *Terra marique: studies in art history and marine archaeology in honor of Anna Marguerite McCann on the receipt of the gold medal of the Archaeological Institute of America.* Oxford; Oakville: 211–31.

Owens, J. (1991) "Aristotle on the pure and applied sciences." In Bowen, A. C. (ed.), *Science and philosophy in classical Greece*. New York: 31–42.

Panofsky, E., and Lavin, I. (1995) *Meaning in the visual arts: views from the outside: a centennial commemoration of Erwin Panofsky (1892–1968)*. Princeton, NJ.

Peachin, M. (2004) *Frontinus and the curae of the curator aquarum*. Stuttgart.

Peirce, C. S. (1931) "Logical tracts No. 2." In Hartshorne, C., Weiss, P., and Burks, A. W. (eds.), *Collected papers* (vol. IV). Cambridge: 418–509.

Perry, E. (2005) *The aesthetics of emulation in the visual arts of ancient Rome*. Cambridge; New York.

Perry, E. E. (2002) "Rhetoric, literary criticism, and the Roman aesthetics of artistic imitation." *Memoirs of the American Academy in Rome. Supplementary Volumes* 1: 153–71.

Petronōtīs, A. (1968) *Bauritzlinien und andere Aufschnürungen am Unterbau griechischer Bauwerke in der Archaik und Klassik, eine Studie zur Baukunst und -technik der Hellenen*. Munich.

Pfeiffer, R. (1968) *History of classical scholarship from the beginnings to the end of the Hellenistic age*. Oxford.

Philo (1974) *Pneumatica: the first treatise on experimental physics, western version and eastern version. Facsimile and transcript of the Latin manuscript, CLM 534, Bayer. Staatsbibliothek, Munich. Translation and illustrations of the Arabic manuscript, A.S. 3713, Aya-Sofya, Istanbul*. Prager, F. D., trans. Wiesbaden.

Pijanowski, B. C., Villanueva-Rivera, L. J., Dumyahn, S. L., Farina, A., Krause, B. L., Napoletano, B. M., Gage, S. H., and Pieretti, N. (2011) "Soundscape ecology: the science of sound in the landscape." *BioScience* 61(3): 203–16.

Platt, V. J. (2007) "'Honour takes wing': unstable images and anxious orators in the Greek tradition." In Newby, Z., and Leader-Newby, R. E. (eds.), *Art and inscriptions in the ancient world*. Cambridge; New York: 247–71.

(2011) *Facing the gods: epiphany and representation in Graeco-Roman art, literature and religion*. Cambridge; New York.

Polanyi, M. (2009) *The tacit dimension*. Chicago; London.

Popper, K. R. (1935) *Logik der Forschung: zur Erkenntnistheorie der modernen Naturwissenschaft*. Vienna.

(1959) *The logic of scientific discovery*. New York.

Ptolemy (2000) *Harmonics*. Solomon, J., trans. Leiden; Boston.

Purcell, N. (1983) "The apparitores: a study in social mobility." *Papers of the British School at Rome* 51: 125–73.

Reay, B. (2005) "Agriculture, writing, and Cato's aristocratic self-fashioning." *Classical Antiquity* 24(2): 331–61.

Rees, R. (1994) "Common sense in Catullus 64." *The American Journal of Philology* 115(1): 75–88.

Reinhardt, T. (2011) "Galen on unsayable properties." *Oxford Studies in Ancient Philosophy* 40: 297–317.

Riccobono, S. (1950) *Il gnomon dell'Idios Logos*. Palermo.

Riggsby, A. (2007) "Guides to the wor(l)d." In König, J., and Whitmarsh, T. (eds.), *Ordering knowledge in the Roman empire*. Cambridge; New York: 88–107.

Rihll, T. E. (2007) *The catapult: a history*. Yardley, PA.

Risselada, R. (1993) *Imperatives and other directive expressions in Latin: a study in the pragmatics of a dead language*. Amsterdam.

Roberts, M. (1989) *The jeweled style: poetry and poetics in late antiquity*. Ithaca, NY.

Robillard, V. K., and Jongeneel, E. (eds.) (1998) *Pictures into words: theoretical and descriptive approaches to ekphrasis*. Amsterdam.

Roby, C. (2014) "Experiencing geometry in Roman surveyors' texts." *Nuncius* 29(1): 9–52.

Rodríguez Almeida, E. (1981) *Forma urbis marmorea: aggiornamento generale 1980*. Rome.

Rosenmeyer, P. A. (2001) *Ancient epistolary fictions: the letter in Greek literature*. Cambridge; New York.

Rosmarin, A. (1985) *The power of genre*. Minneapolis.

Rowland, H. A. (1902) "Screw." In *The physical papers of Henry Augustus Rowland, Johns Hopkins University, 1876–1901: collected for publication by a committee of the faculty of the university*. Baltimore: 506–11.

Rüpke, J. (1992) "Wer las Caesars *Bella* als *Commentarii*?" *Gymnasium* 99: 201–26.

Saastamoinen, A. (2003) "The literary character of Frontinus' De aquaeductu." In Bruun, C., and Saastamoinen, A. (eds.), *Technology, ideology, water: from Frontinus to the Renaissance and beyond: papers from a conference at the Institutum Romanum Finlandiae, May 19–20, 2000*. Rome: 15–39.

Sachs, A. (1974) "Babylonian observational astronomy." *Philosophical Transactions of the Royal Society of London: Series A, Mathematical and Physical Sciences* 276(1257): 43–50.

Sager, J. C., and Dungworth, D. (1980) *English special languages: principles and practice in science and technology*. Wiesbaden.

Saito, K. (2009) "Reading Greek mathematics." In Robson, E., and Stedall, J. A. (eds.), *The Oxford handbook of the history of mathematics*. Oxford; New York: 801–26.

(2012) "Traditions of the diagram, tradition of the text: a case study." *Synthese* 186(1): 7–20.

Sallmann, K. (1984) "Bildungsvorgaben des Fachschriftstellers: Bemerkungen zur Pädagogik Vitruvs." In Wesenberg, B., and Deutscher Archäologen-Verband (eds.), *Vitruv-Kolloquium des deutschen Archäologen-Verbandes e.V., durchgeführt an der Technischen Hochschule Darmstadt, 17. bis 18. Juni 1982*. Darmstadt: 11–26.

Santini, C. (1990) "Le praefationes dei gromatici." In Santini, C., and Scivoletto, N. (eds.), *Prefazioni, prologhi, proemi di opere tecnico-scientifiche latine*. Rome: 135–48.

Scarry, E. (1999) *Dreaming by the book*. New York.

Schaldach, K. (1998) *Römische Sonnenuhren: eine Einführung in die antike Gnomonik*. Thun.

Scheidel, W. (1996) "Finances, figures and fiction." *The Classical Quarterly* 46(1): 222–38.

Schellenberg, H. M. (2008) "Anmerkungen zu Heron von Alexandria und zu seinem Werk über den Geschützbau." In Schellenberg, H. M., Hirschmann, V.-E., and Krieckhaus, A. (eds.), *A Roman miscellany: essays in honour of Anthony R. Birley on his seventieth birthday*. Gdańsk: 92–130.

Schiefsky, M. (2005) "Technical terminology in Greco-Roman treatises on artillery construction." In Fögen, T. (ed.), *Antike Fachtexte = Ancient technical texts*. Berlin; New York: 253–70.

(2007) "Art and nature in ancient mechanics." In Bensaude-Vincent, B., and Newman, W. R. (eds.), *The artificial and the natural: an evolving polarity*. Cambridge, MA: 67–108.

(2008) "Theory and practice in Heron's Mechanics." In Laird, W. R., and Roux, S. (eds.), *Mechanics and natural philosophy before the scientific revolution*. Dordrecht; London: 15–49.

Schironi, F. (2010) "Technical languages: science and medicine." In Bakker, E. J. (ed.), *A companion to the ancient Greek language*. Chichester; Malden, MA: 338–53.

Schlachter, A. (1927) *Der Globus, seine Entstehung und Verwendung in der Antike nach den literarischen Quellen und den Darstellungen in der Kunst*. Leipzig; Berlin.

Schneider, R. (1906) "Herons Cheiroballistra." *Deutsches Archäologisches Institut: Römische Abteilung* 21: 142–68.

Scholz, B. (1998) "*Sub oculos subiectio*: Quintilian on *ekphrasis* and *enargeia*." In Robillard, V. K., and Jongeneel, E. (eds.), *Pictures into words: theoretical and descriptive approaches to ekphrasis*. Amsterdam: 73–99.

Schrijvers, P. H. (1989) "Vitruve et la vie intellectuelle de son temps." In Geertman, H., and Jong, J. J. de (eds.), *Munus non ingratum: proceedings of the international symposium on Vitruvius' De architectura and the Hellenistic and Republican architecture, Leiden, 20–23 January 1987*. Leiden: 13–21.

Schürmann, A. (2002) "Pneumatics on stage in Pompeii: ancient automatic devices and their social context." In Castagnetti, G. (ed.), *Homo faber: studies on nature, technology, and science at the time of Pompeii*. Rome: 35–56.

Schütz, M. (1990) "Zur Sonnenuhr des Augustus auf dem Marsfeld." *Gymnasium* 97: 432–57.

Shapin, S., and Schaffer, S. (1985) *Leviathan and the air-pump: Hobbes, Boyle, and the experimental life: including a translation of Thomas Hobbes, Dialogus physicus de natura aeris by Simon Schaffer*. Princeton, NJ.

Sidoli, N. (2011) "Heron of Alexandria's date." *Centaurus* 53(1): 55–61.

Sidoli, N., and Saito, K. (2012) "Diagrams and arguments in ancient Greek mathematics: lessons drawn from comparisons of the manuscript diagrams with those in modern critical editions." In Chemla, K. (ed.), *The history of mathematical proof in ancient traditions*. Cambridge: 135–62.

Sinding, M. (2002) "After definitions: genre, categories, and cognitive science." *Genre* 35(2): 181–220.

Sorabji, R. (1988) *Matter, space and motion: theories in antiquity and their sequel.* Ithaca, NY.

Spatharakis, I. (2004) *The illustrations of the Cynegetica in Venice: codex Marcianus Graecus Z 139*. Leiden.

Squire, M. (2007) "The motto in the grotto: inscribing illustration and illustrating inscription at Sperlonga." In Newby, Z., and Leader-Newby, R. E. (eds.), *Art and inscriptions in the ancient world.* Cambridge; New York: 102–27.

(2009) *Image and text in Graeco-Roman antiquity*. New York.

(2010) "Making Myron's cow moo? Ecphrastic epigram and the poetics of simulation." *American Journal of Philology* 131(4): 589–634.

Standage, T. (2002) *The mechanical Turk: the true story of the chess-playing machine that fooled the world.* London.

Statius, P. P. (1988) *Statius: Silvae IV*. Coleman, K. M., ed. Oxford.

Steiner, W. (1988) *Pictures of romance: form against context in painting and literature*. Chicago.

Stern, J. P. (1973) *On realism*. London; Boston.

Stewart, A. (1998) "Nuggets: mining the texts again." *American Journal of Archaeology* 102(2): 271–82.

Straub, H. (1949) *Die Geschichte der Bauingenieurkunst: ein Überblick von der Antike bis in die Neuzeit*. Basel.

Stückelberger, A. (1994) *Bild und Wort: das illustrierte Fachbuch in der antiken Naturwissenschaft, Medizin und Technik*. Mainz am Rhein.

(1998) "Vom anatomischen Atlas des Aristoteles zum geographischen Atlas des Ptolemaios: Beobachtungen zu wissenschaftlichen Bilddokumentationen." In Kullmann, W., Althoff, J., and Asper, M. (eds.), *Gattungen wissenschaftlicher Literatur in der Antike*. Tübingen: 287–307.

Swetnam-Burland, M. (2010) "Aegyptus redacta: the Egyptian obelisk in the Augustan Campus Martius." *The Art Bulletin* 92(3): 135.

Tanner, J. (2005) *The invention of art history in ancient Greece: religion, society and artistic rationalisation*. Cambridge.

Taub, L. C. (1993) *Ptolemy's universe: the natural philosophical and ethical foundations of Ptolemy's astronomy*. Chicago.

(2002) "Instruments of Alexandrian astronomy: the uses of the equinoctial rings." In Tuplin, C., and Rihll, T. E. (eds.), *Science and mathematics in ancient Greek culture*. Oxford; New York: 133–49.

(2008) "Eratosthenes sends greetings to King Ptolemy." In Dauben, J. W. (ed.), *Mathematics celestial and terrestrial: Festschrift für Menso Folkerts zum 65. Geburtstag*. Halle (Saale); Stuttgart: 285–302.

Taub, L. C., and Doody, A. (eds.) (2009) *Authorial voices in Greco-Roman technical writing*. Trier.

Thomas, E. (1995) "Zum Zeugniswert griechischer Beischriften auf römischen Wandgemälden der späten Republik und frühen Kaiserzeit." *Mededelingen van het Nederlands Instituut te Rome* 54: 110–23.

Thomas, J. (2001) *Reading images*. Houndsmills, Hants; New York.

Thomasson, A. L. (2009) "Artefacts in metaphysics." In Meijers, A. (ed.), *Philosophy of technology and engineering sciences*. Amsterdam; London; Boston: 191–212.

Toneatto, L. (1994) *Codices artis mensoriae: i manoscritti degli antichi opuscoli latini d'agrimensura (V–XIX sec.)*. Spoleto.

Tueller, M. A. (2008) *Look who's talking: innovations in voice and identity in Hellenistic epigram*. Leuven; Dudley, MA.

Turner, M. (1991) *Reading minds: the study of English in the age of cognitive science*. Princeton, NJ.

Tybjerg, K. (2000) *Doing philosophy with machines: Hero of Alexandria's rhetoric of mechanics in relation to the contemporary philosophy*. Ph.D. diss., University of Cambridge.

(2003) "Wonder-making and philosophical wonder in Hero of Alexandria." *Studies in the History and Philosophy of Science* 34: 443–66.

(2004) "Hero of Alexandria's mechanical geometry." *Apeiron* 37: 29–56.

(2012) "'Senses and hands to the same degree as thought': Ole Rømer's mechanical astronomy." *Centaurus* 54(1): 77–102.

Tybout, R. A. (1989) "Die Perspektive bei Vitruv: zwei Überlieferungen von *scaenographia*." In Geertman, H., and Jong, J. J. de (eds.), *Munus non ingratum: proceedings of the international symposium on Vitruvius' De architectura and the Hellenistic and Republican architecture, Leiden, 20–23 January 1987*. Leiden: 55–68.

Ulrich, R. (2008) "Representations of technical processes." In Oleson, J. P. (ed.), *Oxford handbook of engineering and technology in the classical world*. Oxford; New York: 37–68.

Université Strasbourg and Centre de recherche sur le Proche-Orient et la Grèce antiques (eds.) (1985) *Le dessin d'architecture dans les sociétés antiques: actes du colloque de Strasbourg, 26–28 janvier 1984*. Leiden.

Untermann, M. (1983) "Neues zur Skeuothek des Philon." In Deutsches Archäologisches Institut, A.-R. (ed.), *Bauplanung und Bautheorie der Antike*. Berlin: 81–6.

Usener, S. (1994) *Isokrates, Platon und ihr Publikum: Hörer und Leser von Literatur im 4. Jahrhundert v. Chr*. Tübingen.

Valenzani, R. S. (2007) "Public and private building activity in late antique Rome." In Lavan, L., Zanini, E., and Sarantis, A. C. (eds.), *Technology in transition: A.D. 300–650*. Leiden; Boston: 435–49.

Van der Eijk, P. (1997) "Towards a rhetoric of ancient scientific discourse: some formal characteristics of Greek medical and philosophical texts." In Bakker, E. J. (ed.), *Grammar as interpretation: Greek literature in its linguistic contexts*. Leiden; New York: 72–129.

Van Leeuwen, J. (2014) "Thinking and learning from diagrams in the Aristotelian Mechanics." *Nuncius* 29(1): 53–87.

Vázquez Buján, M. E. (1994) *Tradición e innovación de la medicina latina de la antigüedad y de la alta edad media: actas del IV coloquio internacional sobre los "textos médicos latinos antiguos."* Santiago de Compostela.

Vincenti, W. G. (1990) *What engineers know and how they know it: analytical studies from aeronautical history*. Baltimore.

Vitrac, B. (2010) "Héron d'Alexandrie et le corpus métrologique: état des lieux." In *Géométrie(s), pratiques d'arpentage et enseignement: quels liens et dans quel contexte?* Paris.

Vitruvius (1969) *De l'architecture*. Fleury, P., and Callebat, L., eds. (vol. x). Paris.

Volk, K. (2002) *The poetics of Latin didactic: Lucretius, Vergil, Ovid, Manilius*. Oxford; New York.

Von Staden, H. (1994) "Author and authority: Celsus on the construction of a scientific self." In Vázquez Buján, M. E. (ed.), *Tradición e innovación de la medicina latina de la antigüedad y de la alta edad media: actas del IV coloquio internacional sobre los "textos médicos latinos antiguos."* Santiago de Compostela: 103–17.

(1996) "Body and machine: interactions between medicine, mechanics, and philosophy in early Alexandria." In J. Paul Getty Center for the History of Art and the Humanities (ed.), *Alexandria and Alexandrianism*. Malibu, CA: 85–106.

(1998) "Andréas de Caryste et Philon de Byzance: médecine et mécanique à Alexandrie." In Argoud, G., and Guillaumin, J.-Y. (eds.), *Sciences exactes et sciences appliquées à Alexandrie*. Saint-Étienne: 147–72.

Walker, A. D. (1993) "Enargeia and the spectator in Greek historiography." *Transactions of the American Philological Association* 123: 353–77.

Webb, R. (1999) "Ekphrasis ancient and modern: the invention of a genre." *Word and Image* 15(1): 7–18.

(2006) "The *Imagines* as a fictional text: ekphrasis, apatê, and illusion." In Costantini, M., Graziani, F., and Rolet, S. (eds.), *Le défi de l'art: Philostrate, Callistrate et l'image sophistique*. Rennes: 113–36.

(2009) *Ekphrasis, imagination and persuasion in ancient rhetorical theory and practice*. Farnham, Surrey; Burlington, VT.

Weitzmann, K. (1947) *Illustrations in roll and codex: a study of the origin and method of text illustration*. Princeton, NJ.

Wescher, C. (1867) *Poliorcétique des Grecs: traités theoriques: récits historiques*. Paris.

White, K. D. (1979) "Harvesting machines, Palladius and technology in the later Roman empire." *British Archaeological Reports*: 39–45.

(1984) *Greek and Roman technology*. Ithaca, NY.

White, L., Jr (1980) "Technological development in the transition from antiquity to the Middle Ages." In Gabba, E. (ed.), *Tecnologia, economia e società nel mondo romano: atti del convegno di como, 27/28/29 settembre 1979*. Como: 235–51.

Whitehouse, H. (2007) "Drawing a fine line in Oxyrhynchus." In Bowman, A., Coles, R., Gonis, N., and Parsons, P. (eds.), *Oxyrhynchus: a city and its texts*. London: 296–306.

Wolkenhauer, A. (2011) *Sonne und Mond, Kalender und Uhr: Studien zur Darstellung und poetischen Reflexion der Zeitordnung in der römischen Literatur*. Berlin; New York.

Wright, M. T. (2003) "Epicyclic gearing and the Antikythera Mechanism part I." *Antiquarian Horology* 27: 270–9.

(2005) "Epicyclic gearing and the Antikythera Mechanism part II." *Antiquarian Horology* 29(1): 51–63.

Yacobi, T. (1995) "Pictorial models and narrative ekphrasis." *Poetics Today* 16(4): 599–649.

Zanker, G. (1981) "Enargeia in the ancient criticism of poetry." *Rheinisches Museum für Philologie* 124(3/4): 297–311.

(1987) *Realism in Alexandrian poetry: a literature and its audience.* London; Wolfboro, NH.

(2007) *Modes of viewing in Hellenistic poetry and art.* Madison, WI.

Zhmud, L. (2006) *The origin of the history of science in classical antiquity.* Berlin.

Zimmer, G. (1982) *Römische Berufsdarstellungen.* Berlin.

Texts Consulted

Aelian (1853) "Tactica." In Köchly, H. A. T., and Rüstow, W. (eds.), *Griechische Kriegsschriftsteller: Griechisch und deutsch, mit kritischen und erklärenden Anmerkungen* (vol. II.1). Leipzig.

Aelius Theon (1854) "Progymnasmata." In Spengel, L. von (ed.), *Rhetores graeci, ex recognitione Leonardi Spengel.* Leipzig.

[Anonymous] (1957) *Anthologia graeca.* Beckby, H., ed. Munich.

[Anonymous] (1964) *Rhetorica ad Herennium.* Marx, F., ed. (vol. I). Leipzig.

[Anonymous] (1979) *De rebus bellicis.* Ireland, R. I., ed. Oxford.

Anthemius (1959) *Anthemius of Tralles: a study of later Greek geometry.* Huxley, G. L., ed. Cambridge, MA.

Aphthonius (1926) *Progymnasmata.* Rabe, H., ed. Leipzig.

Apollodoros of Damascus (1908) "Poliorcetica." In Schneider, R. (ed.), *Griechische Poliorketiker.* Berlin.

Apollonius (1974) *Apollonii Pergaei quae Graece exstant cum commentariis antiquis.* Heiberg, J. L., ed. Stuttgart.

Archimedes (1970a) *Archimède.* Mugler, C., ed. Paris.

(1970b) "De conoidibus et sphaeroidibus." In Mugler, C. (ed.), *Archimède.* Paris.

Archimedes and Eutocius (2004) *The works of Archimedes: translated into English, together with Eutocius' commentaries, with commentary and critical edition of the diagrams.* Netz, R., ed. Cambridge.

Aristotle and ps.-Aristotle (1913) *Aristotelis De animalium motione et De animalium incessu; Ps-Aristotelis De spiritu libellus.* Jaeger, W., ed. Leipzig.

Aristotle (1924) *Aristotle's Metaphysics: a revised text with introduction and commentary.* Ross, W. D., ed. Oxford.

(1957) *Politica.* Ross, W. D., ed. New York.

(1960) "Mechanica." In Gigon, O., and Bekker, I. (eds.), *Aristotelis opera ex recensione Immanuelis Bekkeri edidit Academia regia borussica, accedunt fragmenta scholia index Aristotelicus* (vol. II). Berlin.

(1964) *Histoire des animaux.* Louis, P., ed. (vols. I–III). Paris.

(1966) *Aristotelis De arte poetica liber.* Kassel, R., ed. Oxford.

(1978) *Aristotelis Analytica priora et posteriora.* Ross, W. D., ed. Oxford.

(1979) *Aristotelis Ethica nicomachea.* Bywater, I., ed. Oxford.

Athenaeus (1912) "De machinis." In Schneider, R. (ed.), *Griechische Poliorketiker* (vol. I). Berlin.

(2004) *On machines = Peri mēchanēmatōn*. Whitehead, D., and Blyth, P. H., trans. Stuttgart.
Augustine (2001) *De trinitate (Bücher VIII–XI, XIV–XV, Anhang: Buch V): Lateinisch-deutsch*. Kreuzer, J., ed. Hamburg.
Aulus Gellius (1990) *A. Gellii Noctes atticae*. Marshall, P. K., ed. Oxford.
Balbus (1848) "Expositio et ratio omnium mensurarum." In Lachmann, K. (ed.), *Die Schriften der römischen Feldmesser* (vol. 1). Berlin.
(1996) *Présentation systematique de toutes les figures*. Guillaumin, J.-Y., trans. Napoli.
Baldwin, B. (1978) *The De rebus bellicis* (vol. xvi). Prague.
Berthelot, M., and Ruelle, C.-E. (eds.) (1888) *Collection des anciens alchimistes grecs*. Paris.
Callimachus (1949) *Callimachus*. Pfeiffer, R., ed. Oxford.
Cassiodorus (1894) *Cassiodori senatoris variae*. Mommsen, T., ed. Berlin.
Cassius Dio (1895) *Cassii Dionis Cocceiani Historiarum romanarum quae supersunt*. Boissevain, U. P., ed. (vols. i–iii). Berlin.
Cato, Marcus Porcius (1962) *M. Porci Catonis De agri cultura*. Mazzarino, A., ed. Leipzig.
Catullus (1983) *Catullus*. Goold, G. P., ed. London.
Cicero (1922) "Academicorum reliquiae cum Lucullo." In Plasberg, O., Friedrich, W., and Müller, C. F. W. (eds.), *Scripta quae manserunt omnia* (fasc. 42). Stuttgart.
(1942a) *Cicero: in twenty-eight volumes*. Rackham, H., ed. (vol. iv). Cambridge, MA; London.
(1942b) "De partitione oratoria." In Rackham, H. (ed.), *Cicero: in twenty-eight volumes* (vol. iv). Cambridge, MA; London.
(1961a) "De natura deorum." In Ax, W., Müller, C. F. W., and Friedrich, W. (eds.), *Scripta quae manserunt omnia* (fasc. 45). Stuttgart.
(1961b) "De oratore." In Reis, P., Müller, C. F. W., and Friedrich, W. (eds.), *Scripta quae manserunt omnia* (fasc. 5). Stuttgart.
(1961c) "De finibus bonorum et malorum." In Schiche, T. (ed.), *Scripta quae manserunt omnia* (fasc. 43). Stuttgart.
(1995) *De re publica: selections*. Zetzel, J. E. G., ed. Cambridge; New York.
Columella (1897) *Opera quae exstant*. Lundström, V., ed. Uppsala.
Demetrius (1967) *Demetrii Phalerei qui dicitur De elocutione libellus*. Radermacher, L., ed. Stuttgart.
Diodorus (1888) *Diodori Bibliotheca historica*. Fischer, C. T., Dindorf, L. A., Bekker, I., and Vogel, F., eds. Leipzig.
Dionysius (1965a) "De Dinarcho." In Usener, H., and Radermacher, L. (eds.), *Dionysii Halicarnasei quae exstant* (vol. v). Stuttgart.
(1965b) "De Lysia." In Usener, H. and Radermacher, L. (eds.), *Dionysii Halicarnasei quae exstant* (vol. v). Stuttgart.
Epicurus (1967) *Opere*. Arrighetti, G., ed. Torino.
Eutocius (1972) "Commentarii in libros de sphaera et cylindro." In Heiberg, J. L., and Stamatis, E. S. (eds.), *Archimedis opera omnia; cum commentariis Eutocii* (vol. iii). Stuttgart.

Frontinus (1973) *De aquaeductu urbis Romae*. Kunderewicz, C., ed. Leipzig.
(2004) *Aquaeductu urbis Romae*. Rodgers, R. H., ed. Cambridge; New York.
Galen (1821a) "De anatomicis administrationibus." In Kühn, K. G. (ed.), *Klaudiou Galenou Hapanta: Clavdii Galeni opera omnia* (vol. II). Leipzig.
(1821b) "De dignoscendis pulsibus." In Kühn, K. G. (ed.), *Opera omnia* (vol. VIII). Leipzig.
(1821c) "De pulsibus libellus ad tirones." In Kühn, K. G. (ed.), *Klaudiou Galenou Hapanta: Clavdii Galeni opera omnia* (vol. VIII). Leipzig.
(1968) *Galen on the usefulness of the parts of the body*. May, M. T., trans. Ithaca, NY.
(1978) *On the doctrines of Hippocrates and Plato*. De Lacy, P., ed. Berlin.
Geminus (1897) "Fragmenta optica." In Schöne, R. (ed.), *Damianos Schrift über Optik: mit Auszügen aus Geminos: Griechisch und Deutsch*. Berlin.
Gow, A. S. F., and Page, D. L. (eds.) (1965) *The Greek anthology*. London.
Heliodorus, Damianus, and Geminus (1897) *Damianos Schrift über Optik: mit Auszügen aus Geminos: Griechisch und Deutsch*. Schöne, R., ed. Berlin.
Hermogenes (1985) "Peri ideōn logou." In Rabe, H. (ed.), *Hermogenis opera*. Stuttgart.
(1997) *L'art rhétorique*. Patillon, M., trans. Paris.
(2005) *Invention and method: two rhetorical treatises from the Hermogenic corpus*. Kennedy, G. A., trans., Rabe, H., ed. Atlanta.
Hero (1918) "Herons Belopoiika (Schrift vom Geschützbau) Griechisch und Deutsch." In Diels, H., and Schramm, E. (eds.), (vol. II). Berlin.
(1976a) "De automatis (Peri automatopoiētikēs)." In Schmidt, W. (ed.), *Heronis Alexandrini opera quae supersunt omnia* (vol. I). Stuttgart.
(1976b) "Definitiones." In Heiberg, J. L. (ed.), *Heronis Alexandrini opera quae supersunt omnia* (vol. IV). Stuttgart.
(1976c) "Dioptra." In Schöne, H. (ed.), *Heronis Alexandrini opera quae supersunt omnia* (vol. III). Stuttgart.
(1976d) "Mechanicorum fragmenta." In Schmidt, W., and Nix, L. L. M. (eds.), *Heronis Alexandrini opera quae supersunt omnia* (vol. II.I). Stuttgart.
(1976e) "Pneumatica." In Schmidt, W. (ed.), *Heronis Alexandrini opera quae supersunt omnia* (vol. I). Stuttgart.
(1976f) "Rationes dimetiendi et commentatio dioptrica." In Schöne, H. (ed.), *Heronis Alexandrini opera quae supersunt omnia* (vol. III). Stuttgart.
(1988) *Les mécaniques, ou, L'élévateur des corps lourds*. Qusṭā ibn Lūqā and Carra de Vaux, B., trans. Paris.
(1995) "Heron of Alexandria's On automaton-making." Murphy, S., trans. *History of Technology* 17: 1–44.
Heron Byzantius (1909) "De strategematibus." In Schneider, R. (ed.), *Griechische Poliorketiker* (vol. I). Berlin.
Horace (1959) *Q. Horati Flacci opera*. In Klingner, F. ed. Teubner.
Lucian (1913a) "De domo." In Harmon, A. M. (ed.), *Lucian* (vol. I). London; New York.

(1913b) "Quomodo historia conscribenda sit." In Kilburn, K. (ed.), *Lucian* (vol. VI). London; New York.

Lucretius (1992) *T. Lucreti Cari De rerum natura: libri sex.* Martin, J., ed. Stuttgart.

Maximus (1994) *Dissertationes.* Trapp, M. B., ed. Stuttgart.

Nicolaus (1913) *Nicolai Progymnasmata.* Felten, J., ed. Leipzig.

Onasander (1923) "Strategicus." In Oldfather, W. A., Pease, A. S., and Titchener, J. B. (eds.), *Aeneas Tacticus, Asclepiodotus, Onasander.* London; New York.

Optatianus Porfyrius, Publilius (1926) *Carmina.* Kluge, E., ed. Leipzig.

Oribasius (1928) *Oribasii collectionum medicarum reliquiae.* Raeder, H., ed. (vol. I–IV). Leipzig; Berlin.

Ovid (1988) *P. Ovidi Nasonis Fastorum libri sex.* Courtney, E., Alton, E. H., and Wormell, D. E. W., eds. Leipzig.

Palladius, R. T. A. (1898) *Agricultura.* Schmitt, J. K., ed. Leipzig.

Pappus (1876) *Pappi Alexandrini collectionis quae supersunt.* Hultsch, F., ed. Berlin.

Pausanias (1903) *Pausaniou Hellados periēgēsis = Pausaniae Graeciae descriptio.* Spiro, F., ed. Leipzig.

Philo (1919) "Philons Belopoiika (viertes Buch der Mechanik)." In Diels, H., and Schramm, E. A. (eds.), *Philons Belopoiika (viertes Buch der Mechanik): Griechisch und deutsch* (vol. XVI). Berlin.

(1974) *Pneumatica: the first treatise on experimental physics, western version and eastern version. Facsimile and transcript of the Latin manuscript, CLM 534, Bayer. Staatsbibliothek, Munich. Translation and illustrations of the Arabic manuscript, A.S. 3713, Aya-Sofya, Istanbul.* Prager, F. D., trans. Wiesbaden.

Philo, Diels, H., and Schramm, E. A. (1920) *Exzerpte aus Philons Mechanik b. VII und VIII (vulgo fünftes Buch) Griechisch und Deutsch.* Berlin.

Philostratus (1870) "Apollonii vita." In Kayser, C. L. (ed.), *Flavii Philostrati opera* (vol. I). Leipzig.

(1893) *Philostrati maioris Imagines.* Benndorf, O., and Schenkl, K., eds. Leipzig.

Plato (1899a) "Philebus". In Burnet, J. (ed.), *Platonis opera* (vol. II). Oxford.

(1899b) "Timaeus." In Burnet, J. (ed.), *Platonis opera* (vol. IV). Oxford.

Plinius Secundus, Gaius (1967) *Naturalis historia: libri XXXVII.* Mayhoff, K., ed. Stuttgart.

Plinius Caecilius Secundus, Gaius (1963) *Epistularum libri decem.* Mynors, R. A. B., ed. Oxford.

Plutarch (1925) "De gloria Atheniensium." In Nachstädt, W. (ed.), *Plutarchi Moralia* (vol. II.2). Leipzig.

(1964) *Plutarchi Vitae parallelae.* Ziegler, K., ed. Leipzig.

Polybius (1993) *Polybii Historiae.* Dindorf, L. A., and Büttner-Wobst, T., eds. Stuttgart.

Posidippus (2002) *Posidippi Pallaei quae supersunt omnia.* Austin, C., and Bastianini, G., eds. Milan.

(2005) *The new Posidippus: a Hellenistic poetry book.* Gutzwiller, K. J., ed. New York.

Proclus (1873) *Procli Diadochi in primum Euclidis Elementorum librum commentarii*. Friedlein, G., ed. Leipzig.

(1974) *Procli Diadochi Hypotyposis astronomicarum positionum: una cum scholiis antiquis e libris manu scriptis*. Manitius, C., ed. Stuttgart.

ps.-Callisthenes (2005) *Historia Alexandri magni*. Kroll, W., ed. (vol. 1). Hildesheim.

ps.-Hermogenes (1985) "Progymnasmata." In Rabe, H. (ed.), *Hermogenis opera*. Stuttgart.

ps.-Longinus (1964) *On the sublime*. Russell, D. A., ed. Oxford.

Ptolemy (1930) *Die Harmonielehre des Klaudios Ptolemaios*. Düring, I., ed. Göteborg.

Ptolemy (1957) *Opera quae exstant omnia*. Heiberg, J. L., ed. (vols. 1.1, 1.2). Leipzig.

Ptolemy (2000) *Harmonics*. Solomon, J., trans. Leiden; Boston.

Statius (1988) *Statius: Silvae IV*. Coleman, K. M., ed. Oxford.

Straton (1944) "Fragmenta." In Wehrli, F. (ed.), *Die Schule des Aristoteles, Texte und Kommentare* (vol. v). Basel.

Theophrastus (1917) *Theophrastus and the Greek physiological psychology before Aristotle*. Stratton, G. M., trans. London.

Tiberius (1968) *Tiberii De figuris Demosthenicis: libellus cum deperditorum operum fragmentis*. Ballaira, G., ed. Rome.

Varro (1929) *M. Terenti Varronis Rerum rusticarum libri tres*. Goetz, G., ed. Leipzig.

Vegetius Renatus, Flavius (1990) *Epitoma rei militaris*. Stelten, L. F., ed. New York.

Vitruvius (1912) *Vitruvii De architectura libri decem*. Krohn, F., ed. Leipzig.

von Spengel, L. (ed.) (1854) *Rhetores Graeci, ex recognitione Leonardi Spengel*. Leipzig.

General Index

Index Locorum

Printed in the United Kingdom by TJ Clays Ltd.